Image Processing

Image Processing

Principles and Applications

Tinku Acharya
Avisere, Inc.
Tucson, Arizona
and
Department of Electrical Engineering
Arizona State University
Tempe, Arizona

Ajoy K. Ray
Avisere, Inc.
Tucson, Arizona
and
Electronics and Electrical Communication Engineering Department
Indian Institute of Technology
Kharagpur, India

WILEY-INTERSCIENCE

A JOHN WILEY & SONS, INC., PUBLICATION

Published by John Wiley & Sons, Inc., Hoboken, New Jersey.
Published simultaneously in Canada.

Library of Congress Cataloging-in-Publication Data:

Acharya, Tinku.
Image processing : principles and applications / Tinku Acharya, Ajoy K. Ray.
p. cm.
"A Wiley-Interscience Publication."
Includes bibliographical references and index.
ISBN-13 978-0-471-71998-4 (cloth : alk. paper)
ISBN-10 0-471-71998-6 (cloth : alk. paper)
1. Image processing. I. Ray, Ajoy K., 1954- II. Title.
TA1637.A3 2005
621.36'7—dc22 2005005170

Printed in the United States of America.

10 9 8 7 6 5 4 3 2 1

In memory of my father, Prohlad C. Acharya

—Tinku

In memories of my mother, father, and uncle

—Ajoy

Contents

Preface

There is a growing demand of image processing in diverse application areas, such as multimedia computing, secured image data communication, biomedical imaging, biometrics, remote sensing, texture understanding, pattern recognition, content-based image retrieval, compression, and so on. As a result, it has become extremely important to provide a fresh look at the contents of an introductory book on image processing. We attempted to introduce some of these recent developments, while retaining the classical ones.

The first chapter introduces the fundamentals of the image processing techniques, and also provides a window to the overall organization of the book. The second chapter deals with the principles of digital image formation and representation. The third chapter has been devoted to color and color imagery. In addition to the principles behind the perception of color and color space transforation, we have introduced the concept of color interpolation or demosaicing, which is today an integrated part of any color imaging device. We have described various image transformation techniques in Chapter 4. Wavelet transformation has become very popular in recent times for its many salient features. Chapter 5 has been devoted to wavelet transformation.

The importance of understanding the nature of noise prevalent in various types of images cannot be overemphasized. The issues of image enhancement and restoration including noise modeling and filtering have been detailed in Chapter 6. Image segmentation is an important task in image processing and pattern recognition. Various segmentation schemes have been elaborated in Chapter 7. Once an image is appropriately segmented, the next important task involves classification and recognition of the objects in the image. Various pattern classification and object recognition techniques have been presented in Chapter 8. Texture and shape play very important roles in image understanding. A number of texture and shape analysis techniques have been detailed in Chapter 9.

In sharp contrast with the classical crisp image analysis, fuzzy set theoretic approaches provide elegant methodologies for many image processing tasks. Chapter 10 deals with a number of fuzzy set theoretic approaches. We introduce content-based image retrieval and image mining in Chapter 11. Biomedical images like x-Ray, ultrasonography, and CT-Scan images provide sufficient information for medical diagnostics in biomedical engineering. We devote Chapter 12 to biomedical image analysis and interpretation. In this chapter, we also describe some of the biometric algorithms, particularly face recognition, signature verification, etc. In Chapter 13, we present techniques for remotely sensed images and their applications. In Chapter 14, we describe principles and applications of dynamic scene analysis, moving-object detection, and tracking. Image compression plays an important role for image storage and transmission. We devote Chapter 15 to fundamentals of image compression. We describe the JPEG standard for image compression in Chapter 16. In Chapters 17 and 18, we describe the new JPEG2000 standard.

The audience of this book will be undergraduate and graduate students in universities all over the world, as well as the teachers, scientists, engineers and professionals in R&D and research labs, for their ready reference.

We sincerely thank Mr. Chittabrata Mazumdar who was instrumental to bring us together to collaborate in this project. We are indebted to him for his continuous support and encouragement in our endeavors.

We thank our Editor, Val Moliere, and her staff at Wiley, for their assistance in this project. We thank all our colleagues in Avisere and Indian Institute of Technology, Kharagpur, particularly Mr. Roger Undhagen, Dr. Andrew Griffis, Prof. G. S. Sanyal, Prof. N. B. Chakrabarti, and Prof. Arun Majumdar for their continuous support and encouragement. We specially thank Odala Nagaraju, Shyama P. Choudhury, Brojeswar Bhowmick, Ananda Datta, Pawan Baheti, Milind Mushrif, Vinu Thomas, Arindam Samanta, Abhik Das, Abha Jain, Arnab Chakraborti, Sangram Ganguly, Tamalika Chaira, Anindya Moitra, Kaushik Mallick and others who have directly or indirectly helped us in the preparation of this manuscript in different ways. We thank anonymous reviewers of this book for their constructive suggestions.

Finally, we are indebted to our families for their active support throughout this project. Especially, Mrs. Baishali Acharya and Mrs. Supriya Ray stood strongly behind us in all possible ways. We would like to express our sincere appreciation to our children, Arita and Arani, and Aniruddha and Ananya, who were always excited about this work and made us proud.

Tinku Acharya
Ajoy K. Ray

1

Introduction

1.1 FUNDAMENTALS OF IMAGE PROCESSING

We are in the midst of a visually enchanting world, which manifests itself with a variety of forms and shapes, colors and textures, motion and tranquility. The human perception has the capability to acquire, integrate, and interpret all this abundant visual information around us. It is challenging to impart such capabilities to a machine in order to interpret the visual information embedded in still images, graphics, and video or moving images in our sensory world. It is thus important to understand the techniques of storage, processing, transmission, recognition, and finally interpretation of such visual scenes. In this book we attempt to provide glimpses of the diverse areas of visual information analysis techniques.

The first step towards designing an image analysis system is digital image acquisition using sensors in optical or thermal wavelengths. A two-dimensional image that is recorded by these sensors is the mapping of the three-dimensional visual world. The captured two dimensional signals are sampled and quantized to yield digital images.

Sometimes we receive noisy images that are degraded by some degrading mechanism. One common source of image degradation is the optical lens system in a digital camera that acquires the visual information. If the camera is not appropriately focused then we get blurred images. Here the blurring mechanism is the defocused camera. Very often one may come across images of outdoor scenes that were procured in a foggy environment. Thus any outdoor scene captured on a foggy winter morning could invariably result into a blurred image. In this case the degradation is due to the fog and mist in the atmosphere, and this type of degradation is known as atmospheric degradation. In some other cases there may be a relative motion between the object and the camera. Thus if the camera is given an impulsive displacement during the image capturing interval while the object is static, the resulting image will invariably be blurred and noisy. In some of the above cases, we need appropriate techniques of refining the images so that the resultant images are of better visual quality, free from aberrations and noises. Image enhancement, filtering, and restoration have been some of the important applications of image processing since the early days of the field [1]–[4].

Segmentation is the process that subdivides an image into a number of uniformly homogeneous regions. Each homogeneous region is a constituent part or object in the entire scene. In other words, segmentation of an image is defined by a set of regions that are connected and nonoverlapping, so that each pixel in a segment in the image acquires a unique region label that indicates the region it belongs to. Segmentation is one of the most important elements in automated image analysis, mainly because at this step the objects or other entities of interest are extracted from an image for subsequent processing, such as description and recognition. For example, in case of an aerial image containing the ocean and land, the problem is to segment the image initially into two parts–land segment and water body or ocean segment. Thereafter the objects on the land part of the scene need to be appropriately segmented and subsequently classified.

After extracting each segment, the next task is to extract a set of meaningful features such as texture, color, and shape. These are important measurable entities which give measures of various properties of image segments. Some of the texture properties are coarseness, smoothness, regularity, etc., while the common shape descriptors are length, breadth, aspect ratio, area, location, perimeter, compactness, etc. Each segmented region in a scene may be characterized by a set of such features.

Finally based on the set of these extracted features, each segmented object is classified to one of a set of meaningful classes. In a digital image of ocean, these classes may be ships or small boats or even naval vessels and a large class of water body. The problems of scene segmentation and object classification are two integrated areas of studies in machine vision. Expert systems, semantic networks, and neural network-based systems have been found to perform such higher-level vision tasks quite efficiently.

Another aspect of image processing involves compression and coding of the visual information. With growing demand of various imaging applications, storage requirements of digital imagery are growing explosively. Compact representation of image data and their storage and transmission through communication bandwidth is a crucial and active area of development today. Interestingly enough, image data generally contain a significant amount of superfluous and redundant information in their canonical representation. Image compression techniques helps to reduce the redundancies in raw image data in order to reduce the storage and communication bandwidth.

1.2 APPLICATIONS OF IMAGE PROCESSING

There are a large number of applications of image processing in diverse spectrum of human activities–from remotely sensed scene interpretation to biomedical image interpretation. In this section we provide only a cursory glance in some of these applications.

1.2.1 Automatic Visual Inspection System

Automated visual inspection systems are essential to improve the productivity and the quality of the product in manufacturing and allied industries [5]. We briefly present few visual inspection systems here.

- **Automatic inspection of incandescent lamp filaments:** An interesting application of automatic visual inspection involves inspection of the bulb manufacturing process. Often the filament of the bulbs get fused after short duration due to erroneous geometry of the filament, e.g., nonuniformity in the pitch of the wiring in the lamp. Manual inspection is not efficient to detect such aberrations.

  In an automated vision-based inspection system, a binary image slice of the filament is generated, from which the silhouette of the filament is produced. This silhouette is analyzed to identify the non-uniformities in the pitch of the filament geometry inside the bulb. Such a system has been designed and installed by the General Electric Corporation.

- **Faulty component identification:** Automated visual inspection may also be used to identify faulty components in an electronic or electromechanical systems. The faulty components usually generate more thermal energy. The infra-red (IR) images can be generated from the distribution of thermal energies in the assembly. By analyzing these IR images, we can identify the faulty components in the assembly.

- **Automatic surface inspection systems:** Detection of flaws on the surfaces is important requirement in many metal industries. For example, in the hot or cold rolling mills in a steel plant, it is required to detect any aberration on the rolled metal surface. This can be accomplished by using image processing techniques like edge detection, texture identification, fractal analysis, and so on.

1.2.2 Remotely Sensed Scene Interpretation

Information regarding the natural resources, such as agricultural, hydrological, mineral, forest, geological resources, etc., can be extracted based on remotely sensed image analysis. For remotely sensed scene analysis, images of the earth's surface are captured by sensors in remote sensing satellites or by a multi-spectral scanner housed in an aircraft and then transmitted to the Earth Station for further processing [6, 7]. We show examples of two remotely sensed images in Figure 1.1 whose color version has been presented in the color figure pages. Figure 1.1(a) shows the delta of river Ganges in India. The light blue segment represents the sediments in the delta region of the river, the deep blue segment represents the water body, and the deep red regions are mangrove swamps of the adjacent islands. Figure 1.1(b) is the glacier flow in Bhutan Himalayas. The white region shows the stagnated ice with lower basal velocity.

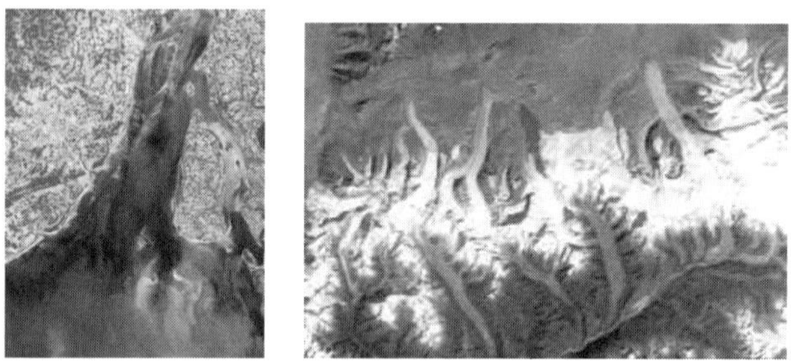

(a) (b)

Fig. 1.1 Example of a remotely sensed image of (a) delta of river Ganges, (b) Glacier flow in Bhutan Himalayas. *Courtesy: NASA/GSFC/METI/ERSDAC/JAROS, and U.S./Japan ASTER Science Team.*

Techniques of interpreting the regions and objects in satellite images are used in city planning, resource mobilization, flood control, agricultural production monitoring, etc.

1.2.3 Biomedical Imaging Techniques

Various types of imaging devices like X-ray, computer aided tomographic (CT) images, ultrasound, etc., are used extensively for the purpose of medical diagnosis [8]–[10]. Examples of biomedical images captured by different image formation modalities such as CT-scan, X-ray, and MRI are shown in Figure 1.2.

(i) localizing the objects of interest, i.e. different organs

(ii) taking the measurements of the extracted objects, e.g. tumors in the image

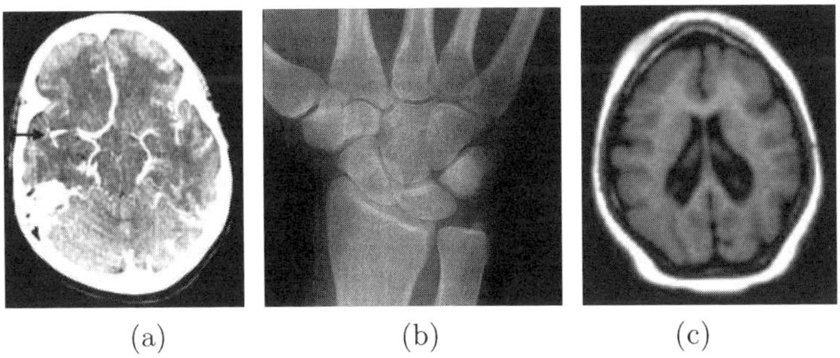

Fig. 1.2 Examples of (a) CT-scan image of brain, (b) X-ray image of wrist, (c) MRI image of brain.

(iii) interpreting the objects for diagnosis.

Some of the biomedical imaging applications are presented below.

(A) *Lung disease identification:* In chest X-rays, the structures containing air appear as dark, while the solid tissues appear lighter. Bones are more radio opaque than soft tissue. The anatomical structures clearly visible on a normal chest X-ray film are the ribs, the thoracic spine, the heart, and the diaphragm separating the chest cavity from the abdominal cavity. These regions in the chest radiographs are examined for abnormality by analyzing the corresponding segments.

(B) *Heart disease identification:* Quantitative measurements such as heart size and shape are important diagnostic features to classify heart diseases. Image analysis techniques may be employed to radiographic images for improved diagnosis of heart diseases.

(C) *Digital mammograms:* Digital mammograms are very useful in detecting features (such as micro-calcification) in order to diagnose breast tumor. Image processing techniques such as contrast enhancement, segmentation, feature extraction, shape analysis, etc. are used to analyze mammograms. The regularity of the shape of the tumor determines whether the tumor is benign or malignant.

1.2.4 Defense surveillance

Application of image processing techniques in defense surveillance is an important area of study. There is a continuous need for monitoring the land and oceans using aerial surveillance techniques.

Suppose we are interested in locating the types and formation of Naval vessels in an aerial image of ocean surface. The primary task here is to segment different objects in the water body part of the image. After extracting the segments, the parameters like area, location, perimeter, compactness, shape, length, breadth, and aspect ratio are found, to classify each of the segmented objects. These objects may range from small boats to massive naval ships. Using the above features it is possible to recognize and localize these objects. To describe all possible formations of the vessels, it is required that we should be able to identify the distribution of these objects in the eight possible directions, namely, north, south, east, west, northeast, northwest, southeast and southwest. From the spatial distribution of these objects it is possible to interpret the entire oceanic scene, which is important for ocean surveillance.

1.2.5 Content-Based Image Retrieval

Retrieval of a query image from a large image archive is an important application in image processing. The advent of large multimedia collection and digital libraries has led to an important requirement for development of search tools for indexing and retrieving information from them. A number of good search engines are available today for retrieving the text in machine readable form, but there are not many fast tools to retrieve intensity and color images. The traditional approaches to searching and indexing images are slow and expensive. Thus there is urgent need for development of algorithms for retrieving the image using the embedded content in them.

The features of a digital image (such as shape, texture, color, topology of the objects, etc.) can be used as index keys for search and retrieval of pictorial information from large image database. Retrieval of images based on such image contents is popularly called the content-based image retrieval [11, 12].

1.2.6 Moving-Object Tracking

Tracking of moving objects, for measuring motion parameters and obtaining a visual record of the moving object, is an important area of application in image processing [13, 14]. In general there are two different approaches to object tracking:

1. Recognition-based tracking

2. Motion-based tracking.

A system for tracking fast targets (e.g., a military aircraft, missile, etc.) is developed based on motion-based predictive techniques such as Kalman filtering, extended Kalman filtering, particle filtering, etc. In automated image processing based object tracking systems, the target objects entering the sensor field of view are acquired automatically without human intervention. In recognition-based tracking, the object pattern is recognized in successive image-frames and tracking is carried-out using its positional information.

1.2.7 Image and Video Compression

Image and video compression is an active application area in image processing [12, 15]. Development of compression technologies for image and video continues to play an important role for success of multimedia communication and applications. Although the cost of storage has decreased significantly over the last two decades, the requirement of image and video data storage is also growing exponentially. A digitized 36 cm $\times$ 44 cm radiograph scanned at 70 μm requires approximately 45 Megabytes of storage. Similarly, the storage requirement of high-definition television of resolution 1280 $\times$ 720 at 60 frames per second is more than 1250 Megabits per second. Direct transmission of these video images without any compression through today's communication channels in real-time is a difficult proposition. Interestingly, both the still and video images have significant amount of visually redundant information in their canonical representation. The redundancy lies in the fact that the neighboring pixels in a smooth homogeneous region of a natural image have very little variation in their values which are not noticeable by a human observer. Similarly, the consecutive frames in a slow moving video sequence are quite similar and have redundancy embedded in them temporally. Image and video compression techniques essentially reduce such visual redundancies in data representation in order to represent the image frames with significantly smaller number of bits and hence reduces the requirements for storage and effective communication bandwidth.

1.3 HUMAN VISUAL PERCEPTION

Electromagnetic radiation in the optical band generated from our visual environment enters the visual system through eyes and are incident upon the sensitive cells of the retina. The activities start in the retina, where the signals from neighboring receivers are compared and a coded message dispatched on the optic nerves to the cortex, behind our ears. An excellent account of human visual perception may be found in [16]. The spatial characteristics of our visual system have been proposed as a nonlinear model in [17, 18].

Although the eyes can detect tranquility and static images, they are essentially motion detectors. The eyes are capable of identification of static objects and can establish spatial relationships among the various objects and regions in a static scene. Their basic functioning depends on comparison of stimuli from neighboring cells, which results in interpretation of motion. When observing a static scene, the eyes perform small repetitive motions called *saccades* that move edges past receptors. The perceptual recognition and interpretation aspects of our vision, however, take place in our brain. The objects and different regions in a scene are recognized in our brain from the edges or boundaries that encapsulate the objects or the regions inside the scene. The maximum information about the object is embedded along these edges or boundaries. The process of recognition is a result of learning that takes place in our neural organization. The orientation of lines and the directions of movements are also used in the process of object recognition.

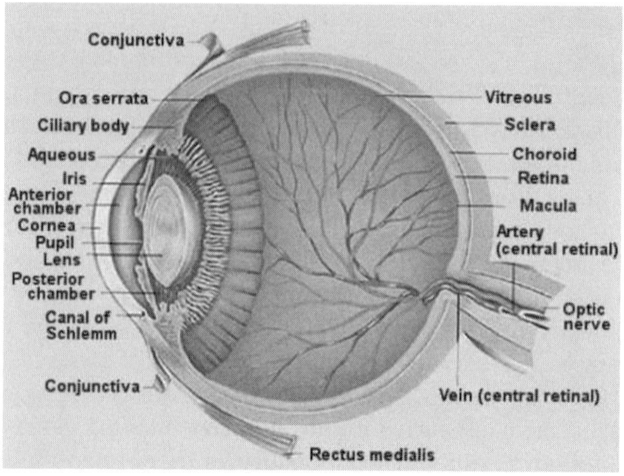

Fig. 1.3 Structure of human eye.

1.3.1 Human Eyes

The structure of an eye is shown in Figure 1.3. The transportation of the visual signal from the retina of the eye to the brain takes place through approximately one and a half million neurons via optic nerves. The retina contains a large number of photo-receptors, compactly located in a more or less regular, hexagonal array. The retinal array contains three types of color sensors, known as *cones* in the central part of the retina named as fovea centralis. The cones are distributed in such a way that they are densely populated near the central part of the retina and the density reduces near the peripheral part of the fovea. There are three different types of cones, namely red, green and blue cones which are responsible for color vision. The three distinct classes of cones contain different photosensitive pigments. The three pigments have maximum absorptions at about 430 nm (violet), 530 nm (blue-green) and 560 nm (yellow-green).

Another type of small receptors fill in the space between the cones. These receptors are called *rods* which are responsible for gray vision. These receptors are more in number than the cones.

Rods are sensitive to very low-levels of illumination and are responsible for our ability to see in dim light (scotopic vision). The cone or photopic system, on the other hand, operates at high illumination levels when lots of photons are available, and maximizes resolution at the cost of reduced sensitivity.

1.3.2 Neural Aspects of the Visual Sense

The optic nerve in our visual system enters the eyeball and connects with rods and cones located at the back of the eye.

The neurons contain dendrites (inputs), and a long axon with an arborization at the end (outputs). The neurons communicate through synapses. The transmission of signals is associated with the diffusion of the chemicals across the interface and the receiving neurons are either stimulated or inhibited by these chemicals, diffusing across the interface. The optic nerves begin as bundles of axons from the ganglion cells on one side of the retina. The rods and cones, on the other side, are connected to the ganglion cells by bipolar cells, and there are also horizontal nerve cells making lateral connections.

The signals from neighboring receptors in the retina are grouped by the horizontal cells to form a receptive field of opposing responses in the center and the periphery, so that a uniform illumination of the field results in no net stimulus. In case of nonuniform illumination, a difference in illumination at the center and the periphery creates stimulations. Some receptive fields use color differences, such as red-green or yellow-blue, so the differencing of stimuli applies to color as well as to brightness. There is further grouping of receptive field responses in the lateral geniculate bodies and the visual cortex for directional edge detection and eye dominance. This is low-level processing preceding the high-level interpretation whose mechanisms are unclear. Nevertheless, it demonstrates the important role of differencing in the senses, which lies at the root of contrast phenomena. If the retina is illuminated evenly in brightness and color, very little nerve activity occurs.

There are 6 to 7 million cones, and 110 to 130 million rods in a normal human retina. Transmission of the optical signals from rods and cones takes place through the fibers in the optic nerves. The optic nerves cross at the optic chiasma, where all signals from the right sides of the two retinas are sent to the right half of the brain, and all signals from the left, to the left half of the brain. Each half of the brain gets half a picture. This ensures that loss of an eye does not disable the visual system. The optical nerves end at the lateral geniculate bodies, halfway back through the brain, and the signals are distributed to the visual cortex from there. The visual cortex still has the topology of the retina, and is merely the first stage in perception, where information is made available. Visual regions in two cerebral hemispheres are connected in the corpus callosum, which unites the halves of the visual field.

1.4 COMPONENTS OF AN IMAGE PROCESSING SYSTEM

There are several components of an image processing system. The first major component of an image processing system is a camera that captures the images of a three-dimensional object.

1.4.1 Digital Camera

The sensors which are used in most of the cameras are either charge coupled device (CCD) or CMOS sensors. The CCD camera comprises a very large number of very small photo diodes, called photosites. The electric charges which are accumulated at each cell in the image are transported and are recorded after appropriate analog to digital conversion.

In CMOS sensors, on the other hand, a number of transistors are used for amplification of the signal at each pixel location. The resultant signal at each pixel location is read individually. Since several transistors are used the light sensitivity is lower. This is because of the fact that some of the photons are incident on these transistors (used for signal amplification), located adjacent to the photo-sensors. The current state-of-the-art CMOS sensors are more noisy compared to the CCD sensors. However, they consume low power and they are less expensive.

In case of bright sunlight the aperture, located behind the camera lens, need not be large since we do not require much light, while on cloudy days when we need more light to create an image the aperture should be enlarged. This is identical to the functioning of our eyes. The shutter speed gives a measure of the amount of time during which the light passes through the aperture. The shutter opens and closes for a time duration which depends on the requirement of light. The focal length of a digital camera is the distance between the focal plane of the lens and the surface of the sensor array. Focal length is the critical information in selecting the amount of required magnification which is desired from the camera.

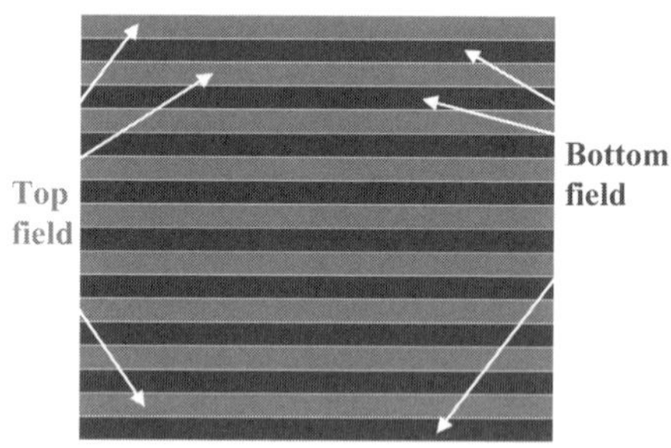

Fig. 1.4 Top and bottom fields in interlace scan.

In an interlaced video camera, each image frame is divided in two fields. Each field contains either the even (top field) or odd (bottom field) horizontal video lines. These two fields are assembled by the video display device. The mode of assembling the top and bottom fields in an interlace camera is shown in Fig. 1.4. In progressive scan cameras on the other hand, the entire frame is output as a single frame. When a moving scene is imaged, such as in robotic vision, it is captured using strobe pulse to illuminate the object in the scene. In such cases of imaging applications, progressive scan cameras are preferable.

Interlaced cameras are not used in such applications because the illumination time may be shorter than the frame time and only one field will be illuminated and captured if interlaced scanning is used.

A digital camera can capture images in various resolutions, e.g., 320×240, or 352×288, or 640×480 pixels on the low to medium resolution range to 1216×912 or 1600×1200 pixels on the high resolution size. The cameras that we normally use can produce about 16 million colors, i.e., at each pixel we can have one of 16 million colors.

The spatial resolution of an image refers to the image size in pixels, which corresponds to the size of the CCD array in the camera. The process of zooming an image involves performing interpolation between pixels to produce a zoomed or expanded form of the image. Zooming does not increase the information content in addition to what the imaging system provides. The resolution, however, may be decreased by subsampling which may be useful when system bandwidth is limited. Sensor resolution depends on the smallest feature size of the objects in a scene that we need our imaging system to distinguish, which is a measure of the object resolution. For example in an OCR system, the minimum object detail that needs to be discerned is the minimum width of line segments that constitute the pattern. In case of a line drawing, the minimum feature size may be chosen as two pixels wide. The sensor resolution of a camera is the number of rows and columns of the CCD array, while the field of view FOV is the area of the scene that the camera can capture. The FOV is chosen as the horizontal dimension of the inspection region that includes all the objects of interest. The sensor resolution of the camera = 2FOV/object resolution. The sensor resolution or sensor size is thus inversely proportional to the object resolution. The resolution of quantization refers to the number of quantization levels used in analog to digital (A/D) conversions. Higher resolution in this sense implies improved capability of analyzing low-contrast images.

Line scan cameras use a sensor that has just a row of CCD elements. An image may be captured by either moving the camera or by moving the image being captured by the camera. The number of elements in a line scan camera ranges from 32 to 8096. Even a single detector moved in a scanning pattern over an area can also be used to produce a video signal. A number of features, such as shutter control, focus control, exposure time control along with various triggering features are supported in cameras.

1.4.1.1 Capturing colors in a digital camera There are several ways in which a digital camera can capture colors. In one approach, one uses red, green, and blue filters and spins them in front of each single sensor sequentially one after another and records three separate images in three colors at a very fast rate. Thus the camera captures all the three color components at each pixel location. While using this strategy an automatic assumption is that during the process of spinning the three filters, the colors in the image must not change (i.e., they must remain stationary). This may not be a very practical solution.

A practical solution is based on the concept of color interpolation or demosaicing, which is a more economical way to record the three primary colors of an image. In this method, we permanently place only one type of filter over each individual photo-site. Usually the sensor placements are carried out in accordance to a pattern. The most popular pattern is called the Bayer's pattern [19], where each pixel is indicated by only one color–red, blue, or green pixel. It is possible to make very accurate guesses about the missing color component in each pixel location by a method called *color interpolation* or *demosaicing* [20, 21]. We cover different methods of color interpolation in Chapter 3.

In high-quality cameras, however, three different sensors with the three filters are used and light is directed to the different sensors by using a beam splitter. Each sensor responds only to small wavelength band of color. Thus the camera captures each of the three colors at each pixel location. These cameras will have more weight and they are costly.

1.5 ORGANIZATION OF THE BOOK

In this chapter, we introduced some fundamental concepts and a brief introduction to digital image processing. We have also presented few interesting applications of image processing in this chapter.

Chapter 2 deals with the principles of image formation and their digital representation in order to process the images by a digital computer. In this chapter, we also review the concepts of sampling and quantization, as well as the various image representation and formatting techniques.

In Chapter 3, we present the basics of color imagery, the color spaces and their transformation techniques. In this chapter, we also present a novel concept of color interpolation to reconstruct full color imagery from sub-sampled colors prevalent in low-cost digital camera type image processing devices.

Chapter 4 has been devoted to discuss various image transformation techniques and their underlying theory. Some of the popular image transformation techniques such as Discrete Fourier Transform, Discrete Cosine Transform, Karhaunen-Loeve Transform, Singular Value decomposition, Walsh-Hadamard transform and their salient properties are discussed here.

Wavelet transformation has become very popular in image processing applications in recent times for its many salient features. Chapter 5 has been devoted to wavelet transformation. We discuss both the convolution and lifting based algorithms for implementation of the DWT.

The importance of understanding the nature of noise and imprecision prevalent in various types of images cannot be overemphasized. This issue has been detailed in Chapter 6. We present a number of algorithms for enhancement, restoration, and filtering of images in this chapter.

Image segmentation is possibly one of the most important tasks in image processing. Various edge detection schemes have been elaborated in Chapter 7. Region based segmentation strategies such as thresholding, region growing, and clustering strategies have been discussed in this chapter.

Once an image is appropriately segmented, the next important task involves classification and recognition of the objects in the image. The various supervised and unsupervised pattern classification and object recognition techniques have been presented in Chapter 8. Several neural network architectures namely multilayer perceptron, Kohonen's Self Organizing feature map, and counterpropagation networks have been discussed in this chapter.

Texture and shape of objects play a very important role in image understanding. A number of different texture representation and analysis techniques have been detailed in Chapter 9. In this chapter, we have also discussed various shape discrimination strategies with examples.

In sharp contrast with the classical crisp image analysis techniques, fuzzy set theoretic approaches provide elegant methodologies which yield better results in many image processing tasks. We describe a number of image processing algorithms based on fuzzy set theoretic approaches in Chapter 10.

In today's world dealing with Internet, the application on content based image retrieval became important because of image search and other multimedia applications. We introduce the concepts of *content-based image retrieval* and *image mining*in Chapter 11.

Biomedical images like x-Ray, ultrasonography, and CT-scan images provide sufficient information for medical diagnostics in biomedical engineering. We devote Chapter 12 to biomedical image analysis and interpretation. In this chapter, we also describe two important applications of biometric recognition, viz., face recognition and signature verification.

Remote sensing is one of the most important applications in image processing. We discuss various satellite based remotely sensed image processing applications in Chapter 13.

In Chapter 14, we describe principles and applications of dynamic scene analysis, moving-object detection, and tracking. We also included recent developments such as condensation algorithm and particle filtering for object tracking.

Image compression plays an important role for image storage and transmission. We devote Chapter 15 to describe the fundamentals of image compression and principles behind it. There are many image compression techniques in the literature. However, adhering to image compression standards is important for interoperability and exchange of image data in today's networked world. The international standard organization, defined the algorithms and formats for image compression towards this goal. We describe the JPEG standard for image compression in Chapter 16.

In this era of internet and multimedia communication, it is necessary to incorporate new features and functionalities in image compression standards in order to serve diverse application requirements in the market place.

JPEG2000 is the new image compression standard to achieve this goal. In Chapters 17 and 18, we elaborate on the JPEG2000 standard, its applications and implementation issues.

1.6 HOW IS THIS BOOK DIFFERENT?

With the growth of diverse applications, it became a necessity to provide a fresh look at the contents of an introductory image processing book. In our knowledge there is no other book that covers the following aspects in detail.

- We present a set of *advanced topics*, in this book, retaining the classical ones.

- We cover several applications such as *biomedical* and *biometric* image processing, *Content based image retrieval*, *remote sensing*, *dynamic scene analysis*, *pattern recognition*, *shape* and *texture* analysis, etc.

- We include new concepts in *color interpolation* to produce full color from sub-sampled Bayer pattern color prevalent in today's digital camera and other imaging devices [21].

- The concepts of Discrete Wavelet Transform and its efficient implementation by *lifting* approach have been presented in great detail.

- In this era of internet and multimedia communication, there is necessity to incorporate many new features and functionalities in *image compression standards* to serve diverse application. *JPEG2000* is the new image compression standard to achieve this goal [15]. We devote two chapters on the *JPEG2000 standard* in great detail.

- We present the concepts and techniques of *Content based image retrieval* and *image mining* [11].

- The principles of *moving-object detection* and *tracking*, including recent developments such as *condensation algorithm* and *particle filtering* for object tracking [14] have been discussed in this book.

- Applications of *dental* and *mammogram* image analysis in biomedical image processing [9, 10] have been presented here.

- Both the *soft* and *hard* computing approaches have been dealt in greater length with respect to the major image processing tasks [11].

- The *fuzzy set theoretic* approaches are rich to solve many image processing tasks, but not much discussions are present in the classical image processing books [22, 23].

- We present the direction and development of current *research* in certain areas of image processing.
- We have provided *extensive bibliography* in the unified framework of this book.

1.7 SUMMARY

In this chapter, we have introduced the concepts, underlying principles, and applications of image processing. We have visited the role of eyes as the most important visual sensor in the human and animal world. The components constituting a computer vision system are presented briefly here. The organization of book and how this book is different from other image processing books currently in the market have also been discussed.

REFERENCES

1. A. Rosenfeld and A. C. Kak, *Digital Picture Processing*, Second Edition, Volume 1, Academic Press, 1982.
2. W. K. Pratt, *Digital Image Processing*, Second Edition, Wiley, New York, 1991.
3. R. C. Gonzalez and R. E. Woods, *Digital Image Processing*, Addison-Wesley, Reading, MA, 1992.
4. R. N. Bracewell, *Two-Dimensional Imaging*, Prentice Hall, Englewood Cliffs, NJ, 1995.
5. D. T. Pham and R. Alcock, *Smart Inspection Systems: Technicuqes and Applications of Intelligent Vision*, Academic Press, Oxford, 2003.
6. T. M. Lissesand and R. W. Kiefer, *Remote Sensing and Image Interpretation*, 4th Edition, John Wiley and Sons, 1999.
7. J. R. Jensen, *Remote Sensing of the Environment: An Earth Resource Perspective*, Prentice Hall, 2000.
8. P. Suetens, *Fundamentals of Medical Imaging* Cambridge University Press, 2002.
9. P. F. Van Der stelt and Qwil G.M.Geraets, "Computer aided interpretation and quantication of angular periodontal Bone defects on dental radiographs", *IEEE Transactions on Biomedical engineering*, 38(4), April 1998, 334–338.

10. M. A. Kupinski and M. Giger, "Automated Seeded Lesion Segmentation on Digital Mammograms," IEEE Trans. Med. Imag., Vol. 17, 1998, 510–517.

11. S. Mitra and T. Acharya, *Data Mining: Multimedia, Soft Computing, and Bioinformatics*, Wiley, Hoboken, NJ, 2003.

12. A. K. Ray and T. Acharya. *Information Technology: Principles and Applications.* Prentice Hall of India, New Delhi, India, 2004.

13. D. Reid, "An algorithm for tracking multiple targets," *IEEE Trans. on Automation and Control*, Vol. AC-24, December 1979, 84–90.

14. R. Cucchiara, C. Grana, G. Neri, M. Piccardi, and A. Prati, "The Sakbot System for Moving Object Detection and Tracking," *Video-Based Surveillance Systems-Computer Vision and Distributed Processing*, 2001, 145–157.

15. T. Acharya and P. S. Tsai. *JPEG2000 Standard for Image Compression: Concepts, Algorithms, and VLSI Architectures*, Wiley, Hoboken, NJ, 2004.

16. G. Wyszecki and W. S. Stiles, *Color Science*, Second Edition, McGraw-Hill, NY, 1982.

17. T. G. Stockham, Jr., "Image Processing in the context of a Visual Model," *Proceedings of IEEE*, 60(7), July 1972, 828–842.

18. C. F. Hall, and E. L. Hall, "A Nonlinear Model for the Spatial Characteristics of the Human Visual System," *IEEE Trans. Systems, Man, and Cybernetics*, SMC-7(3), March 1977, 161–170.

19. B. E. Bayer, "Color Imaging Array," *US Patent 3,971,065*, Eastman Kodak Company, 1976.

20. T. Sakamoto, C. Nakanishi, and T. Hase, "Software Pixel Interpolation for Digital Still Cameras Suitable for A 32-bit MCU," *IEEE Transactions on Consumer Electronics*, 44(4), November 1998, 1342–1352.

21. P. Tsai, T. Acharya, and A. K. Ray, "Adaptive Fuzzy Color Interpolation," *Journal of Electronic Imaging*, 11(3), July 2002, 293–305.

22. L. A. Zadeh, "Fuzzy Sets," *Information and Control*, 8, 1965, 338–353.

23. C. V. Jawahar and A. K. Ray, "Fuzzy Statistics of Digital Images," *IEEE Signal Processing Letter*, 3, 1996, 225–227.

2

Image Formation and Representation

2.1 INTRODUCTION

There are three basic components of image formation, i.e., the illumination, the reflectance models of surfaces which are imaged, and the process of image formation at the retina of human eyes or at the sensor plane of the camera. Once the images are formed (which is a two-dimensional analog signal), the next process involves sampling and digitization of the analog image. The digital images so formed after all these processes need to be represented in appropriate format so that they may be processed and manipulated by a digital computer for various applications. In this chapter, we discuss the principles of image formation and the various representation schemes.

2.2 IMAGE FORMATION

Understanding of physics of illumination is the first step of understanding of image formation. We start our discussion with the physics of illumination.

2.2.1 Illumination

Illumination is a fundamental component in the image formation process, which generates sense in our visual organ. Light produces the psychological sensation when it impinges on our eyes and excites our visual sense. The strength of this sensation, which is the sensation of brightness, can be quantified by averaging the responses of many human observers. The average response, i.e., the psychovisual sensation is determined at different spectral wavelengths. The peak spectral sensitivity of a human observer happens at 555 nm wavelength. If this sensitivity is normalized to one, then the sensitivity drops down to 0.0004 at the two ends of the optical spectrum (i.e., at 400 nm and 735 nm).

It may be noted here that equal amounts of luminous flux produce equal brightness, which is proportional to the logarithm of the luminous flux. Fechner's Law defines the brightness by the relation

$$B = k \log(\frac{F}{F_0}),$$

where F_0 is a reference luminous flux, measured in lumane (lm). The above relation shows that doubling the luminous flux does not double the apparent brightness.

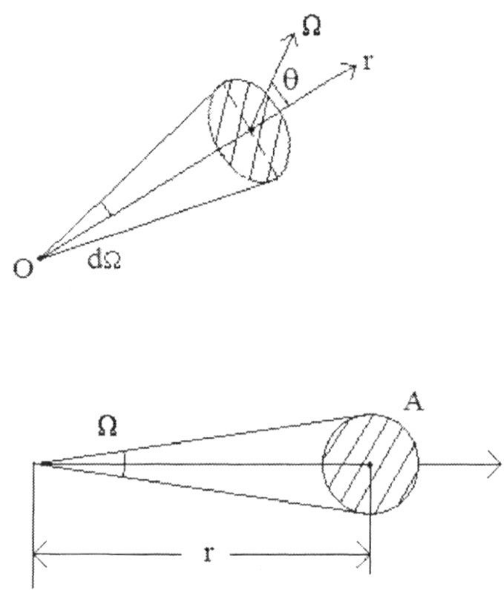

Fig. 2.1 Differential solid angle formation.

Let us consider a point source which emits luminous flux along radial lines. This point source of illumination may be anisotropic. A finite amount of radiation is emitted from the anisotropic point source in a finite cone. This cone has its vertex at the point source O, and its base of area dA at a distance r from the point source O, the normal to dA making an angle θ with the radius. Then, this cone is measured by the differential solid angle

$$d\Omega = \frac{dA \cos\theta}{r^2}$$

measured in steradians as shown in Figure 2.1. It is positive or negative as the normal to dA points outwards or inwards.

It is clear that the total solid angle surrounding a point is 4π. The luminous intensity I of a point source is the ratio $\frac{dF}{d\Omega}$, where F is the luminance flux. The luminous intensity is in general a function of direction and it is measured in candela (cd). If 1 lm (lumane) is emitted per steradian, the intensity is 1 cd. An isotropic point source of intensity I candela emits $4\pi I$ lumane.

The luminous flux incident on area dA from a source of intensity I is

$$dF = I\frac{dA\cos\theta}{r^2},$$

as shown in Figure 2.1. This follows directly from the definition of I as luminous flux per unit solid angle and the definition of solid angle. If the source is an extended one, then this must be integrated over the source area. The luminous flux per unit area falling on a surface is called the illumination E of the surface, and is measured in lm/m^2 (lumane per square meter), which is called a lux. For a point source,

$$E = \frac{dF}{dA} = I\frac{\cos\theta}{r^2}.$$

When a surface is illuminated, the response to incident light differs quite significantly, depending upon the nature of the surface. There are different types of surfaces with different characteristics. Some surfaces may be perfectly absorbing (e.g., black absorbing surfaces), which absorb the entire incident luminous flux and do not reflect any light. Other surfaces reflect the light incident on them.

2.2.2 Reflectance Models

Depending on the nature of reflection we group them in three categories– *Lambertian*, *Specular*, and *Hybrid* surfaces.

- **Lambertian Reflectance:** The *Lambertian* surfaces are those surfaces from which light is reflected in all directions. The nature of such reflectance is a diffused one. The reflection from the wall painted with flat paints, papers, fabrics, ground surfaces are some of the examples of *Lambertian reflection*. The illuminated region of the surface emits the entire incident light in all directions covering solid angle 2π radians. The Lambertian surface appears equally bright from all directions (i.e., equal projected areas radiate equal amounts of luminous flux). Many real surfaces approach to be nearly Lambertian. The reflectance map of the Lambertian surface may be modelled as

$$I_L = E_0 A\cos\theta,$$

where E_0 is the strength of the incident light source, A is the surface area of the Lambertian patch and θ is the angle of incidence. Such a model applies better when the angle of incidence as well as the angle of reflection are both small.

- **Specular Reflectance:** A specularly reflecting surface, such as that of a metal or mirror reflects the light according to the laws of reflection (i.e., the angle of reflection is equal to the angle of incidence). The reflectance from such a surface is known as *specular* reflection.

- **Hybrid Reflectance Model:** There exists another type of reflection, which are mostly found in display devices. These are known as *Hazes.* In the real world most of the surfaces we come across are neither Lambertian nor specular. They possess the combination of both the properties and are termed as *hybrid* surfaces. For example, the cathode-ray oscilloscopes may be considered as having considerable specular reflection and very low to moderate Lambertian reflection. The specular components of reflection from these surfaces may be reduced by using antireflection coatings on these surfaces. The reflectance models from such surfaces may be described as

$$I = wI_S + (1 - w)I_L,$$

where w is the weight of the specular component of the hybrid surface, and I_S and I_L are the specular and Lambertian intensities of the hybrid surface.

The problem of sun glint and glare assumes importance while working with optical imagery of water, snow, or even roads. This problem increases as the sun angle increases. This is due to the specular reflection of light from the object surface. In scenes containing water body the glint increases at high sun angle. At high sun angle much of the sunlight reaches the bottom of the water body and as a result the bottom of the water body gets illuminated and the potential for glint increases. The effect of glint depends on the sun angle, and also on the focal length and the field of view of the imaging devices. The glare is much more common over water, which has a much higher natural reflectance than vegetation. This can be seen on the waters, where the glare appears grayish-silver.

2.2.3 Point Spread Function

The basis of image formation can be explained by the *point spread function* (PSF). The PSF indicates how a point source of light results in a spread image in the spatial dimension.

Let us assume that we want to find the image of a single point at (x, y). If the imaging system is perfectly focused and without any stochastic disturbance, then all the photons from the point source will strike the detector focal plane at the same point and will produce a point image. However, the resultant image of this point source will interestingly not be a point, or a perfect copy of the point; but a blurred version of it. Usually the intensity at the center will be maximum and it will progressively reduce away from the center, resulting in a Gaussian distribution function. The blurring results from several factors – the blurring may be due to inappropriate focusing, imperfection of the lens, scatter of photons or the interaction of photons with the detector array. The resultant image is described in terms of its point spread function (PSF), as defined below

$$I_{res}(x, y) = I_{id}(x, y) \otimes P(x, y)$$

where $\otimes$ is the convolution operation, and I_{res} is the resultant image when the input image I_{id} is convolved with the point spread function $P(x, y)$ at location $(x,\ y)$. The width of the PSF decides the nature of the resultant image.

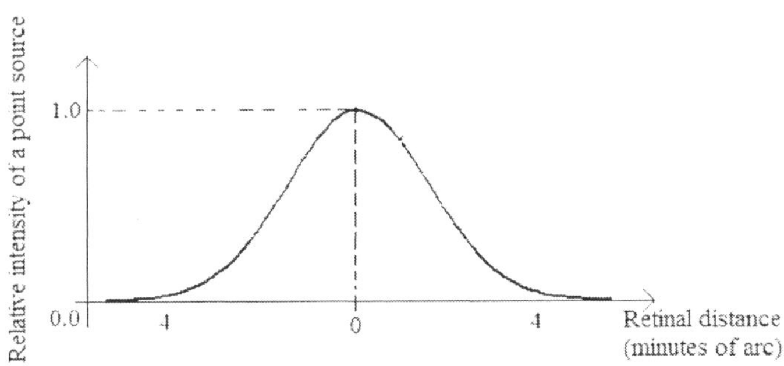

Fig. 2.2 Example of point spread function.

Thus if we know the point spread function, it is possible to restore the image by deconvolution. We know that the convolution in the time domain is analogous to the multiplication in the frequency domain. In the Fourier Transform domain

$$F\left(I_{res}(x, y)\right) = F\left(I_{id}(x, y)\right) \cdot F\left(P(x, y)\right)$$

or

$$F\left(I_{id}(x, y)\right) = \frac{F\left(I_{res}(x, y)\right)}{F\left(P(x, y)\right)},$$

where $F\left(f(x, y)\right)$ represents the Fourier transform of the two-dimensional image function $f(x, y)$. Thus given the Fourier transform of the resultant image along with the Fourier transform of the point spread function, we can reconstruct the original point object by taking the inverse transform of $F\left(I_{id}(x, y)\right)$.

Figure 2.2 shows the PSF of a typical imaging device. The width within which the PSF drops to half on both the sides of the center point is known as *full width half maximum* (FWHM). If now there are two points which are separated by a distance of FWHM or more, then the two points can be distinguished in the image. Otherwise the points will be indistinguishable in the image plane. This is shown in Figure 2.3. The PSF is not necessarily symmetrical and it may have different spreads in different directions.

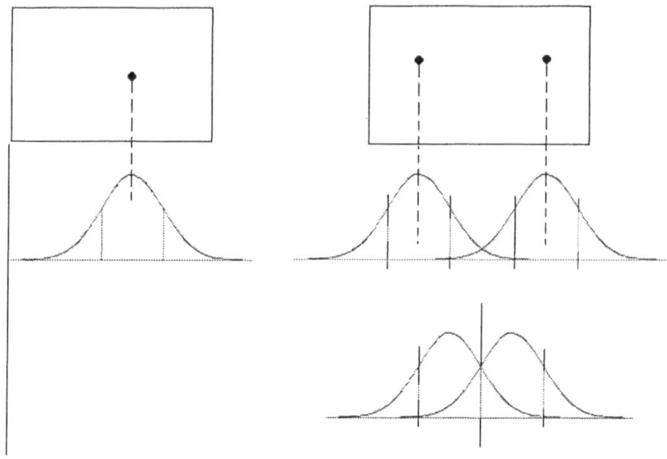

Fig. 2.3 Indistinguishability of point sources.

Often it is difficult to produce a perfect point source to measure the point spread function. In this case a line or edge is often used instead, giving the line spread function (LSF), or edge response function (ERF). The line spread function is a simple extension of the concept of PSF. As in case of a PSF, profiles can be generated orthogonally through the line image, and as in case of PSF, FWHM is used for defining the resolution.

2.3 SAMPLING AND QUANTIZATION

Understanding the process of *sampling and quantization* is one of the key areas in image processing. A comprehensive and detailed description of the theory may be found in [1, 2]. The phenomenal research of Shannon on the diverse aspects of communications in a noisy environment has led to the understanding of the process of sampling continuous signals [3]. The theories of image sampling and quantization have been investigated from two viewpoints. The two-dimensional images may be viewed as deterministic systems, where a continuous-valued image, representing the intensity or luminance at each point of the image, is sampled by an array of *Dirac-Delta* functions of infinite size. The results of sampling and reconstruction of such a deterministic image field may be found in [4]. In an alternative view images have been considered as samples of two-dimensional random processes. In this approach an image is viewed as a two-dimensional stationary random process with a certain mean and autocorrelation function. The practical images may always be viewed as the ideal image with additive noise, which is modelled as a random field. Sampling of such a two-dimensional random field model of images has been discussed in [5].

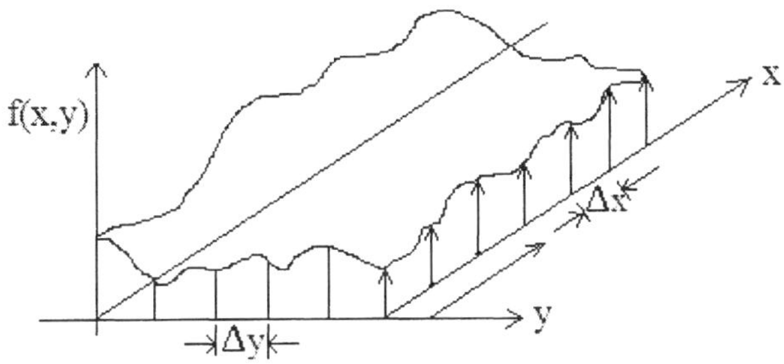

Fig. 2.4 Two-dimensional sampling array.

Let $f(x, y)$ be a continuous-valued intensity image and let $s(x, y)$ be a two-dimensional sampling function of the form

$$s(x,y) = \sum_{j=-\infty}^{\infty} \sum_{k=-\infty}^{\infty} \delta(x - j\Delta x, y - k\Delta y.)$$

The two-dimensional sampling function is an infinite array of dirac delta functions as shown in Figure 2.4. The sampling function, also known as a comb function, is arranged in a regular grid of spacing Δx and Δy along X- and Y axes respectively. The sampled image may be represented as

$$\begin{aligned} f_s(x,y) &= f(x,y)s(x,y) \\ &= \sum_{j=-\infty}^{\infty} \sum_{k=-\infty}^{\infty} f(j\Delta x, k\Delta y) \cdot \delta(x - j\Delta x, y - k\Delta y) \end{aligned}$$

The sampled image $f_s(x, y)$ is an array of image intensity values at the sample points $(j\Delta x, k\Delta y)$ in a regular two-dimensional grid. Images may be sampled using rectangular and hexagonal lattice structures as shown in Figure 2.5. One of the important questions is how small Δx and Δy should be, so that we will be able to reconstruct the original image from the sampled image. The answer to this question lies in the Nyquist theorem, which states that a time varying signal should be sampled at a frequency which is at least twice of the maximum frequency component present in the signal. Comprehensive discussions may be found in [1, 2, 4, 6].

2.3.1 Image Sampling

A static image is a two-dimensional spatially varying signal. The sampling period, according to Nyquist criterion, should be smaller than or at the most

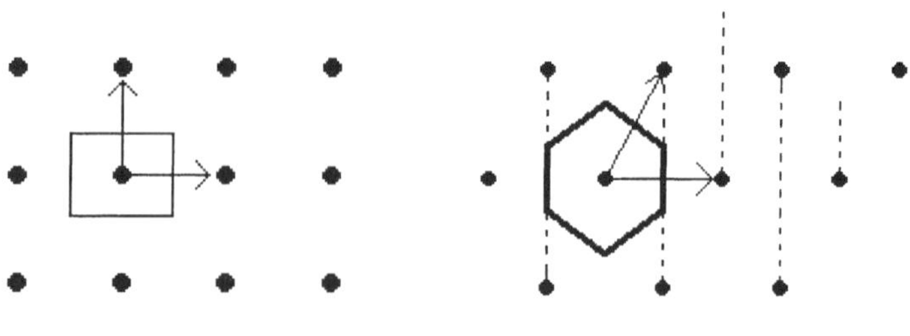

Fig. 2.5 (a) Rectangular and (b) hexagonal lattice structure of the sampling grid.

equal to half of the period of the finest detail present within an image. This implies that the sampling frequency along x axis $w_{xs} \geq 2w_x^L$ and along y axis $w_{ys} \geq 2w_y^L$, where w_x^L and w_y^L are the limiting factors of sampling along x and y directions. Since we have chosen sampling of Δx along X-axis and Δy along Y-axis, $\Delta x \leq \frac{\pi}{w_x^L}$ and $\Delta y \leq \frac{\pi}{w_y^L}$. The values of Δx and Δy should be chosen in such a way that the image is sampled at Nyquist frequency. If Δx and Δy values are smaller, the image is called oversampled, while if we choose large values of Δx and Δy the image will be undersampled. If the image is oversampled or exactly sampled, it is possible to reconstruct the bandlimited image. If the image is undersampled, then there will be spectral overlapping, which results in *aliasing effect*. We have shown images sampled at different spatial resolutions in Figure 2.6 to demonstrate that the aliasing effect increases as the sampling resolution decreases.

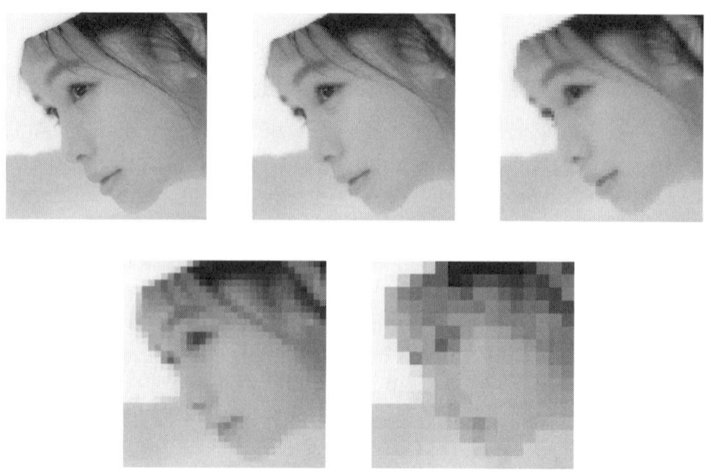

Fig. 2.6 Images sampled at 256 × 256, 128 × 128 , 64 × 64, 32 × 32, and 16 × 16 rectangular sampling grids.

If the original image is bandlimited in Fourier domain, and if the sampling is performed at the Nyquist rate, then it is possible to reconstruct the original image using appropriate sample interpolation. This property is valid in both the cases (i.e., for deterministic and random image fields). The theory of sampling in a lattice of two or more dimensions has been well documented in [7]. The aliasing errors caused by undersampling of the image has been discussed in [6]. The aliasing error can be reduced substantially by using a presampling filter. Thus by choosing a filter which attenuates the high spatial frequencies, the errors get reduced. However, if there is any attenuation in the pass band of the reconstruction filter, it results in a loss of resolution of the sampled image [2]. Reports on the results of sampling errors using Fourier and optimal sampling have been presented in [8].

2.3.2 Image Quantization

Conversion of the sampled analog pixel intensities to discrete valued integer numbers is the process of *quantization.* Quantization involves assigning a single value to each sample in such a way that the image reconstructed from the quantized sample values are of good quality and the error introduced because of quantization is small. The dynamic range of values that the samples of the image can assume is divided into a finite number of intervals, and each interval is assigned a single level. Early work on quantization may be found in [9]–[11].

Some of the interesting questions are as follows:

- How many quantized levels are sufficient to represent each sample?
- How do we choose the quantization levels?

As the number of quantization levels increases, obviously the quantized image will approximate the original continuous-valued image in a better way with less quantization error. When the quantization levels are chosen equally spaced at equal interval, it is known as uniform quantization. When the sample intensity values are equally likely to occur at different intervals, uniform quantization is always preferred. In many situations, however, the image samples assume values in a small range quite frequently and other values infrequently. In such a situation, it is preferable to use nonuniform quantization. The quantization in such cases should be such that they will be finely spaced in the small regions in which the sample values occur frequently, and coarsely spaced in other regions. The uniform and nonuniform quantization levels are shown in Figures 2.7(a) and 2.7(b) respectively. The process of nonuniform quantization is implemented by the process of companding, in which each sample is first processed by a nonlinear compressor, then quantized uniformly and finally again processed by an expander before reconstruction of the original image [11].

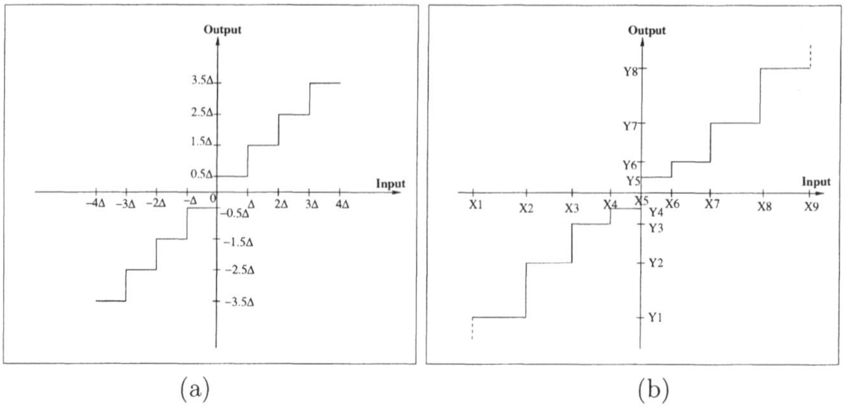

Fig. 2.7 Two-dimensional (a) uniform quantization (b) nonuniform quantization.

2.3.2.1 Monochrome and Color Image Quantization In monochrome image quantization, we assign uniform length code to each image sample. If n is the number of code-bits assigned to each sample, then the number of amplitude quantization levels $M = 2^n$. This kind of code assignment is known as *pulse code modulation* (PCM) coding. The number of levels M is so chosen that the resultant image quality is acceptable to the human observers. The eye is able to discriminate the absolute brightness of only around 15 shades of gray values, however, it is more sensitive to the difference in the brightness of adjacent gray shades. If we choose a reduced number of gray levels, the noticeable artifacts is a gray scale contouring. This contouring artifact occurs in an image where in some regions, the analog change in brightness is very slow. In the quantized image, however, this change in brightness appears as a step jump. This is shown in Figure 2.8, where the effect of reduction of the number of quantized levels is prominently observed, specially in those regions of the image where the brightness changes very slowly.

A color image, represented by red, green, and blue components, is quantized in individual color bands. When each color component is linearly quantized over a maximum range into 2^n levels, then each color sampled pixel is quantized in $3n$ bits, because it requires n bits for each color component.

2.4 BINARY IMAGE

In light of the above discussion we may consider that images may be binary, gray or color. The pixels in a binary image can assume only two values, 0 or 1; a gray image may be quantized to a number of intensity levels, depending on the application, while a color image may be quantized in different color bands. As the number of intensity levels increases, the image is represented to a better approximation, although the storage requirements also grow propor-

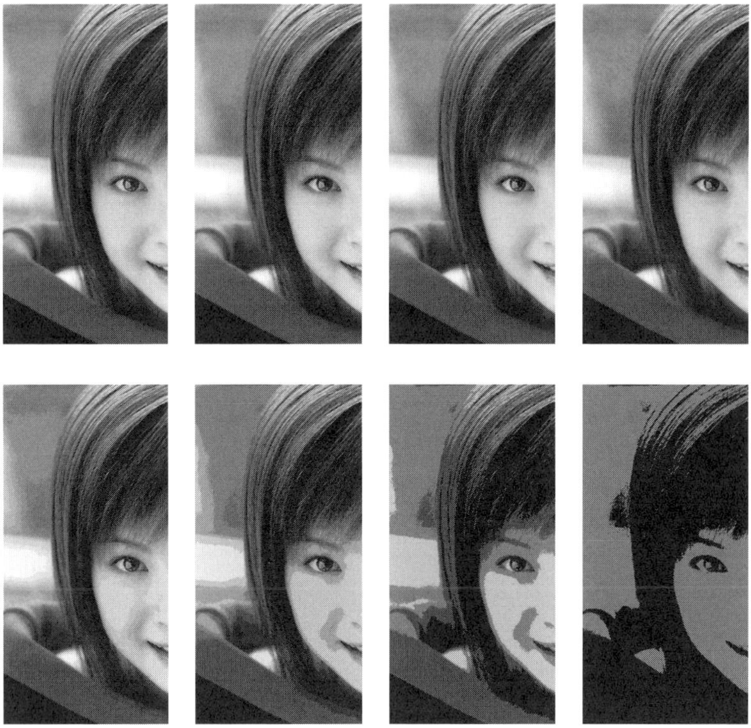

Fig. 2.8 Image quantization: results of finer to coarser quantization.

tionately. The binary images are thus least expensive, since the storage and also processing requirement is the least in case of binary images. Examples of binary images are line drawings, printed text on a white page, or silhouette. These images contain enough information about the objects in the image and we can recognize them easily. There are a number of applications in computer vision where binary images are used for object recognition, tracking, and so on. The applicability of binary images is, however, limited because the overall information content in such images is limited. A gray level image may be converted to a binary image by thresholding process. In the next section we will review some of the interesting properties and operations of a binary image [12].

2.4.1 Geometric Properties

Geometric properties of a binary image such as connectivity, projection, area, and perimeter are important components in binary image processing.

An object in a binary image is a **connected** set of 1 pixels. The following definitions related to connectivity of pixels in a binary image are important.

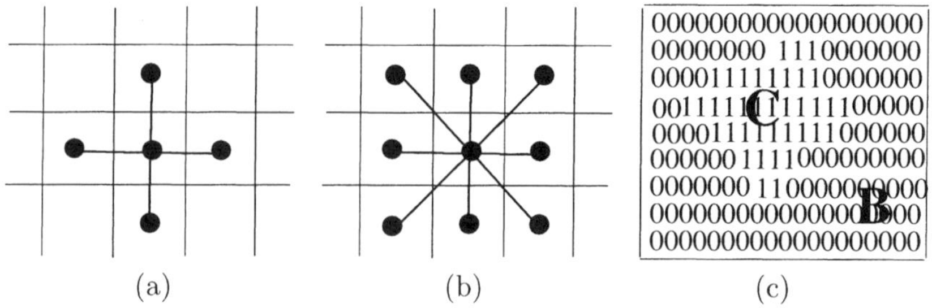

Fig. 2.9 Examples of (a) 4-connected pixels, (b) 8-connected pixels, and (c) connected component and background.

- **Connected Pixels:** A pixel P_0 at (i_0, j_0) is *connected* to another pixel P_n at (i_n, j_n) if and only if there exists a path from P_0 to P_n, which is a sequence of points $(i_0, j_0), (i_1, j_1), \cdots, (i_n, j_n)$, such that the pixel at (i_k, j_k) is a neighbor of the pixel at (i_{k+1}, j_{k+1}) and $P_k = P_{k+1}$ for all k, $0 < k < n - 1$.

- **4-connected:** When a pixel at location (i, j) has four immediate neighbors at $(i + 1, j)$, $(i - 1, j)$, $(i, j + 1)$, and $(i, j - 1)$, they are known as *4-connected.* Two four connected pixels share a common boundary as shown in Figure 2.9(a).

- **8-connected:** When the pixel at location (i, j) has, in addition to above four immediate neighbors, a set of four corner neighbors at $(i+1, j+1)$, $(i + 1, j - 1)$, $(i - 1, j + 1)$, and $(i - 1, j - 1)$, they are known as *8-connected.* Thus two pixels are eight neighbors if they share a common corner. This is shown in Figure 2.9(b).

- **Connected component:** A set of connected pixels (4 or 8 connected) forms a *connected component.* Such a connected component represents an object in a scene as shown in Figure 2.9(c).

- **Background:** The set of connected components of 0 pixels forms the *background* as shown in Figure 2.9(c).

Once an object is identified, some of the attributes of the object may be defined as follows.

- *Area of an object*: The area of a binary object is given by

$$A = \sum_i \sum_j O[i, j],$$

where $O[i, j]$ represents the object pixels (binary 1). The area is thus computed as the total number of object pixels in the object.

- *Location of object*: The location of the object is usually given by the center of mass and is given as

$$x_c = \frac{\sum_i \sum_j [i\, O(i,j)]}{A}, \qquad y_c = \frac{\sum_i \sum_j [j\, O(i,j)]}{A}$$

  where x_c and y_c are the coordinates of the centroid of the object and A is the area of the object. In effect thus the location of the object is given by the first-order moment.

- *Orientation of an object*: When the objects have elongated shape, the axis of elongation is the orientation of the object. The axis of elongation is a straight line so that the sum of the squared distances of all the object points from this straight line is minimum. The distance here implies the perpendicular distance from the object point to the line.

- *Perimeter of an object*: To compute the perimeter of an object, we identify the boundary object pixels covering an area. Perimeter is defined by the sum of these boundary object pixels.

- *Projection of an object onto a line*: The projections of a binary image provide good information about the image. The projections may be computed along horizontal, vertical, or diagonal lines. The horizontal projection is obtained by counting the number of object pixels in each column of the binary image, while the total number of object pixels in each row yields the vertical projection as follows:

$$P_{hor} = \sum_j O(i,j), \quad P_{ver} = \sum_i O(i,j)$$

  In Figure 2.10, we show the histograms of the binary object along the horizontal and vertical axes in terms of number of object pixels projected in the corresponding axis. Each square in the grid in the object represents an object pixel. Sum of the counts of projected pixels in either axis gives the area of the binary object. From Figure 2.10, it can be validated that $P_{hor} = P_{ver} = 59$ which is the area of the object. If there exists only one object in the scene, either of P_{hor} or P_{ver} yields the area of the object.

It should be noted that different images may have the same projection and hence the projection is not a unique attribute of an object.

2.4.2 Chain code representation of a binary object

One efficient way to represent an object in a binary image is by *chain codes*. Each boundary pixel of an object has at least one adjacent neighboring boundary pixel such that the direction of the next boundary pixel from the current

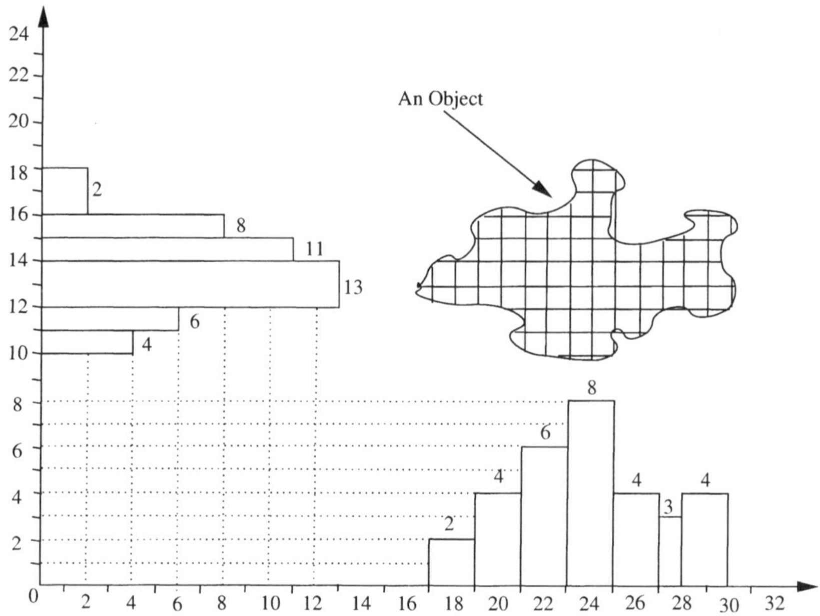

Fig. 2.10 Histogram of a binary object along horizontal and vertical axes.

one is specified by a unique number between 1 and 8. Each number represents one of eight possible directions as shown in Fig 2.11(a). The representation is efficient because each direction can be encoded by only 3 bits. The partial chain code of the boundary for the head-and-shoulder binary image in Fig 2.11(b) is ".......7 7 7 7 7 7 8 7 7 7 8 77 77 7 7 7 8 7 7 8 7 7 7 7 7 7 7 7 8 7 7 8 8 7 8 7 8 8 8 1 8 8 8 1 8 8 1 8 1 8 1 8 1 8 1 8 1 8 1 8 8 8 1 1 8 8 1 8 1 8 1 8 1 7 7 1 7 1 7 7 7 7 7 7 7 7 7 7 7 6 7 6 6 6 7 6 6 7 7 7 7 ....". Figure 2.11(c) shows the dominant vertices along the head-and-shoulder contour of the binary image.

(a) (b) (c)

Fig. 2.11 (a) chain code, (b) binary image, (c) dominant vertices along the contour.

Now given two shape patterns of two objects in a binary image and their chain codes, we can get a similarity measure to determine similarity in shape and orientation between the objects.

Let C' and C'' be two chain codes of length m and n respectively and $m \leq n$, the cross correlation between the two chain codes is given by

$$R_{C',C''}(j) = \frac{1}{m} \sum_{j=1}^{n} \cos\left[C'(i) - C''((i+j)mod\, n)\pi/4\right]$$

The value of j at which $R_{C',C''}(j)$ assumes maximum value, yields the chain code segment of C'' that best matches C' It is easy to note that when C' and C'' are exactly identical the cross correlation between them is 1.

There are other interesting features which provide adequate information about a binary image. Some of these are *moments*, *compactness*, *Euler number*, *medial axis* and so on. Also a number of interesting operations, known as morphological operations provide rich information about the shape of a binary image. These find number of applications in image processing as discussed in Chapters 7 and 9.

2.5 THREE-DIMENSIONAL IMAGING

Three-dimensional imaging using stereo vision finds a number of applications (e.g., industrial inspection of three-dimensional objects, biomedical imaging, creating three-dimensional database of city, regional planning, image based rendering, etc.).

2.5.1 Stereo Images

The human eye has a remarkable property of three-dimensional perception of the world around us. We use two images of the same object captured by both our eyes and combine them to get a three-dimensional perception. There are several techniques to extract three-dimensional information using various sensors like radar sensors, acoustic sensors, laser range finders, etc. In the use of stereo vision, we can use two images of the object using two cameras and combine them to get the depth perception as in case of human vision. There exist a number of algorithms which yield the depth information from such stereo images. The principle involved in these algorithms is that of finding the same feature from both the images captured from left and right cameras and then measuring the depth using *triangulation* technique.

The process of triangulation involves intersecting the lines of sight from each camera to the object. As a matter of fact, identifying the same point in the original image from two stereo images (i.e., establishing correspondence between two image points), which are the projections of the same point in the original image is an interesting problem in computer vision. Figure 2.12 shows

Fig. 2.12 Stereo image pair (a) from left camera and (b) from right camera.

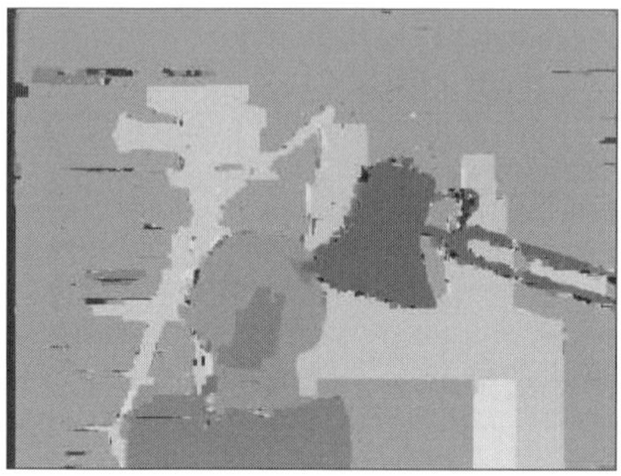

Fig. 2.13 The depth map image from a range camera.

the pair of a stereo image (left and right). Once we determine the distance maps of the scene, a number of shape features or volumetric features can be extracted from the object.

2.5.2 Range Image Aquisition

In contrast to the intensity capturing cameras, e.g., CCD or CMOS sensor based cameras that we have described so far, there is a distinctly different type of camera known as the range camera, which captures a raster image of depth information about a three-dimensional object. Using the range camera we do not record the intensity or the color information of the object to be imaged. On the contrary, a range camera yields an image, where each image location gives a measure of the range or depth or distance of the object point from the camera plane. As in the case of quantization of intensity images here also the range or depth values are quantized in say 256 levels. The quantized distance or the range values are displayed as a gray scale image, such that each intensity value gives a measure of the distance. These cameras provide two-and-a-half-dimensional image, in contrast to the three-dimensional one, which provides complete three-dimensional information of the image. It may be noted here that MRI images yield complete three-dimensional information. The range cameras are again of several types. The laser range finder uses a single optical beam and computes the depth information from the time delay between the incident and the reflected pulse of laser beam. In continuous scan case, the phase shift is measured. The laser beam sweeps across the entire field of view in equiangular increments, say 60 degree $\times$ 60 degree, and for achieving this objective two mirrors are used.

In another version of range cameras, using structured lights, two optical paths are used. The depth information is computed in such scanners using the method of triangulation. The example of depth map image (depth map of the image in Figure 2.12) from a range camera is shown in Figure 2.13. The color version of this figure is provided in the color pages to show different segments in distinct colors.

2.6 IMAGE FILE FORMATS

There are a number of file formats in which one may store the images in files and retrieve them from files. These are known as image file format standards. Here we will present some of the most popularly used Image file format standards.

Tagged Image Format (.tif, .tiff): The .tif format is a very broad format, which can handle anything from bitmaps to compressed color palette images. The *tiff* format supports several compression schemes, but is often used for uncompressed images as well. This format is popular, relatively simple, and allows color.

Portable Network Graphics (.png): This is an extensible file format that provides lossless, well-compressed storage of raster images. This simple format covers the major functionalities of .tiff. Gray scale, color palette, and full-color (true-color) images are supported by this file format. It supports an optional alpha channel, and depths from 1 to 16 bits per channel.

JPEG (.jpg): It is the most widely used standard for transmission of pictorial information and includes a variable lossy encoding as part of the standard, set by a quality parameter.

MPEG (.mpg): This format is extensively used throughout the Web and is used only for motion images. This uses compression, yielding only lossy videos.

Graphics Interchange Format (.gif): This format supports 8-bit color palette images and is not very popular among the image processing researchers.

RGB (.rgb): This is an image file standard from Silicon Graphics for color images.

RAS (.ras) This is an uncompressed scan-out of three color bands for Sun Raster images.

Postscript (.ps, .eps, .epsf): This image format is mainly used while introducing images or figures in a book or note and for printing. In postscript format, gray level images are represented by decimal or hex numerals encoded in ASCII.

Portable Image File Formats: Some of the most commonly used image file formats are *Portable Image Formats*, which include Portable Bitmap, Portable Graymap, Portable Pixmap, and Portable network map. The default suffixes for these formats are .pbm, .pgm, .ppm, and .pnm. These formats are a convenient method of saving and reading the image data. These are some of the image formats which support all kinds of images of increasing complexity–from bits to gray levels to color pixmaps of various sorts.

PPM: A PPM file consists of two parts, a header and the image data. The header consists of at least three parts. The first part is a magic PPM identifier. The PPM identifier can be either *P3* (for ASCII format image data) or *P6* (data in binary format). The next part consists of the width and height of the image as ASCII numbers. The last part of the header gives the maximum value of the color components for the pixels. In addition to the above, a comment can be placed anywhere with a # character; the comment extends to the end of the line.

PGM: This format is identical to the above except it stores gray scale information, that is, one value per pixel instead of three (r, g, b). The only difference in the header section is the magic identifiers which are *P2* and *P5*; these correspond to the ASCII and binary form of the data respectively.

PBM: PBM stores single-bit pixel image as a series of ASCII 0 or 1's. Traditionally 0 refers to white while 1 refers to black. The header is identical to PPM and PGM format except there is no third header line (the maximum pixel value doesn't have any meaning). The magic identifier for PBM is *P1*.

2.7 SOME IMPORTANT NOTES

There are some important notes which need to be remembered while designing an image processing system. It is important to ensure that the appropriate imaging system is set up before capturing an image. If the illumination level is low, it may lead to underexposure of the sensor while too much illumination can lead to overexposure. Both under and overexposed images need preprocessing for detecting objects in a scene. It is also important to choose correct sensor resolution so that the captured images have adequate object resolution which help in automated recognition.

An image is represented by a two-dimensional matrix of size $M \times N$. It is convenient to choose the topmost and leftmost point as the origin (0, 0) and coordinates of all other pixels are assigned accordingly. Quite frequently, we perform local operation inside a 3×3 or sometimes larger neighborhoods. Every pixel in an image, except those at the borders has two horizontal two vertical and four diagonal neighbors, totaling 8 neighbors. Only the pixels on the borders have five neighbors. In such situations, there are three possible options:

1. We may pad up the entire image on all the sides with zero gray value pixels, making it a $(M + 1) \times (N + 1)$ image.

2. We may ignore the pixels on the boundary and perform the operations on the inner pixels only.

3. The image may be considered as cyclically closed, which means that the last column is adjacent to the first column and last row is adjacent to the first row.

We categorize the low level image processing operations into following three different types.

1. *Type 0 operation*: If the output intensity level at a certain pixel is strictly dependent on only the input intensity level at that point, such an operation is known as *type 0* or a *point* operation. Point operations are quite frequently used in image segmentation, pixel classification, image summing, differencing, etc.

2. *Type 1 operations*: If the output intensity level at a pixel depends on the input intensity levels of the neighboring pixels as well, then such operations are termed *type 1* or *local* operations. Examples of local operations are Edge detection, image filtering, etc.

3. *Type 2 operations*: If the operations are such that the output level at a point is dependent on some geometrical transformation, these operations are termed *type 2* or *Geometrical* operations.

2.8 SUMMARY

In this chapter we have discussed the theory of digital Image formation, their representation and various image formats in which the digital images can be stored in the computer. We have discussed the role of illumination and the different reflectance models, which are the two most important components of image formation. The principal ideas behind formation of digital images are based on the concepts of sampling and quantization. We have discussed these concepts in this chapter. Some of the important properties of binary images have also been discussed here.

REFERENCES

1. A. Rosenfeld and A. C. Kak. *Digital Picture Processing*, 2nd Ed., Vol. 1, Academic Press, 1982.

2. W. K. Pratt. *Digital Image Processing*, 2nd Ed., Wiley, New York, 1991.

3. C. E. Shannon, "Communications in the Presence of Noise," *Proceedings of IRE*, 37(1), 10–21, 1949.

4. H. J. Landa, "Sampling, Data Transmission, and the Nyquist rate," *Proceedings of IEEE*, 55(10), 1701–1706, 1967.

5. H. S. Shapiro and R. A. Siverman, "Alias Free Sampling of Random Noise," *Journal of SIAM*, 8(2), 225–248, 1960.

6. D. G. Childers, "Study and Experimental Investigation of Sampling Rate and Aliasing in Time Division Telemetry Systems," *IRE Trans. Space Electronics and Telemetry*, SET-8, 267–283, 1962.

7. D. P. Peterson and D. Middleton, "Sampling and Reconstruction of Wave number Limited functions in N-Dimensional Eucledian Space," *Information & Control*, Vol.5, 279–323, 1962.

8. A. Habibi and P. A. Wintz, "Image Coding by Linear Transformation & Block Quantization," *IEEE Trans Communications*, Com-19, 1971, 50–62.

9. L. H. Harper, "PCM Picture Transmission," *IEEE Spectrum*, 3(6), 1966.

10. J. Max, "Quantizing for Minimum Distortion," *IRE Trans. Information Theory*, IT-6(1) 1960, 7–12.

11. B. Smith, "Instantaneous Companding of Quantized Signals," BSTJ36, 1957, 653–709.

12. R. Jain, R. Kasturi, and B. G. Schunk, *Machine Vision*, McGraw-Hill, New York, 1995.

3

Color and Color Imagery

3.1 INTRODUCTION

Color is important visual information which keeps humans fascinated since birth. Light (chromatic information) reflected from an object is absorbed by the cone cells of our visual system and ultimately leads to a perception of color. There are three cone classes in the human vision system to perceive color. The light reflected from the object leads to different amounts of absorptions in these three classes of cones in the human eye. The interpretation of these cone absorptions by our nervous system is the basis of our color perception. So color is a perceptual representation of the surface reflectance of an object.

In addition to shape, texture, and other low-level image features color information is an important feature which has been successfully used in many image processing applications such as object recognition, image matching, content-based image retrieval, computer vision, color image compression, etc. Color science still remains a challenging field of study in the computer vision and digital image processing community today [1]–[3]. In this chapter, we briefly describe the principles behind the perception of colors by the human visual system and then describe the important color space transformation techniques suitable for digital image processing. We describe the interpolation of color to restore full color from subsampled color information that is common in most of the color digital cameras and other imaging devices.

3.2 PERCEPTION OF COLORS

The human eye is sensitive to electromagnetic radiation in wavelengths ranging from 400 nm (violet) to 770 nm (red) [2]. The spectral response of the human eye is shown in Figure 3.1. It may be seen from Figure 3.1 that the eye responds to only a small spectral wavelength in the optical band of the entire electromagnetic spectrum. The machine vision systems today, however, can respond to a much larger extended spectrum ranging, from 780 nm to 1400 nm in the near-infrared range, 1400 nm to 3300 nm in mid-infrared, 3 to 10 μm in far-infrared and also 100 nm to 380 nm in ultraviolet light range. This has been possible because of the availability of a large range of special sensors available for recording these signals.

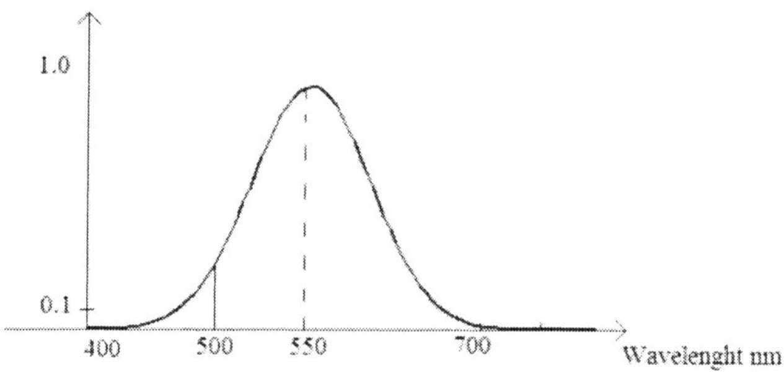

Fig. 3.1 Spectral response of human eye.

A color image can be represented as a function $C(x, y, t, \lambda)$. It is thus a function of the location (x, y), wavelength λ of the reflected light, and also of time in case of a dynamic image. When an image is captured in a fixed wavelength λ, it is called a monochromatic image.

The existence of three spectral sensitivity functions $V_R(\lambda)$, $V_G(\lambda)$, and $V_B(\lambda)$ provides a basis for color vision. In fact the monochromatic lights at 430, 530, and 560 nm at which the eye responses are maximum are not exactly blue, green, and red respectively, and some researchers prefer to use the nomenclature of short-, medium-, and long-wavelength cones instead of R, G, and B cones. The cones provide us with color vision (photopic vision) that can distinguish remarkably fine wavelength changes. The relative spectral sensitivity of the eye has been measured as the function of the wavelength. The measurements reveal that human eyes have peak scotopic spectral sensitivity at wavelength 507 nm and peak photopic spectral sensitivity at wavelength 555 nm as shown in Figure 3.1.

The object in the scene as perceived by human eyes or the camera system is characterized by its radiance $R(\lambda, x, y, t)$, where λ is the wavelength of the electromagnetic radiation at position (x, y) and at time t for a particular color. There exists a direct relationship between the physical stimulii from the object, say the luminance of a display device, and the subjective human perception, say the response of a human being. The relationship was first formulated by Weber in his law, which states that

$$\frac{W_L}{L} = k$$

where W_L is the just noticeable difference in the brightness or luminance which is required to perceive a difference between L and $L + W_L$. The constant k assumes an approximate value 0.015. It is obvious that as the brightness L increases, W_L also increases, for k to be constant. This implies that for large values of luminance L, a larger increase in luminance W_L is required for distinguishing two objects of luminance L and $L + W_L$ respectively. The just noticeable difference is, however, much smaller at the lower values of luminance. The relationship between the stimulus and the observer's perception is not linear. Experimental investigations show that Weber's law is valid only for intermediate luminance values and not for very high or low luminance values. The relationship between the perceived brightness B and the luminance L is logarithmic (i.e., $B \propto \log L$).

3.3 COLOR SPACE QUANTIZATION AND JUST NOTICEABLE DIFFERENCE (JND)

The availability of a large number of colors in the images causes problems in many image processing applications. Large redundancy, nonlinearity, and huge color space dimensionality in the real-life color image data reduces the efficiency of color image processing, coding, and analysis algorithms. Reduction in the number of colors reduces the search space and hence improves the convergence time considerably. After removing the redundancy and reducing the number of colors in a digital color image, the efficiency of the algorithms may considerably increase. The process of representing the complete color space using a comparatively few representative colors without compromising much on the image quality is widely known as quantization or sampling the RGB color space. The RGB color space may be nonlinearly sampled to reduce the redundancy in colors and to reduce the color space dimensionality [4, 5].

The modern computer graphics systems may be able to generate millions of colors while a sound vision observer (as defined by CIE) can respond to only 17,000 colors at the maximum intensity. Thus the huge space containing millions of colors may be mapped on to a new space containing approximately 17,000 colors by maintaining the same perceptual quality of the image, simulating the human color vision performance. The red color cones have minimum spectral sensitivity, green color cones have the maximum sensitivity, while blue color cones have moderate but near to green sensitivity. One may use 24 samples in each basic color shade for color rendition experiments. Taking into consideration the spectral sensitivity of the color cones, nonuniform sampling of the color space yields better results for accommodating approximately 17,000 colors in the RGB space. Thus 24 quantization levels on R-axis, 26 levels on B-axis and 28 levels on G-axis yields better sampling.

Use of a sampling theorem in this direction has not yet been fully explored. New approaches have been suggested which modifies the color space itself considering the physiological limitations of human color vision. Thus the RGB space is transformed into a new color space with much less dimensionality.

According to the three color theory of Thomas Young, all the colors are perceived by our visual system as linear combination of the basic colors. The response curves of the three cones (i.e., red, green, and blue cones) against the wavelength are presented in [4]. A human eye can discriminate between two colors if they are at least one just noticeable difference away from each other. Actual value of the *just noticeable difference* (JND) in terms of coordinates may not be constant over the RGB space due to nonlinearity of human vision and the nonuniformity of the RGB space. At higher illumination, the color sensing power of the eye is more if it has not been saturated. The constants in Buchsbaum's nonlinearity equation take care of the eye adaptation conditions and the illumination conditions. Buchsbaum started his work with the findings of Weber and analytically derived the visual nonlinearity in the logarithmic form. Though his predicted logarithmic behavior is supported by physiological literature, it does not consider the saturation behavior shown by the human vision at high intensities [5, 6].

3.4 COLOR SPACE AND TRANSFORMATION

A number of color spaces or color models have been suggested and each one of them has a specific color coordinate system and each point in the color space represents only one specific color [2, 3, 7]. Each color model may be useful for specific applications. Typical color images, particularly those generated by a digital imaging system, are represented as red, green, blue and are normally called RGB images. They are useful for color monitors, and video cameras. An RGB color image, represented by 8 bits of R, G, and B pixels has 256^3 or 16,777,216 colors. There are a number of such color spaces like CMYK, HSV, HIS, or LUV, etc.

3.4.1 CMYK space

Another interesting color model utilizes CMYK (cyan, magenta, yellow, and black) and this model finds utility in color printers. Most of the output devices including color printers or copiers use CMY color model. Just as the primary additive colors are red, green and blue, the primary colors of pigments on the other hand are magenta, cyan and yellow and the corresponding secondary colors are red, green and blue. The conversion from RGB to CMY may be performed as

$$\begin{bmatrix} C \\ M \\ Y \end{bmatrix} = \begin{bmatrix} 1 \\ 1 \\ 1 \end{bmatrix} - \begin{bmatrix} R \\ G \\ B \end{bmatrix}. \tag{3.1}$$

where R, G, B represent the normalized color values in the range 0 to 1.

It may be easily verified from the above that a cyan coated surface does not contain red, or a surface pigmented by magenta is devoid of green. It may also be noted that equal amounts of pigments primaries (e.g. cyan, magenta, and yellow) produces black. Thus a four-color system cyan (C), magenta (M), yellow (Y), and black (B) forms a four-color model.

3.4.2 NTSC or YIQ Color Space

In this color space (also known as YIQ color space), the luminance information Y represents the gray scale information, while hue (I) and saturation (Q) carry the color information. The conversion from RGB to YIQ is

$$\begin{bmatrix} Y \\ I \\ Q \end{bmatrix} = \begin{bmatrix} .299 & .587 & .114 \\ .596 & -.274 & -.322 \\ .211 & -.523 & .312 \end{bmatrix} \begin{bmatrix} R \\ G \\ B \end{bmatrix}. \tag{3.2}$$

The elements of the first row when added become unity and the elements in the second and third row sum to 0. Thus in a gray scale image, where $R = G = B$, the color components I and Q are zero. The NTSC color space is used in television.

3.4.3 YC_bC_r Color Space

In this color space, Y is the luminous component while C_b and C_r provide the color information. The color information is stored as two color difference components C_b and C_r. This color space is used in digital video. The information from RGB to YC_bC_r is as follows:

$$\begin{bmatrix} Y \\ C_b \\ C_r \end{bmatrix} = \begin{bmatrix} 16 \\ 128 \\ 128 \end{bmatrix} + \begin{bmatrix} 65.481 & 128.553 & 24.966 \\ -37.797 & -74.203 & 112.00 \\ 112.00 & -93.786 & -18.214 \end{bmatrix} \begin{bmatrix} R \\ G \\ B \end{bmatrix}. \tag{3.3}$$

3.4.4 Perceptually Uniform Color Space

Although both RGB and CMYK color models are extremely useful for color representation, color processing, and also for development of hardware, these models are in no way similar to the human vision model.

One of the major limitations of the RGB color space is that it is a nonuniform one. Uniform color space is one in which the Euclidean color distance between two color points at any part of the color space corresponds to the perceptual difference between the two colors by the human vision system. In nonuniform color spaces, on the other hand, two different colors at a distance d apart in one part of the color space does not exhibit the same degree of perceptual difference as two other colors at the same distance apart in another part of the color space. In imaging applications, perceptually uniform color spaces are of great importance. Based on the physiological knowledge of human vision, the nonuniform RGB space needs to be mapped into new perceptually uniform spaces.

3.4.4.1 Colors and Hues Color is an attribute of visual perception and can be described by many color names like red, green, yellow, white, gray, black, and so on. Like colors, hue is also an attribute of human perception and can be described as red, green, blue, purple, and yellow as primary hues or any intermediate combinations of the primary hues. Interestingly, although black, white, and gray are considered as colors, according to CIE (Commission International de l'Eclairage) they are not hues. From the above discussions thus we can consider there are two classes of perceived colors: (1) chromatic colors (i.e., the hues which do not include black, white and gray) and (2) achromatic colors, which are not hues (i.e., black, white, and gray).

3.4.4.2 HSV Çolor Space HSL or HSI is one color space, which describes colors as perceived by human beings. HSI (or HSV) stands for hue (H), saturation (S) and intensity (I) (or value V).

Hue has been already mentioned as a color property of light. It may also be conceived as a property of the surface reflecting or transmitting the light. For example, a blue car reflects blue hue. Moreover it is also an attribute of the human perception.

The hue which is essentially the chromatic component of our perception may again be considered as weak hue or strong hue. The colorfulness of a color is described by the saturation component. For example, the color from a single monochromatic source of light, which produces colors of a single wavelength only, is highly saturated, while the the colors comprising hues of different wavelengths have little chroma and have less saturation. The gray colors do not have any hues and hence they are zero saturation or unsaturated. Saturation is thus a measure of colorfulness or whiteness in the color perceived. The *lightness* (L) or *intensity* (I) or *value* (V) essentially provides a measure of the brightness of colors. This gives a measure of how much light is reflected from the object or how much light is emitted from a region. It is proportional to the electromagnetic energy radiated by the object. The luminosity (or intensity) essentially helps human eye to perceive color. A colorful object in dark doesn't appear colorful at all.

Another method of expressing the colors in an image is *Principal Component Transform* (PCT), which examines all the RGB vectors of an image and finds the linear transformation that aligns the coordinate axes, so that most of the information is along one axis, called the *principal* axis. PCT color spaces help in image compression as we can get about 90% or more of the information into one band. However, in computer vision applications, the images obtained by the systems are readily in RGB form. Thus one may use either RGB format or any other formats, so that the work can be easily ported to other color representation systems using appropriate mathematical transformations.

3.4.4.3 RGB to HSV Color Space Transformation The HSV image may be computed from RGB using different forms of transformations. Some of them are as follows:

- The simplest form of HSV transformation is

$$H = \tan\left[\frac{3(G-B)}{(R-G)+(R-B)}\right], S = 1 - \frac{\min(R,G,B)}{V}, V = \frac{R+G+B}{3}.$$

However, the hue (H) becomes undefined when saturation $S = 0$.

- The most popular form of HSV transformation is shown next, where the r, g, b values are first obtained by normalizing each pixel such that

$$r = \frac{R}{R+G+B}, \quad g = \frac{G}{R+G+B}, \quad b = \frac{B}{R+G+B}.$$

Accordingly, the H, S, and V values can be computed as

$$V = \max(r, g, b), \tag{3.4}$$

$$S = \begin{cases} 0 & \text{if } V = 0 \\ V - \frac{\min(r,g,b)}{V} & \text{if } V > 0, \end{cases} \tag{3.5}$$

$$H = \begin{cases} 0 & \text{if } S = 0 \\ \frac{60*(g-b)}{S*V} & \text{if } V = r \\ 60 * \left[2 + \frac{(b-r)}{S*V}\right] & \text{if } V = g \\ 60 * \left[4 + \frac{(r-g)}{S*V}\right] & \text{if } V = b, \end{cases} \tag{3.6}$$

$$H = \quad H + 360 \quad \text{if } H < 0. \tag{3.7}$$

The results obtained by using either of the above transformations yield reasonably good results.

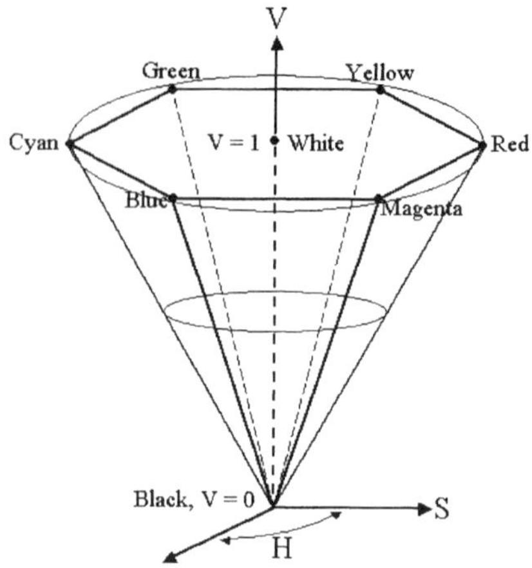

Fig. 3.2 Perceptual representation of HSV color space.

The perceptual representation of the HSV color space has a conical shape, as shown in Figure 3.2. The *Value* (V) varies along the vertical axis of the cone, the *Hue* (H) varies along the periphery of the circle of the cone and is represented as an angle about the vertical axis, and the *Saturation* (S) varies along the radial distance as shown in Figure 3.2. We have shown the vertices a hexagon on the periphery of the circle in Figure 3.2 to show six colors separated by 60° angles. Red is at 0° (coincides with 300°), Yellow at 60°, Green at 120°, Cyan at 180°, Blue at 240°, and Magenta at 300°. The complementary colors are 180° apart. For example, Blue and Yellow are the complementary colors. Apex of the cone ($V = 0$) represents *Black*. Again, $V = 1$ and $S = 0$ represents *White*. The colors have their maximum luminosity ($V = 1$) at the periphery of the circle. $V = 1$ and $S = 1$ represent the *pure* hues for any color.

3.4.5 CIELAB color Space

The CIELAB color space, which was adopted as an international standard in the 1970s, by CIE is indeed a perceptually uniform space. The Euclidean distance between two color points in the CIELAB color space corresponds to the perceptual difference between the two colors by the human vision system. This property of the CIELAB color space has made it particularly attractive and useful for color analysis, and the superiority of the CIELAB color space over other color spaces has been demonstrated in many color image applications. For example the CIELAB color space has been successfully used for color clustering. In this approach the color difference in the CIELAB color space has been used in the computation of the dissimilarities between colors and this has been used for color clustering. The color difference formula in the CIELAB color space is used in the computation of the dissimilarities between colors and the formulation of the color membership function.

3.4.5.1 RGB to CIELAB Color Space Transformation The transformation from RGB to CIELAB is as follows [2, 7]:

$$\begin{aligned} X &= 0.412453R + 0.357580G + 0.180423B \\ Y &= 0.212671R + 0.715160G + 0.072169B \\ Z &= 0.019334R + 0.119193G + 0.950227B \end{aligned}$$

Based on this definition, $L^*a^*b^*$ is defined as

$$\begin{aligned} L^* &= 116f(Y/Y_n) - 16 \\ a^* &= 500[f(X/X_n) - f(Y/Y_n)] \\ b^* &= 200[f(Y/Y_n) - f(Z/Z_n)] \end{aligned}$$

where

$$f(q) = \begin{cases} q^{1/3} & \text{if } q > 0.008856 \\ 7.787q + 16/116 & \text{otherwise} \end{cases} .$$

X_n,Y_n, and Z_n represent a reference white as defined by a CIE standard illuminant, D_{65} in this case, and are obtained by setting $R = G = B = 100$ $(q \in \{\frac{X}{X_n}, \frac{Y}{Y_n}, \frac{Z}{Z_n}\})$.

3.5 COLOR INTERPOLATION OR DEMOSAICING

Due to the cost and packaging consideration, in digital imaging devices such as a digital camera, the image color is captured in a subsampled pattern. Typically each pixel in the captured raw image contains only one of the three primary color components, R (Red), G (Green), or B (Blue). This subsampled color image is generated using certain pattern of a Color Filter Array (CFA). This CFA is realized by coating the surface of the electronic sensor array using some optical material that acts as a band-pass filter. This coating allows the photons corresponding to only one color component to be transmitted to the sensor. A typical and widely used CFA pattern is called Bayer Pattern [8]. In Figure 3.3, we have shown a Bayer pattern image of dimension 8×8. Each cell in Figure 3.3 represents a pixel with only one color component as indicated by either R or G or B.

A full-color image needs the information of all the three colors in each pixel location. As a result, it is essential to interpolate the missing two colors in each pixel location using the information of the neighboring pixels.

	1	2	3	4	5	6	7	8
1	G	R	G	R	G	R	G	R
2	B	G	B	G	B	G	B	G
3	G	R	G	R	G	R	G	R
4	B	G	B	G	B	G	B	G
5	G	R	G	R	G	R	G	R
6	B	G	B	G	B	G	B	G
7	G	R	G	R	G	R	G	R
8	B	G	B	G	B	G	B	G

Fig. 3.3 Bayer pattern.

The methodology to recover or interpolate these missing color components is known as *color interpolation* or *color demosaicing* [9]–[11]. Color interpolation algorithms can be broadly classified into two categories:

1. **Nonadaptive algorithms:** In nonadaptive color interpolation algorithms, a fixed pattern of computation is applied in every pixel location in the subsampled color image to recover the missing two color components.

2. **Adaptive algorithms:** In adaptive color interpolation algorithms, intelligent processing is applied in every pixel location based on the characteristics of the image in order to recover the missing color components. This type of algorithm yields better results in terms of visual image quality as compared with the nonadaptive algorithms.

3.5.1 Nonadaptive Color Interpolation Algorithms

In this section, we review some nonadaptive color interpolation algorithms reported in the literature. These algorithms are simple to implement and their computational requirements are much lower than the nonadaptive algorithms.

3.5.1.1 Nearest Neighbor Replication In this simple method [11, 12], each missing color is approximated by nearest pixel representing that color in the input image. The nearest neighbor can be any one of the upper, lower, left, and right pixels. The only advantage of this approach is that the computational requirement is very small and suitable for applications where speed is very crucial. However, the significant color errors make it unacceptable for a still imaging system, such as high-resolution digital cameras.

3.5.1.2 Bilinear Interpolation In this algorithm [11], a missing color component is interpolated by linear average of the adjacent pixels representing the missing color. As an example, the pixel coordinate (4, 7) in Figure 3.3 contains BLUE component only. Hence the missing GREEN component at pixel coordinate (4, 7) can be estimated as average of the left, right, top, and bottom GREEN pixels at pixel coordinates (4, 6), (4, 8), (3, 7), and (5, 7) respectively. The missing RED component can be estimated as average of the four diagonally adjacent corner RED pixels at coordinates (3, 6), (3, 8), (5, 6), and (5, 8) respectively.

This method is very simple, but the results suffer from pixel artifacts (e.g., zipper effect) introduced in the neighborhood of the interpolated pixels. This may be acceptable in a moderate quality video application because the artifact may not be immediately visible by the human eye due to the effect of motion blur, but is not acceptable in still imaging applications.

3.5.1.3 Median Interpolation Median interpolation [13] allocates the missing color component with the "median" value of the color components in the adjacent pixels, as opposed to the linear average used in bilinear interpolation. This provides a slightly better result in terms of visual quality as compared with the bilinear interpolation. However, the resulting images are still blurry and not acceptable for high-quality imagery.

3.5.1.4 Smooth Hue Transition Interpolation The key problem in both bilinear and median interpolation is that the hue values of adjacent pixels change suddenly because of the nature of these algorithms. On the other hand, the Bayer pattern can be considered as a combination of a luminance channel (green pixels, because it green contains mostly luminous information) and two chrominance channels (red and blue pixels). The smooth hue transition interpolation algorithm [14] treats these channels differently. The missing GREEN component in every RED and BLUE pixel in the Bayer pattern can first be interpolated using bilinear interpolation. The idea of chrominance channel interpolation is to impose a smooth transition in hue value from pixel to pixel. In order to do so, it defines "blue hue" as B/G, and "red hue" as R/G. To interpolate the missing BLUE component $B_{m,n}$ in pixel location (m, n) in the Bayer pattern, the following three cases may arise:

Case 1: The pixel at location (m, n) is GREEN and the adjacent left and right pixels are BLUE color. Pixel at location (2, 2) in Figure 3.3 is such an example. The missing BLUE component in location (m, n) can be estimated as

$$B_{m,n} = \frac{G_{m,n}}{2}\left[\frac{B_{m,n-1}}{G_{m,n-1}} + \frac{B_{m,n+1}}{G_{m,n+1}}\right].$$

Case 2: The pixel at (m, n) is GREEN and the adjacent top and bottom pixels are BLUE. The pixel at (3, 3) in Figure 3.3 is such an example.

The missing BLUE component can be estimated as

$$B_{m,n} = \frac{G_{m,n}}{2}\left[\frac{B_{m-1,n}}{G_{m-1,n}} + \frac{B_{m+1,n}}{G_{m+1,n}}\right].$$

Case 3: The pixel at (m, n) is RED and four diagonally neighboring corner pixels are BLUE. The pixel at location (3, 2) in Figure 3.3 is such an example. The missing BLUE component in location (m, n) can be estimated as

$$B_{m,n} = \frac{G_{m,n}}{4}\left[\frac{B_{m-1,n-1}}{G_{m-1,n-1}} + \frac{B_{m-1,n+1}}{G_{m-1,n+1}} + \frac{B_{m+1,n-1}}{G_{m+1,n-1}} + \frac{B_{m+1,n+1}}{G_{m+1,n+1}}\right].$$

We can interpolate the missing RED component in each location in a similar fashion. In any color image processing system, after the raw image is captured by an electronic sensor it goes through several image processing steps before it can be displayed or stored for usage. As explained in [11], if the captured image is transformed into logarithmic exposure space from linear space before interpolation, instead of B/G or R/G, one can now define the "hue value" as $B - G$ or $R - G$, because of the logarithmic property $\log(X/Y) = \log(X) - \log(Y) = X' - Y'$, where $X' = \log(X)$ and $Y' = \log(Y)$. Since all the division for calculating hue value is replaced by subtraction, this helps reduce computational complexity for implementation.

3.5.2 Adaptive algorithms

3.5.2.1 Pattern Matching-Based Interpolation Algorithm In the Bayer pattern, a BLUE or RED pixel has four neighboring GREEN pixels. A simple pattern matching technique for reconstructing the missing color components based on the pixel contexts was proposed in [15]. This pattern matching algorithm defines a *green pattern* for the pixel at location (m, n) containing a non-GREEN color component as a four-dimensional integer-valued vector:

$$g_{m,n} = (G_{m-1,n}, G_{m+1,n}, G_{m,n-1}, G_{m,n+1}).$$

The similarity (or difference) between two green patterns g_1 and g_2 is defined as the vector 1-norm

$$\|g_1 - g_2\| = \sum_{i=0}^{3} |g_{1i} - g_{2i}|.$$

When the difference between two green patterns is small, it is likely that the two pixel locations where the two green patterns are defined will have similar RED and BLUE color components.

A weighted average proportional to degree of similarity of the green patterns is used to calculate the missing color component. For example if the pixel at location (m, n) is RED, the missing BLUE component $B_{m,n}$ is estimated by comparing the green pattern $g_{m,n}$ with the four neighboring green patterns $g_{m-1,n-1}$, $g_{m+1,n-1}$, $g_{m-1,n+1}$, and $g_{m+1,n+1}$. If the difference between $g_{m,n}$ and each of the four neighboring green patterns is uniformly small (below a certain threshold), then a simple average is used to estimated the missing BLUE color component,

$$B_{m,n} = \frac{B_{m-1,n-1} + B_{m+1,n-1} + B_{m-1,n+1} + B_{m+1,n+1}}{4}.$$

Otherwise, only the top two best-matched green pattern information items are used. For example, if $\|g(m,n) - g(m-1,n-1)\|$ and $\|g(m,n) - g(m+1,n-1)\|$ are the two smallest differences, then the missing BLUE color is estimated as

$$B_{m,n} = \frac{B_{m-1,n-1}\Delta_1 + B_{m+1,n-1}\Delta_2}{\Delta_1 + \Delta_2},$$

where $\Delta_1 = \|g(m,n) - g(m+1,n-1)\|$ and $\Delta_2 = \|g(m,n) - g(m-1,n-1).\|$

The missing RED components can be interpolated in a similar fashion.

3.5.2.2 A Block Matching Algorithm Acharya, et al. [9] defined a block matching algorithm for color interpolation based on a concept of *Color Block*. The *color block* of a non-green pixel is defined as a set $x = (x_1, x_2, x_3, x_4)$ formed by the four neighboring green pixels, say, x_1, x_2, x_3, and x_4. We define a new metric *Color Gravity* as the mean $\bar{x} = (x_1 + x_2 + x_3 + x_4)/4$. The similarity between two color blocks is defined as the absolute difference of their color gravities. The block matching algorithm is developed based on the selection of a neighboring color block whose color gravity is closest to the color gravity of the color block under consideration.

For any non-green pixel in the Bayer pattern image, there are four neighboring green pixels G_N (the North neighbor), G_S (South), G_E (East), and G_W (the West neighbor), which form the *color block* $g = (G_N, G_S, G_E, G_W)$ and its *color gravity* is $\bar{g} = (G_N + G_S + G_E + G_W)/4$. The missing GREEN value is simply computed by the median of G_N, G_S, G_E, and G_W. If the pixel at (m, n) is BLUE, it will have four diagonally RED pixels R_{NE} (the Northeast neighbor), R_{SE} (the Southeast neighbor), R_{SW} (the Southwest neighbor), and R_{NW} (the Northwest neighbor) whose color blocks are g_{NE}, g_{SE}, g_{SW}, and g_{NW} and the corresponding color gravities are $\bar{g}_{NE}$, $\bar{g}_{SE}$, $\bar{g}_{SW}$, and $\bar{g}_{NW}$ respectively. The missing RED component is assumed to be one of these four diagonal red pixels based on best match of their color gravities. The best match (or minimal difference $\Delta_{\min}$) is the minimum of $\Delta_1 = |\bar{g} - \bar{g}_{NE}|$, $\Delta_2 = |\bar{g} - \bar{g}_{SE}|$, $\Delta_3 = |\bar{g} - \bar{g}_{SW}|$, and $\Delta_4 = |\bar{g} - \bar{g}_{NW}|$. Similarly, we can estimate the missing BLUE component in a RED pixel location due to the symmetry of red and blue sampling position in a Bayer pattern image. For the green pixel location, only two color blocks (either up-bottom or left-right positions) are considered for the missing RED or BLUE color. The algorithm can be described as:

```
Begin
    for each pixel in the Bayer pattern image do
    {
      if the pixel at location (m, n) is not GREEN then
      {
        G_{m,n} ← median{G_N, G_S, G_E, G_W};
        Δ1 = |ḡ − ḡ_NE|;
        Δ2 = |ḡ − ḡ_SE|;
        Δ3 = |ḡ − ḡ_SW|;
        Δ4 = |ḡ − ḡ_NW|;

        Δmin = min{Δ1, Δ2, Δ3, Δ4};

        if the pixel at location (m, n) is RED then
        {
          if (Δmin = Δ1 ) then B_{m,n} ← B_NE;
          if (Δmin = Δ2 ) then B_{m,n} ← B_SE;
          if (Δmin = Δ3 ) then B_{m,n} ← B_SW;
          if (Δmin = Δ4 ) then B_{m,n} ← B_NW;
        }

        if the pixel at location (m, n) is BLUE then
        {
          if (Δmin = Δ1 ) then R_{m,n} ← R_NE;
          if (Δmin = Δ2 ) then R_{m,n} ← R_SE;
          if (Δmin = Δ3 ) then R_{m,n} ← R_SW;
          if (Δmin = Δ4 ) then R_{m,n} ← R_NW;
        }
      }

      if the pixel at location (m, n) is GREEN then
      {
        Δu = |G − ḡ_u| , Δb = |G − ḡ_b| , Δl = |G − ḡ_l|, Δr = |G − ḡ_r|;

        if (Δu < Δb) then B_{m,n} ← B_u else B_{m,n} ← B_b;
        if (Δl < Δr) then R_{m,n} ← R_l else R_{m,n} ← R_r;
      }
    }
End.
```

This algorithm provides a much sharper image as compared to the median, or bilinear interpolation, or the simple pattern matching. However, since it does not consider smooth hue transition, the color bleeding artifacts still remains a problem for some images containing sharp edges, such as the image of a Zebra.

3.5.2.3 Edge Sensing Interpolation In this algorithm, different predictors are used for the missing GREEN values depending on the luminance gradients [11, 16]. For each pixel containing only the RED or the BLUE component, two gradients (one in the horizontal direction, the other in the vertical direction) are defined as

$$\delta H = |G_{m,n-1} - G_{m,n+1}| \text{ and } \delta V = |G_{m-1,n} - G_{m+1,n}|,$$

where $|x|$ denotes absolute value of x. Based on these gradients and a certain threshold (T), the interpolation algorithm can be described as follows:

begin
  if $\delta H < T$ and $\delta V > T$ (i.e. smoother in horizontal direction) **then**
    $G_{m,n} \leftarrow (G_{m,n-1} + G_{m,n+1})/2;$

  else if $\delta H > T$ and $\delta V < T$ (i.e. smoother in vertical direction) **then**
    $G_{m,n} \leftarrow (G_{m-1,n} + G_{m+1,n})/2;$

  else
    $G_{m,n} \leftarrow (G_{m,n-1} + G_{m,n+1} + G_{m-1,n} + G_{m+1,n})/4;$
end

A slightly different edge sensing interpolation algorithm is described in [17]. Instead of luminance gradients, chrominance gradients are used. For a pixel location containing only the BLUE component, the two gradients are

$$\delta H = \left| B_{m,n} - \frac{B_{m,n-2} + B_{m,n+2}}{2} \right| \text{ and } \delta V = \left| B_{m,n} - \frac{B_{m-2,n} + B_{m+2,n}}{2} \right|.$$

Similar treatment can be done for the RED pixels as well.

3.5.2.4 Linear Interpolation with Laplacian Second-Order Corrections This algorithm [18] was developed with the goal of enhanced visual quality of the interpolated image when applied on images with sharp edges. Missing color components are estimated by the following steps.

Estimation of GREEN component: Consider estimating the missing GREEN component $(G_{m,n})$ at a BLUE pixel $(B_{m,n})$ at location (m, n), as an example. Interpolation at a RED pixel location can be done in the similar fashion. We define horizontal and vertical gradients in this pixel location as follows:

$$\delta H = |G_{m,n-1} - G_{m,n+1}| + |(B_{m,n} - B_{m,n-2}) - (B_{m,n+2} - B_{m,n})|$$

and

$$\delta V = |G_{m-1,n} - G_{m+1,n}| + |(B_{m,n} - B_{m-2,n}) - (B_{m+2,n} - B_{m,n})|.$$

Intuitively, we can consider δH and δV as combination of the luminance gradient and the chrominance gradient as described in the edge sensing interpolation algorithm in the previous section. In the expression of δH, the first term $|G_{m,n-1} - G_{m,n+1}|$ is the first-order difference of the neighboring green pixels, considered to be the luminance gradient and the second term $|(B_{m,n} - B_{m,n-2}) - (B_{m,n+2} - B_{m,n})|$ is the second-order derivative of the neighboring blue pixels, considered as the chrominance gradient. Using these two gradients, the missing green component $G_{m,n}$ at location (m, n) is estimated as follows.

if $\delta H < \delta V$ **then**
$G_{m,n} \leftarrow (G_{m,n-1} + G_{m,n+1})/2 + (2B_{m,n} - B_{m,n-2} - B_{m,n+2})/4;$

else if $\delta H > \delta V$ **then**
$G_{m,n} = (G_{m-1,n} + G_{m+1,n})/2 + (2B_{m,n} - B_{m-2,n} - B_{m+2,n})/4;$

else
$$\begin{aligned} G_{m,n} = {} & (G_{m,n-1} + G_{m,n+1} + G_{m-1,n} + G_{m+1,n})/4 + \\ & (4B_{m,n} - B_{m,n-2} - B_{m,n+2} - B_{m-2,n} - B_{m+2,n})/8; \end{aligned}$$
endif

Hence, the interpolation for missing GREEN component has two parts. The first part is the linear average of the neighboring green values, and the second part can be considered as a second-order correction term based on the neighboring blue (red) components.

Estimation of RED (or BLUE) component: The missing RED (or BLUE) color components are estimated in every pixel location after estimation of the missing green components is complete. Depending on the position, we have three cases:

1. Estimate the missing RED (BLUE) component at a GREEN pixel ($G_{m,n}$), where nearest neighbors of RED (BLUE) pixels are in the same column, e.g., pixel location (4, 4) as shown in Figure 3.3. The missing RED (BLUE) component is estimated as

$$R_{m,n} \leftarrow \left[\frac{R_{m-1,n} + R_{m+1,n}}{2} + \frac{2G_{m,n} - G_{m-1,n} - G_{m+1,n}}{4} \right].$$

2. Estimate the missing RED (BLUE) component at a GREEN pixel ($G_{m,n}$), where nearest neighbors of RED (BLUE) pixels are in the same row, e.g., pixel location (3, 3) as shown in Figure 3.3. The missing RED (BLUE) component is estimated as

$$R_{m,n} \leftarrow \left[\frac{R_{m,n-1} + R_{m,n+1}}{2} + \frac{2G_{m,n} - G_{m,n-1} - G_{m,n+1}}{4} \right].$$

3. Estimate RED (BLUE) color component at a BLUE (RED) pixel, e.g. the RED pixel at (3, 4) as shown in Figure 3.3. Here we first define two diagonal gradients as

$$\delta D_1 = |R_{m-1,n-1} - R_{m+1,n+1}| + |(G_{m,n} - G_{m-1,n-1}) + (G_{m,n} - G_{m+1,n+1})|,$$

and

$$\delta D_2 = |R_{m-1,n+1} - R_{m+1,n-1}| + |(G_{m,n} - G_{m-1,n+1}) + (G_{m,n} - G_{m+1,n-1})|$$

Using these diagonal gradients, the algorithm for estimating the missing color components is described as :

if $\delta D_1 < \delta D_2$ **then**

$$R_{m,n} \leftarrow \frac{R_{m-1,n-1}+R_{m+1,n+1}}{2} + \frac{2G_{m,n}-G_{m-1,n-1}-G_{m+1,n+1}}{2};$$

else if $\delta D_1 > \delta D_2$ **then**

$$R_{m,n} \leftarrow \frac{R_{m-1,n+1}+R_{m+1,n-1}}{2} + \frac{2G_{m,n}-G_{m-1,n+1}-G_{m+1,n-1}}{2};$$

else

$$R_{m,n} \leftarrow \frac{R_{m-1,n-1}+R_{m+1,n+1}+R_{m-1,n+1}+R_{m+1,n-1}}{4} + \frac{4G_{m,n}-G_{m-1,n-1}-G_{m+1,n+1}-G_{m-1,n+1}-G_{m+1,n-1}}{4};$$

end.

This algorithm provides much better visual quality of the reconstructed image containing a lot of sharp edges. However, the second-order derivative for calculating the gradients makes the algorithm quite sensitive to noise. Since only the color information in the same direction (vertical, horizontal, or one of the diagonal directions based on the gradient information) is used for interpolation, we believe that it is still possible to further improve the visual quality of the reconstructed image.

3.5.3 A Novel Adaptive Color Interpolation Algorithm

In fuzzy membership assignment strategy, we assign different membership values to members of a set to reflect their degrees of belongingness within that set [19]. Depending upon the correlation among the surrounding pixels, a fuzzy membership assignment to the surrounding horizontal and vertical pixels has been formulated in [10]. The membership grades have been experimentally derived through exhaustive subjective visual inspection, taking into consideration the exhaustive set of images having possible edges in the horizontal and vertical directions.

Let us now consider the estimation of the GREEN component ($G_{m,n}$) at the pixel location (m, n) containing the RED component ($R_{m,n}$) only. The following four cases may arise, where there is a possible edge along the horizontal direction:

1. $|G_{m,n-1} - G_{m,n+1}|$ is small while $|G_{m-1,n} - G_{m+1,n}|$ is arbitrarily large. Here we can assume existence of a horizontal edge.

2. $|G_{m,n-1} - G_{m,n+1}|$ is small and $|G_{m-1,n} - G_{m+1,n}|$ is arbitrary and $G_1m, n-1 \approx G_{m,n+1} \approx G_{m+1,n}$. In this case, there is clearly a possible edge at the pixel location $R_{m,n}$, and the intensity of this edge depends on the surrounding pixel values $G_{m-1,n}$ and $G_{m+1,n}$.

3. This case is similar to case 2 with a difference $G_1m, n-1 \approx G_{m,n+1} \approx G_{m-1,n}$.

4. In this case we consider all the four connecting neighboring pixels $G_{m,n-1}$, $G_{m,n+1}$, $G_{m-1,n}$ and $G_{m+1,n}$ that are all different subject to the condition $|G_{m-1,n} - G_{m+1,n}| - |G_{m,n-1} - G_{m,n+1}| \gg 0$.

In each of these cases, the fuzzy membership value of 0.5 has been assigned to the horizontal green pixels $G_{m,n-1}$ and $G_{m,n+1}$ and 0.1 has been assigned to vertical green pixels $G_{m-1,n}$ and $G_{m+1,n}$. Based on this strategy, the missing green component ($G_{m,n}$) can be interpolated as follows:

$$G_{m,n} = \frac{0.5G_{m,n-1} + 0.5G_{m,n+1} + 0.1G_{m-1,n} + 0.1G_{m+1,n}}{0.5 + 0.5 + 0.1 + 0.1},$$

that is

$$G_{m,n} = 0.8333\frac{G_{m,n-1} + G_{m,n+1}}{2} + 0.1667\frac{G_{m-1,n} + G_{m+1,n}}{2}.$$

Using this fuzzy membership assignment as a weighted-average tool for missing color interpolation, we can fully utilize all the neighboring information for estimating the missing color information.

3.5.3.1 Three Steps of the Interpolation Algorithm The proposed interpolation algorithm is a three-step algorithm as summarized here:

1. Estimation of all missing GREEN components. After completion of this step, each and every pixel location has a green component.

2. Estimation of missing BLUE (or RED) component at each RED (or BLUE) pixel. The GREEN components estimated in the previous step are used in this step. The decision is based on the change of hue values.

3. Estimation of missing RED and BLUE at GREEN pixels in the Bayer pattern. The estimated RED (or BLUE) at BLUE (or RED) pixels in step 2 have been utilized for interpolation.

Estimation of all missing GREEN components: Let us consider estimation of the missing GREEN component ($G_{m,n}$) at a RED pixel ($R_{m,n}$) location (m, n), e.g., pixel coordinate (3, 4) in Figure 3.3. First, we estimate two parameters in terms of changes in the Hue values, one in the horizontal direction and the other in the vertical direction. The change in hue value in the horizontal direction can be estimated as

$$C_{hor} \equiv (R_{m,n+1} - G_{m,n+1}) - (R_{m,n-1} - G_{m,n-1}).$$

Since the RED components $R_{m,n+1}$ and $R_{m,n-1}$ are missing in pixel coordinates (m, $n+1$) and (m, $n-1$), we approximate them as

$$R_{m,n+1} \approx \frac{R_{m,n} + R_{m,n+2}}{2}, \quad \text{and } R_{m,n-1} \approx \frac{R_{m,n} + R_{m,n-2}}{2}.$$

Hence C_{hor} can be approximated as

$$C_{hor} \approx \left(\frac{R_{m,n} + R_{m,n+2}}{2} - G_{m,n+1}\right) - \left(\frac{R_{m,n} + R_{m,n-2}}{2} - G_{m,n-1}\right)$$

and hence

$$C_{hor} \approx \frac{1}{2}(-R_{m,n-2} + 2G_{m,n-1} - 2G_{m,n+1} + R_{m,n+2}).$$

In a similar fashion the change in hue value in the vertical direction can be estimated as

$$C_{ver} \approx \frac{1}{2}(-R_{m-2,n} + 2G_{m-1,n} - 2G_{m+1,n} + R_{m+2,n}).$$

Depending on the values of C_{hor} and C_{ver}, different fuzzy membership numbers are used as weighting factors to estimate the missing GREEN components as described here:

if ($|C_{hor}| < |C_{ver}|$) **then**

$$G_{m,n} = 0.8333\, X_{hor} + 0.1667\, X_{ver};$$

else if ($|C_{hor}| > |C_{ver}|$) **then**

$$G_{m,n} = 0.1667\, X_{hor} + 0.8333\, X_{ver};$$

else

$$G_{m,n} = 0.5\, X_{hor} + 0.5\, X_{ver};$$

endif

where

$$X_{hor} = \left(\frac{G_{m-1,n} + G_{m+1,n}}{2}\right) + 0.5\left(\frac{-R_{m-2,n} + 2R_{m,n} - R_{m+2,n}}{4}\right),$$

$$X_{ver} = \left(\frac{G_{m,n-1} + G_{m,n+1}}{2}\right) + 0.5\left(\frac{-R_{m,n-2} + 2R_{m,n} - R_{m,n+2}}{4}\right).$$

It should be noted that X_{hor} and X_{ver} above are estimated based on the assumption of smooth hue transition as discussed earlier with a weighting factor (0.5) in the second term to reduce the sensitivity of noise in the image.

The similar strategy can be applied for interpolation of the missing GREEN component at the pixel coordinates containing the BLUE component only.

Estimation of BLUE/RED components at the RED/BLUE pixels: Let us consider estimation of the missing BLUE component ($B_{m,n}$) at RED pixel ($R_{m,n}$), e.g., coordinate (3, 4) in Figure 3.3. Since the green components have already been interpolated in step 1, the hue values of the four corner BLUE pixels at $(m-1,\ n-1)$, $(m-1,\ n+1)$, $(m+1,\ n-1)$, $(m+1,\ n+1)$ are

$$H_{nw} = B_{m-1,n-1} - G_{m-1,n-1}, \quad H_{sw} = B_{m+1,n-1} - G_{m+1,n-1},$$

$$H_{ne} = B_{m-1,n+1} - G_{m-1,n+1}, \quad H_{se} = B_{m+1,n+1} - G_{m+1,n+1}$$

respectively and the differences of the hues along the diagonals are

$$\Delta H_1 = H_{nw} - H_{se}, \text{ and } \Delta H_2 = H_{ne} - H_{sw}$$

The procedure for estimation of the missing BLUE component in the RED pixel location in the Bayer pattern is as follows.

if $(|\Delta H_1| < |\Delta H_2|)$ **then**
  $B_{m,n} \leftarrow G_{m,n} + 0.8333X_{Diag1} + 0.1667X_{Diag2};$
else
  $B_{m,n} \leftarrow G_{m,n} + 0.1667X_{Diag1} + 0.8333X_{Diag2};$
endif

where

$$X_{Diag1} = \frac{B_{m-1,n-1} - G_{m-1,n-1} + B_{m+1,n+1} - G_{m+1,n+1}}{2}$$

and

$$X_{Diag2} = \frac{B_{m-1,n+1} - G_{m-1,n+1} + B_{m+1,n-1} - G_{m+1,n-1}}{2}$$

The RED component at a BLUE pixel location is interpolated similarly.

Estimation of missing BLUE/RED components at GREEN pixels: Let us now consider estimation of the missing BLUE component ($B_{m,n}$) at GREEN pixel ($G_{m,n}$), e.g., at coordinate (3, 3) in Figure 3.3. We already interpolated the BLUE components in the pixel coordinates containing the RED component only and vice versa. The blue hue of the north, south, east, and west neighbors of the GREEN pixel at $(m,\ n)$ are

$$H_n = B_{m-1,n} - G_{m-1,n}, \quad H_s = B_{m+1,n} - G_{m+1,n},$$

$$H_e = B_{m,n+1} - G_{m,n+1}, \quad H_w = B_{m,n-1} - G_{m,n-1}$$

respectively. Difference of the hues in horizontal and vertical directions are

$$\Delta H_{hor} = H_e - H_w, \text{ and } \Delta H_{ver} = H_n - H_s$$

The procedure to interpolate the missing BLUE component at the pixel coordinate containing only the GREEN component in the Bayer pattern is as follows:

if $(|\Delta H_{hor}| < |\Delta H_{ver}|)$ **then**

$$B_{m,n} = G_{m,n} + 0.8333\,(B_{m,n-1} - G_{m,n-1} + B_{m,n+1} - G_{m,n+1})/2$$

$$+0.1667\,(B_{m-1,n} - G_{m-1,n} + B_{m+1,n} - G_{m+1,n})/2;$$

else

$$B_{m,n} = G_{m,n} + 0.1667\,(B_{m,n-1} - G_{m,n-1} + B_{m,n+1} - G_{m,n+1})/2$$

$$+0.8333\,(B_{m-1,n} - G_{m-1,n} + B_{m+1,n} - G_{m+1,n})/2;$$

endif

The missing RED component can be interpolated in a similar fashion.

3.5.4 Experimental Results

To show the effectiveness of the above color interpolation algorithm, we can synthetically generate the Bayer pattern from a full-color RGB image by simply dropping two color components in each pixel coordinate (if the original Bayer pattern image cannot be grabbed from a digital camera directly). We show the results of the above color interpolation in Figure 3.4. the color version of the figure is provided in the color pages section. Figure 3.4(a) is a full color image. The image contains lots of high-frequency patterns in the form of black-and-white sharp edges in different angles. We generated the Bayer pattern image from the original image in Figure 3.4(a) by simply dropping two color components in each pixel to show the results. This synthetically generated Bayer pattern image is then interpolated to the full-color image using the adaptive algorithm discussed above. The interpolated result is shown in Figure 3.4(b). We have chosen this particular image because of its high-frequency nature and other interesting image characteristics. The comparative studies of different color interpolation algorithms with number of different adaptive and non-adaptive color interpolation algorithms have been presented in [10]. In all different types of images, the adaptive algorithm discussed here provides better image quality compared to others.

(a)

(b)

Fig. 3.4 (a) Original Image and (b) result of fuzzy assignment-based adaptive interpolation.

3.6 SUMMARY

In this chapter, we have presented a concise description of the human color perception. The need for color space transformation has been explained and some of the popular color spaces such as RGB, HSV, CIELAB, etc. and the formulation for their transformations have been presented.

Due to the cost and packaging consideration, in digital imaging devices such as a digital camera, only a single electronic sensor is used and the need for color interpolation or demosaicing will remain critical until other technologies such as multi-channel color moir free sensor is mature. In this chapter, we have introduced the concept of color interpolation, defined the problem and reviewed different types of color interpolation algorithms proposed in the literature. We discussed an interesting technique based on fuzzy membership assignment strategy along with the concept of smooth hue transition for estimating the missing colors in each pixel. This algorithm significantly improves the overall visual quality of the reconstructed color images.

REFERENCES

1. E. H. Land, "Color vision and the natural image," Part I, *Proceedings of the National Academy of Sciences, USA*, 45, 1959, 116–129.

2. G. Wyszecki and W. S. Stiles, *Color Science*, 2nd ed., McGraw-Hill, New York, 1982.

3. B. A. Wandell, *Foundations of Vision*, Sinauer Associates, Inc., Sunderland, MA, 1995.

4. G. Buchsbaum,"An analytical derivation of visual nonlinearity," IEEE Transactions on biomedical engineering. vol-BME-27, 5, 237–242, May 1980.

5. K. M. Bhurchandi, A. K. Ray and P. M. Nawghare, "An Analytical Approach for Sampling the RGB Color Space Considering Physiological Limitations of Human Vision and its Application for Color Image Analysis," *Proc. Indian Conference on Computer Vision, Graphics and Image Processing*, Bangalore, December 2000, 44–49.

6. G. Sharma, "Digital color imaging," *IEEE Transactions on image processing*, 6(7), July 1997, 901–932.

7. H. R. Kang, *Color Technology for Electronic Imaging Devices*, SPIE Optical Engineering Press, 1997.

8. B. E. Bayer, "Color Imaging Array," *US Patent 3,971,065*, Eastman Kodak Company, 1976.

9. T. Acharya and P. Tsai, "A New Block Matching Based Color Interpolation Algorithm," *Proceedings of the SPIE Electronic Imaging Conference, Color Imaging: Device-Independent Color, Color Hardcopy, and Graphic Arts IV*, Vol. 3648, January 1999, 60-65.

10. P. Tsai, T. Acharya, and A. K. Ray, "Adaptive Fuzzy Color Interpolation," *Journal of Electronic Imaging*, 11(3), July 2002, 293-305.

11. J. E. Adams, Jr., "Interactions between color plane interpolation and other image processing functions in electronic photography," *Proceedings of the SPIE Electronic Imaging Conference*, Vol. 2416, 144–151, 1995.

12. T. Sakamoto, C. Nakanishi, and T. Hase, "Software Pixel Interpolation for Digital Still Cameras Suitable for A 32-bit MCU," *IEEE Transactions on Consumer Electronics*, 44(4), November 1998, 1342-1352.

13. W. T. Freeman, "Method and apparatus for reconstructing missing color samples," *U.S. Patent 4,663,655*, Polaroid Corporation, 1987.

14. D. R. Cok, "Signal processing method and apparatus for producing interpolated chrominance values in a sampled color image signal," *U.S. Patent 4,642,678*, Eastman Kodak Company, 1987.

15. X. Wu, W. K. Choi, and P. Bao, "Color Restoration from Digital Camera Data by Pattern Matching," *Proceedings of the SPIE's Electronic Imaging Conference, Color Imaging: Device-Independent Color, Color Hardcopy, and Graphic Arts II*, Vol. 3018, 12–17, 1997.

16. R. H. Hibbard, "Apparatus and method for adaptively interpolating a full color image utilize luminance gradients," *U.S. Patent 5,382,976*, Eastman Kodak Company, 1995.

17. G. A. Laroche and M. A. Prescott, "Apparatus and method for adaptively interpolating a full color image utilize chrominance gradients," *U.S. Patent 5,373,322*, Eastman Kodak Company, 1994.

18. J. F. Hamilton, Jr. and J. E. Adams, Jr., "Adaptive color plan interpolation in single sensor color electronic camera," *U.S. Patent 5,629,734*, Eastman Kodak Company, 1997.

19. L. A. Zadeh, "Fuzzy Sets," *Information and Control*, 8, 1965, 338–353.

4

Image Transformation

4.1 INTRODUCTION

Two-dimensional image transforms are extremely important areas of study in image processing [1, 2]. The image output in the transformed space may be analyzed, interpreted, and further processed for implementing diverse image processing tasks. These transformations are widely used, since by using these transformations, it is possible to express an image as a combination of a set of basic signals, known as the basis functions. In case of Fourier transform of an image these basis signals are sinusoidal signals with different periods which describe the spatial frequencies in an image. This implies that an image is decomposed into its constituent sinusoids, using the Fourier transform, and the amplitudes of various frequencies constitute the frequency spectrum of the image. The process of inverse Fourier transform operation involves synthesizing the image by adding up its constituent frequencies. The notion of frequency, more specifically spatial frequency, is not a mere mathematical abstraction. On the other hand, interestingly the human vision system, which is a biological system, essentially performs the frequency analysis of the image incident on the retina of our eyes. Thus such transforms, such as the Fourier transform, reveal spectral structures embedded in the image that may be used to characterize the image.

4.2 FOURIER TRANSFORMS

The understanding of the formation and analysis of two-dimensional signals, viz., images, has been possible because of the advent of various orthogonal transforms. Fourier transform is one of the most important tools which have been extensively used not only for understanding the nature of an image and its formation but also for processing the image. We have already mentioned that an image is a two-dimensional signal and can be viewed as a surface in two-dimensional space. Using Fourier transform, it has been possible to analyze an image as a set of spatial sinusoids in various directions, each sinusoid having a precise frequency. But before we venture into understanding Fourier transform on an image, let us first look into the Fourier transform of a continuous valued one-dimensional signal.

4.2.1 One-Dimensional Fourier Transform

The one-dimensional *continuous Fourier transform* (CFT) of a continuous function $f(x)$ is

$$F(\omega) = \int_{-\infty}^{+\infty} f(x) \exp\left[-j\,2\pi\,\omega x\right]\,dx. \tag{4.1}$$

The corresponding inverse Fourier transform is

$$f(x) = \int_{-\infty}^{+\infty} F(\omega) \exp\left[j\,2\pi\,\omega x\right]\,d\omega. \tag{4.2}$$

Eq. 4.1 can be decomposed into a real component $R(\omega)$ and an imaginary component $I(\omega)$ as

$$F(\omega) = R(\omega) + jI(\omega). \tag{4.3}$$

The magnitude function $|F(\omega)|$ is called the *Fourier Spectrum* of the function $f(x)$ and is denoted as

$$|F(\omega)| = \sqrt{R^2(\omega) + I^2(\omega)}. \tag{4.4}$$

The multiplication of a function $f(x)$ with $\exp\left[-j\,2\pi\,\omega x\right]\,dx$ and integrating the product over the entire x results in a function of the parameter ω.

The phase angle $\Phi(\omega)$ of the function $f(x)$ is denoted by

$$\Phi(\omega) = \tan^{-1}\left[\frac{I(\omega)}{R(\omega)}\right]. \tag{4.5}$$

Example: Let us consider a simple one-dimensional rectangular function

$$f(x) = rect\left(\frac{x}{a}\right) = \begin{cases} 1 & \text{when } x = 0 \\ 0 & \text{otherwise} \end{cases}.$$

The Fourier transform of the above signal (rectangular function) may be computed from Eq. 4.1 as

$$F(\omega) = \int_{-\infty}^{+\infty} rect(\frac{x}{a}) \exp\left[-j\, 2\pi\, \omega x\right] dx = \frac{\sin(a)}{a},$$

which is popularly known as a *Sync function.*

4.2.2 Two-Dimensional Fourier Transform

Extending the concept of one-dimensional Fourier transform, the two-dimensional Fourier transform of a continuous function $f(x, y)$ is denoted by

$$F(\omega, \psi) = \int_{-\infty}^{+\infty} \int_{-\infty}^{+\infty} f(x, y) \exp\left[-j\, 2\pi\, (\omega x + \psi y)\right] dy\, dx. \tag{4.6}$$

The operation of multiplication of a two-dimensional function $f(x, y)$ with $\exp\left[-j\, 2\pi\, (\omega x + \psi y)\right]$ results in some interesting observations. Using Euler's formula the exponential function can be decomposed as

$$\exp\left[-j\, 2\pi\, (\omega x + \psi y)\right] = \cos\left(2\, \pi(\omega x + \psi y)\right) \; - j\, \sin\left(2\, \pi(\omega x + \psi y)\right).$$

This implies that the function $f(x, y)$ is essentially multiplied by the terms $\cos(2\,\pi\,\omega x)\cos(2\,\pi\,\psi y)$, $\sin(2\,\pi\,\omega x)\sin(2\,\pi\,\psi y)$, $\sin(2\,\pi\,\omega x)\cos(2\,\pi\,\psi y)$, and $\sin(2\,\pi\,\omega x)\sin(2\,\pi\,\psi y)$. If the function $f(x, y)$ is a doubly symmetric function along both the X and Y directions, then the Fourier transform of $f(x, y)$ involves only the multiplication of the $\cos(2\,\pi\,\omega x)\cos(2\,\pi\,\psi y)$ term. For a general nonsymmetric two-dimensional function $f(x, y)$ as in most of the real-life images, the multiplication will involve all four terms.

The integral $F(\omega, \psi)$ thus yields the results of limit summation of an infinite number of sin and cos terms. The variable ω in Eq. 4.6 indicates the frequency, i.e., the number of waves per unit length in X direction, and ψ indicates the number of waves along the Y direction. For a certain pair of values of these frequency components the integral yields just the amplitude of the chosen component.

Accordingly, the magnitude spectrum and phase angle of the two-dimensional function $f(x, y)$ are

$$|F(\omega, \psi)| = \sqrt{R^2(\omega, \psi) + I^2(\omega, \psi)} \tag{4.7}$$

and

$$\Phi(\omega, \psi) = \tan^{-1}\left[\frac{I(\omega, \psi)}{R(\omega, \psi)}\right] \tag{4.8}$$

respectively. It should be noted that the power spectrum of $f(x, y)$ may be denoted as

$$P(\omega, \psi) = |F(\omega, \psi)|^2 = R^2(\omega, \psi) + I^2(\omega, \psi). \tag{4.9}$$

The corresponding inverse two-dimensional Fourier transform is

$$f(x,y) = \int_{-\infty}^{+\infty} \int_{-\infty}^{+\infty} F(\omega,\psi) \exp\left[j\,2\pi\left(\omega x + \psi y\right)\right] d\psi\, d\omega. \tag{4.10}$$

4.2.3 Discrete Fourier Transform (DFT)

When the function or signal is represented in discrete form using a sequence of discrete samples such as $f(x) = \{f(0), f(1), \cdots, f(N-1)\}$, the corresponding Fourier Transform of the discrete signal is the *Discrete Fourier Transform* (DFT). Since the signal is discretized, the operation of integration in *continuous Fourier transform* (CFT) is replaced by summation operations in DFT. We present the one-dimensional DFT and also the two-dimensional DFT in the following subsections.

4.2.3.1 One-Dimensional DFT The one-dimensional discrete Fourier transform of a function $f(x)$ of size N with integer index x running from 0 to $N-1$, is represented by

$$F(u) = \frac{1}{N} \sum_{y=0}^{N-1} f(x) \exp\left[-j\,\frac{2\pi\,ux}{N}\right]. \tag{4.11}$$

The corresponding one-dimensional inverse DFT is

$$f(x) = \sum_{u=0}^{N-1} F(u) \exp\left[j\,\frac{2\pi\,ux}{M}\right]. \tag{4.12}$$

4.2.3.2 Two-Dimensional DFT The two-dimensional discrete Fourier transform of a two-dimensional signal $f(x,y)$ of dimension $M \times N$ with integer indices x and y running from 0 to $M-1$ and 0 to $N-1$, is represented by

$$F(u,v) = \frac{1}{MN} \sum_{x=0}^{M-1} \sum_{y=0}^{N-1} f(x,y) \exp\left[-j\,2\pi\left(\frac{ux}{M} + \frac{vy}{N}\right)\right]. \tag{4.13}$$

The equivalent two-dimensional inverse DFT is

$$f(x,y) = \sum_{u=0}^{M-1} \sum_{v=0}^{N-1} F(u,v) \exp\left[j\,2\pi\left(\frac{ux}{M} + \frac{vy}{N}\right)\right]. \tag{4.14}$$

4.2.4 Transformation Kernels

As we have already noted the general forward and inverse Fourier transformation can be expressed as

$$F(u,v) = \frac{1}{MN} \sum_{x=0}^{M-1} \sum_{y=0}^{N-1} f(x,y).g(x,y,u,v) \tag{4.15}$$

and

$$f(x,y) = \sum_{u=0}^{M-1} \sum_{v=0}^{N-1} F(u,v).h(x,y,u,v). \tag{4.16}$$

In the above equations $g(x, y, u, v)$ is known as forward transformation kernel and $h(x, y, u, v)$ is the inverse transformation kernel. Here u and v assume values in the range $\{0, 1, \cdots, N-1\}$. As can be further observed, these kernels are dependent only on the spatial positions and frequencies along X and Y directions, and are independent of either $f(x, y)$ or $F(u, v)$. These kernels are like a set of basis functions. If the kernel $g(x, y, u, v) = g_1(x, u)g_2(y, v)$, we say that the kernel is separable. If in addition the functional forms of g_1 and g_2 are identical, then the kernel is said to be symmetric.

It easy to show that the kernel $g(x, y, u, v)$ is separable and symmetric since

$$g(x,y,u,v) = \exp\left[-j\,2\pi\left(\frac{ux}{M} + \frac{vy}{N}\right)\right] = \exp\left[-j\,2\pi\left(\frac{ux}{M}\right)\right]\exp\left[-j\,2\pi\left(\frac{vy}{N}\right)\right]$$

In a likewise fashion it may be easily proved that the inverse kernel $g(x, y, u, v)$ is also symmetric and separable.

4.2.5 Matrix Form Representation

In case the kernel is observed to be symmetric and separable it may be expressed as

$$F = A\,f(x,y)\,A \tag{4.17}$$

where $f(x, y)$ is an image matrix of dimension $N \times N$ and A is a symmetric transformation matrix of dimension $N \times N$ with elements $a_{i,j} = g(i, j)$. Multiplying both the sides of Eq. 4.17 by a matrix B, we get

$$B\,F\,B = B\,[A\,f(x,y)\,A]\,B. \tag{4.18}$$

To recover the original image $f(x, y)$, we need to choose $B = A^{-1} = g^{-1}(i, j)$. In that case, Eq. 4.18 reduces to

$$B\,F\,B = B\,[A\,f(x,y)\,A]\,B = A^{-1}\,A\,f(x,y)\,A\,A^{-1} = f(x,y).$$

Thus the original image $f(x, y)$ can be reconstructed.

The Figure 4.1 shows the results of two-dimensional DFT. Figure 4.1(a) is a two-dimensional sync function which is the result of DFT of the square image shown in Figure 4.1(a). Similarly, Figure 4.1(d) is the DFT of the two-dimensional DC function shown in Figure 4.1(c).

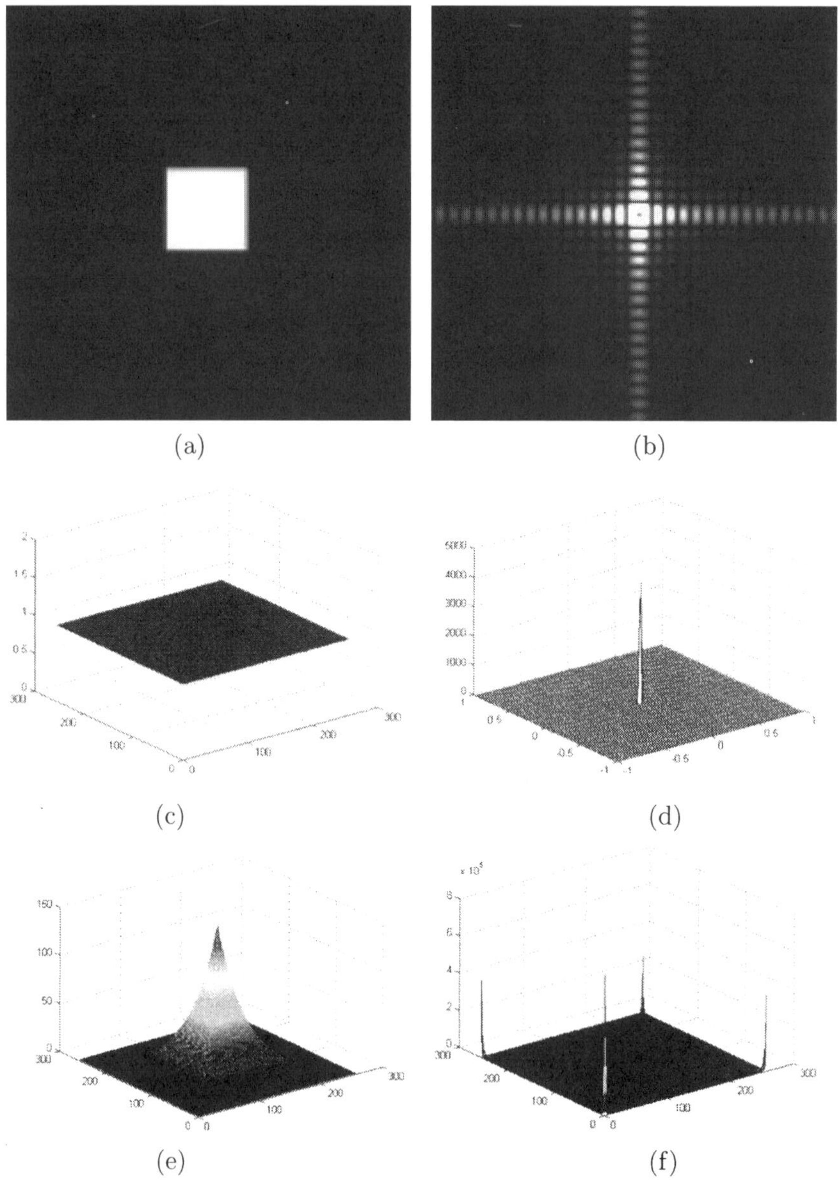

Fig. 4.1 Examples: (a) square image, (b) DFT magnitude of the square image, (c) two-dimensional DC signal, (d) DFT magnitude of the DC signal, (e) two-dimensional exponential function, (f) DFT magnitude of the exponential function.

4.2.6 Properties

- **Translation**: The translation of a Fourier transform pair is

$$f(x, y) \exp [j\, 2\pi\, (px + qy).] \Leftrightarrow F(u-p, v-q) \tag{4.19}$$

The above implication indicates the correspondence between a two-dimensional image function and its Fourier transform.

- **Rotation**: Assuming that the function $f(x, y)$ undergoes a rotation of α, the corresponding function $f(x, y)$ in polar coordinates will then be represented as $f(r, \alpha)$, where $x = r \cos \alpha$ and $y = r \sin \alpha$. The corresponding DFT $F(u, v)$ in polar coordinates will be represented as $F(\beta, \gamma)$, where $u = \beta \cos \gamma$ and $v = \beta \sin \gamma$.

  The above implies that if $f(x, y)$ is rotated by α_0, then $F(u, v)$ will be rotated by the same angle α_0 and hence we can imply that $f(r, \alpha + \alpha_0)$ corresponds to $F(\beta, \gamma + \alpha_0)$ in the DFT domain and vice versa.

- **Separability:** The separability property of a two-dimensional transform and its inverse ensures that such computations can be performed by decomposing the two-dimensional transforms into two one-dimensional transforms. From Eqs. 4.13 and 4.14 describing the DFT and inverse DFT of a two-dimensional function $f(x, y)$, we can express them in separable form as follows.

$$F(u, v) = \left[\frac{1}{M} \sum_{x=0}^{M-1} f(x, y) \exp \left[\frac{-j\, 2\pi\, ux}{M}\right]\right] \left[\frac{1}{N} \sum_{y=0}^{N-1} f(x, y) \exp \left[\frac{-j\, 2\pi\, vy}{N}\right]\right]$$

$$f(x, y) = \left[\sum_{u=0}^{M-1} F(u, v) \exp \left[\frac{j\, 2\pi\, ux}{M}\right]\right] \left[\sum_{v=0}^{N-1} F(u, v) \exp \left[\frac{j\, 2\pi\, vy}{N}\right]\right].$$

Hence the two-dimensional DFT (as well as the inverse DFT) can be computed by the taking the one-dimensional DFT row-wise in the two-dimensional image and the result is again transformed column-wise by the same one-dimensional DFT.

- **Distributive property:** The DFT of sum of two functions $f_1(x, y)$ and $f_2(x, y)$ is identical to the sum of the DFT of these two functions, i.e.,

$$F\{f_1(x, y) + f_2(x, y)\} = F\{f_1(x, y)\} + F\{f_2(x, y)\},$$

where $F\{f(x, y)\}$ is the DFT of $f_1(x, y)$. It should be noted that the distributive property for product of two functions does not hold, i.e.,

$$F\{f_1(x, y).f_2(x, y)\} \neq F\{f_1(x, y)\}.F\{f_2(x, y)\}.$$

- **Scaling property:** The DFT of a function $f(x, y)$ multiplied by a scalar(k) is identical to the multiplication of the scalar with the DFT of the function $f(x, y)$, i.e.,

$$F\{kf(x, y)\} = k\,F(u, v).$$

- **Convolution:** The DFT of convolution of two functions is equal to the product of the DFT of these two functions, i.e.,

$$F\{f_1(x, y) \otimes f_1(x, y)\} = F_1(u, v).F_2(u, v).$$

- **Correlation:** The correlation between two functions $f_1(x, y)$ and $f_2(x, y)$ in continuous domain is denoted as

$$f_1(x, y) \bullet f_2(x, y) = \int_{-\infty}^{+\infty} \int_{-\infty}^{+\infty} f_1^*(\alpha, \beta) f_2(x + \alpha, y + \beta) d\alpha\, d\beta,$$

whereas the correlation in the discrete domain is denoted as

$$f_1(x, y) \bullet f_2(x, y) = \sum_{m=0}^{M-1} \sum_{n=0}^{N-1} f_1^*(m, n) f_2(x + m, y + n).$$

The DFT of the correlation of two functions $f_1(x, y)$ and $f_2(x, y)$ is the product of the complex conjugate of the DFT of the first function and the DFT of the second function, i.e.,

$$F\{f_1(x, y) \bullet f_2(x, y)\} = F_1^*(u, v).F_2(u, v).$$

- **Periodicity:** The DFT of a two-dimensional function $f(x, y)$ and its inverse are both periodic with period τ, i.e.,

$$F(u, v) = F(u + \tau, v) = F(u, v + \tau) = F(u + \tau, v + \tau)$$

All the properties presented in this section are valid for *continuous Fourier transform* (CFT) cases as well.

4.2.7 Fast Fourier Transform

The number of complex multiplications and additions to compute Eq. 4.11 for DFT is $O(N^2)$. However, we can adopt a divide-and-conquer approach to reduce the computational complexity of the algorithm to $O(N \log_2 N)$. This algorithm is popularly known as the *Fast Fourier Transform* (FFT). There are many implementations of the FFT proposed in the literature. Here we present a general idea of the divide-and-conquer approach toward implementation of the FFT. The general principles is based on successive division method using divide-and-conquer approach as explained below.

As shown in Eq. 4.11, the one-dimensional DFT of a one-dimensional signal $f(x)$ is computed

$$F(u) = \frac{1}{N} \sum_{x=0}^{N-1} f(x) \kappa_N^{ux} \tag{4.20}$$

where

$$\kappa_N = \exp\left[-j\frac{2\pi}{N}\right]. \tag{4.21}$$

For ease of explanation, we assume that N is a power of 2 and hence N can be expressed as $M = 2^n = 2M$, where n is a positive integer. Hence, Eq. 4.20 can be written as

$$\begin{aligned} F(u) &= \tfrac{1}{2M} \textstyle\sum_{x=0}^{2M-1} f(x) \kappa_{2M}^{ux} \\ &= \tfrac{1}{2}\left\{ \tfrac{1}{M} \textstyle\sum_{x=0}^{M-1} f(2x) \kappa_{2M}^{u(2x)} + \tfrac{1}{M} \sum_{x=0}^{M-1} f(2x+1) \kappa_{2M}^{u\,(2x+1)} \right\} \\ &= \tfrac{1}{2}\left\{ \tfrac{1}{M} \textstyle\sum_{x=0}^{M-1} f(2x) \kappa_{2M}^{u(2x)} + \tfrac{1}{M} \sum_{x=0}^{M-1} f(2x+1) \kappa_{2M}^{2ux} \kappa_{2M}^{u} \right\} \end{aligned} \tag{4.22}$$

As per Eq. 4.21, $\kappa_{2M}^{2ux} = \kappa_M^{ux}$ and hence Eq. 4.22 can be written as

$$F(u) = \frac{1}{2}\left\{ \frac{1}{M} \sum_{x=0}^{M-1} f(2x) \kappa_M^{ux} + \frac{1}{M} \sum_{x=0}^{M-1} f(2x+1) \kappa_M^{ux} \kappa_{2M}^{u} \right\}. \tag{4.23}$$

Let us define the first component in Eq. 4.23 as $F_{even}(u)$ and the second component as $F_{odd}(u)$. Hence,

$$F(u) = \frac{1}{2}\left\{ F_{even}(u) + \kappa_{2M}^{u} F_{odd}(u) \right\}. \tag{4.24}$$

It may be pointed out that $F_{even}(u)$ is the DFT of the sequence composed of the even samples $f(2x)$ (i.e., $f(0), f(2), f(4), \cdots, f(2M-2)$) of the original discrete signal $f(x)$, whereas $F_{odd}(u)$ is the DFT of the sequence composed of all the odd samples $f(2x+1)$ (i.e., $f(1), f(3), f(5), \cdots, f(2M-1)$) of the original discrete signal $f(x)$. Size of both the even and odd sequences $f(2x)$ and $f(2x+1)$ is $\frac{N}{2}$. Hence computation of both $F_{even}(u)$ and $F_{odd}(u)$ are essentially $\frac{N}{2}$-point DFT each. As a result, the N-point DFT $F(u)$ can be computed as two $\frac{N}{2}$-point DFT operations $F_{even}(u)$ and $F_{odd}(u)$ followed by addition of F_{even} with F_{odd} scaled by κ_{2M}^{u}. In the similar fashion, the $\frac{N}{2}$ point DFT computations of each of $F_{even}(u)$ and $F_{odd}(u)$ may further be decomposed into two $\frac{N}{4}$ point DFT computations for each as shown in Figure 4.2. We can continue this in $\log_2 N$ iterations until each become a 1-point computation as shown in Figure 4.2.

Thus in the algorithm discussed above, each N point transform has been computed as two $N/2$ point transforms. The algorithm for computation of

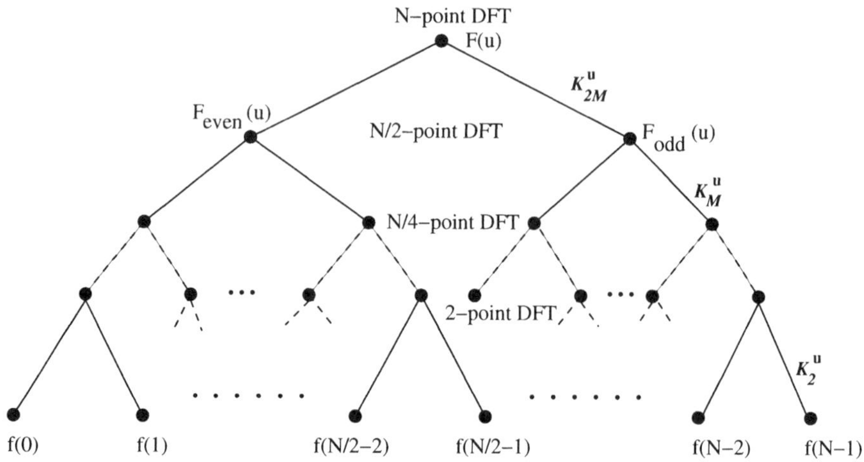

Fig. 4.2 FFT computation by Successive Decomposition.

the N point FFT using successive division approach is shown in Figure 4.2. The computational complexity of the algorithm is $O(N \log_2 N)$.

4.3 DISCRETE COSINE TRANSFORM

Discrete Cosine Transform (DCT) is the basis for many image and video compression algorithms, especially the baseline JPEG and MPEG standards for compression of still and video images respectively.

The one-dimensional *forward discrete Cosine transform* (1D FDCT) of N samples is formulated by

$$F(u) = \sqrt{\frac{2}{N}} C(u) \sum_{x=0}^{N-1} f(x) \cos\left[\frac{\pi(2x+1)u}{2N}\right]$$

for $u = 0, 1, \ldots, N-1$, where

$$C(u) = \begin{cases} \frac{1}{\sqrt{2}} & \text{for } u = 0 \\ 1 & \text{otherwise.} \end{cases}$$

The function $f(x)$ represents the value of the xth sample of the input signal. $F(u)$ represents a DCT coefficient for $u = 0, 1, \cdots, N-1$.

The one-dimensional *inverse discrete Cosine transform* (1D IDCT) is formulated in a similar fashion as follows,

$$f(x) = \sqrt{\frac{2}{N}} \sum_{u=0}^{N-1} C(u) F(u) \cos\left[\frac{\pi(2x+1)u}{2N}\right]$$

for $x = 0, 1, \ldots, N-1$.

The two-dimensional DCT can be computed using the one-dimensional DCT horizontally and then vertically across the signal because DCT is a separable function. The *two-dimensional forward discrete Cosine transform* (2D FDCT) of a block of $M \times N$ samples of a two-dimensional signal $F(x, y)$ is formulated as

$$F(u,v) = \frac{2}{\sqrt{MN}} C(u)C(v) \sum_{x=0}^{N-1} \sum_{y=0}^{M-1} f(x,y) \cos\left[\frac{\pi(2x+1)u}{2N}\right] \cos\left[\frac{\pi(2y+1)u}{2M}\right]$$

for $u = 0, 1, \ldots, N-1$ and $v = 0, 1, \ldots, M-1$, where

$$C(k) = \begin{cases} \frac{1}{\sqrt{2}} & \text{for } k = 0 \\ 1 & \text{otherwise.} \end{cases}$$

The function $f(x, y)$ represents the value of the xth sample in the yth row of a two-dimensional signal. $F(u, v)$ is a two-dimensional transformed coefficient for $u = 0, 1, \ldots, N-1$ and $v = 0, 1, \ldots, M-1$.

The above expression for 2D FDCT is clearly a separable function because we can express the formula as follows:

$$F(u,v) = \sqrt{\frac{2}{M}} C(v) \sum_{y=0}^{M-1} \left\{ \sqrt{\frac{2}{N}} C(u) \sum_{x=0}^{N-1} f(x,y) \cos\left[\frac{\pi(2x+1)u}{2N}\right] \right\} \cos\left[\frac{\pi(2y+1)u}{2M}\right].$$

As a result we can accomplish the 2D FDCT of a two-dimensional signal by applying 1D FDCT first row-wise followed by 1D FDCT column-wise in two steps. First, the 1D FDCT is applied row-wise in all the rows independently to obtain $F(u, y)$, where

$$F(u,y) = \sqrt{\frac{2}{N}} C(u) \sum_{x=0}^{N-1} f(x,y) \cos\left[\frac{\pi(2x+1)u}{2N}\right]$$

for $u = 0, 1, \ldots, N-1$.

In the second step, the same 1D FDCT is applied column-wise in all the columns of $F(u, y)$ to obtain the result $F(u, v)$, where

$$F(u,v) = \sqrt{\frac{2}{M}} C(v) \sum_{y=0}^{M-1} F(u,y) \cos\left[\frac{\pi(2y+1)v}{2M}\right]$$

for $v = 0, 1, \ldots, M-1$.

The *two-dimensional inverse discrete Cosine transform* (2D IDCT) is computed in a similar fashion. The 2D IDCT of $F(u, v)$ is formulated as

$$f(x,y) = \frac{2}{\sqrt{MN}} \sum_{u=0}^{N-1} \sum_{v=0}^{M-1} C(u)C(v)F(u,v) \cos\left[\frac{\pi(2x+1)u}{2N}\right] \cos\left[\frac{\pi(2y+1)u}{2M}\right]$$

for $x = 0, 1, \ldots, N-1$ and $y = 0, 1, \ldots, M-1$.

The above function is again a separable function similar to what we have shown for the 2D FDCT. As a result, the 2D IDCT can be computed in exactly the opposite way of the 2D FDCT. The 2D IDCT is computed in two steps: by first applying 1D IDCT column-wise followed by the 1D IDCT row-wise. After column-wise computation of 1D IDCT in every column of the input signal $F(u, v)$, we obtain $F(u, y)$, where

$$F(u, y) = \sqrt{\frac{2}{M}} \sum_{v=0}^{M-1} C(v) F(u, v) \cos\left[\frac{\pi(2y+1)v}{2M}\right]$$

for $v = 0, 1, \ldots, M-1$.

In the second step, the same 1D IDCT is applied row-wise in all the rows of $F(u, y)$ to obtain the two-dimensional signal $f(x, y)$, where

$$f(x, y) = \sqrt{\frac{2}{N}} \sum_{u=0}^{N-1} C(u) F(u, y) \cos\left[\frac{\pi(2x+1)u}{2N}\right]$$

for $u = 0, 1, \ldots, N-1$.

The two-dimensional DCT kernel is a separable function and hence the 2D DCT computation can be done in two steps, by applying one-dimensional DCT row-wise and then column-wise instead of the direct computation.

Since DCT belongs to the family of DFT, there are fast DCT algorithms of computational complexity $O(N \log_2 N)$ similar to the *Fast Fourier Transform* (FFT). There are many fast DCT algorithms proposed in the literature [3].

4.4 WALSH-HADAMARD TRANSFORM (WHT)

Like the Fourier transform, the discrete *Walsh Hadamard Transform* (WHT) also has a separable symmetric kernel. The discrete Walsh Hadamard transform of a function $f(x)$ is denoted by

$$W(u) = \frac{1}{N} \sum_{x=0}^{N-1} f(x) g(x, u), \tag{4.25}$$

where

$$g(x, u) = \frac{1}{N} \left[\prod_{i=0}^{n-1} (-1)^{b_i(x) b_{n-1-i}(u)}\right]$$

is the kernel of WHT, where $n = \log_2 N$ and $b_i(z)$ is the ith bit in binary representation of z. The WHT kernel $g(x, u)$ is a symmetric matrix having a set of N orthogonal rows and columns. The symmetric WHT kernel for $N =$ 1, 2, 4 and 8 are shown in Eqs. 4.26–4.29 respectively:

$$H_1 = [1] \tag{4.26}$$

$$H_2 = \frac{1}{\sqrt{2}} \begin{bmatrix} H_1 & H_1 \\ H_1 & -H_1 \end{bmatrix} = \frac{1}{\sqrt{2}} \begin{bmatrix} 1 & 1 \\ 1 & -1 \end{bmatrix} \tag{4.27}$$

$$H_4 = \frac{1}{\sqrt{2}} \begin{bmatrix} H_2 & H_2 \\ H_2 & -H_2 \end{bmatrix} = \frac{1}{\sqrt{4}} \begin{bmatrix} 1 & 1 & 1 & 1 \\ 1 & -1 & 1 & -1 \\ 1 & 1 & -1 & -1 \\ 1 & -1 & -1 & 1 \end{bmatrix} \tag{4.28}$$

$$H_8 = \frac{1}{\sqrt{2}} \begin{bmatrix} H_4 & H_4 \\ H_4 & -H_4 \end{bmatrix} = \frac{1}{\sqrt{8}} \begin{bmatrix} 1 & 1 & 1 & 1 & 1 & 1 & 1 & 1 \\ 1 & -1 & 1 & -1 & 1 & -1 & 1 & -1 \\ 1 & 1 & -1 & -1 & 1 & 1 & -1 & -1 \\ 1 & -1 & -1 & 1 & 1 & -1 & -1 & 1 \\ 1 & 1 & 1 & 1 & -1 & -1 & -1 & -1 \\ 1 & -1 & 1 & -1 & -1 & 1 & -1 & 1 \\ 1 & 1 & -1 & -1 & -1 & -1 & 1 & 1 \\ 1 & -1 & -1 & 1 & -1 & 1 & 1 & -1 \end{bmatrix}. \tag{4.29}$$

Hence the recursive relation to generate a Walsh-Hadamard Transform kernel can be represented as

$$H_N = \frac{1}{\sqrt{2}} \begin{bmatrix} H_{\frac{N}{2}} & H_{\frac{N}{2}} \\ H_{\frac{N}{2}} & -H_{\frac{N}{2}} \end{bmatrix}. \tag{4.30}$$

The advantage of using the WHT is the simplicity in its computation in view of the binary nature of the transform kernel. The WHT has been used for shape analysis, and other signal and image processing applications.

4.5 KARHAUNEN-LOEVE TRANSFORM OR PRINCIPAL COMPONENT ANALYSIS

Karhaunen-Loeve Transform or *Principal Component Analysis* (PCA) has been a popular technique for many image processing and pattern recognition applications. This transform which is also known as *Hotelling Transform* is based on the concepts of statistical properties of image pixels or pattern features [1, 2].

Principal component analysis (PCA) forms the basis of the Karhunen-Loeve (KL) transform for compact representation of data [1]. The KL transform and the theory behind the principal component analysis are of fundamental importance in signal and image processing. The principle has also found its place in data mining for reduction of large-dimensional datasets. It has been successfully applied to text analysis and retrieval for text mining as well [4].

One of the major problems in pattern recognition and image processing is the dimensionality reduction. In practical pattern recognition problems, quite often the features that we choose are correlated with each other and a number of them are useless so far as their discriminability is concerned. If we can reduce the number of features, i.e., reduce the dimensionality of the feature space, then we will achieve better accuracy with lesser storage and computational complexities.

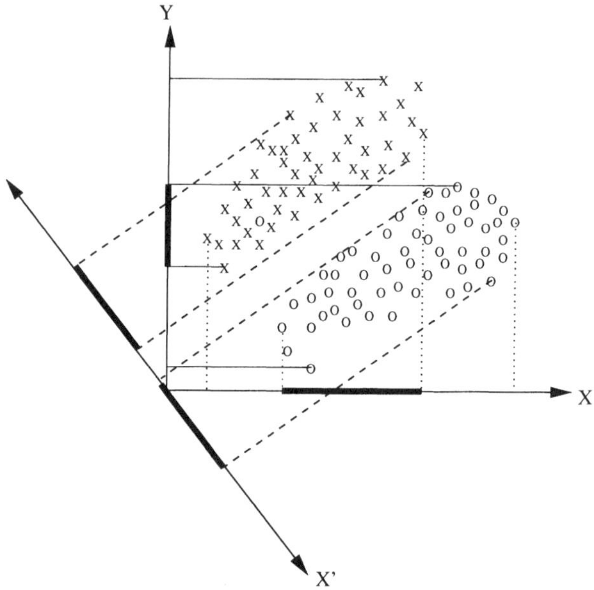

Fig. 4.3 Dimensionality Reduction.

In Figure 4.3 there are a number of two-dimensional pattern points, belonging to two different pattern classes (shown by X and O symbols), where each pattern is described by only two features X and Y. It may be observed that the projection of the pattern points both on X and Y axis are overlapping. As a result, the two features X and Y do not exhibit good discriminability. It is possible to find a reduced set of features that may result in better discrimination between the two classes. This is shown by the nonoverlapping projections of the patterns belonging to two classes on the new feature axis (X') as shown in Figure 4.3. PCA is one such tool which yields an extremely powerful technique for dimensionality reduction and many image processing applications such as compression, classification, feature selection, etc. Before describing the PCA, we would briefly present the concepts of covariance matrix.

4.5.1 Covariance Matrix

In practical pattern recognition problems there are usually more than one feature. During the process of statistical analysis of these data, we have to find out whether these features are independent of one another. Otherwise there exists a relationship between each pair of features. For example, while extracting the features of human face, one may choose two features such as (1) X to denote the distance between the centers of the two irises, and (2) Y to denote the distance between the centers of the left and right eyebrows. From a large set of human faces, we can determine the mean and the standard deviation of the above two features. The standard deviation for each of the above two dimensions of the face data set may be computed independently of each other. To understand whether there exists any relationship between these two features, we have to compute how much the first feature X of each of the patterns in our data set varies from the mean of the second feature Y. This measure, which is computed similar to variance, is always measured between two features. The covariance is computed as follows:

$$Cov(X,Y) = \sum_{i=1}^{n}(X_i - \bar{X})(Y_i - \bar{Y}), \tag{4.31}$$

where n is the number of facial patterns, and $\bar{X}$ and $\bar{Y}$ are the mean of feature X and Y respectively.

If the covariance value is positive, it implies that when one feature (X) increases, the other feature (Y) also increases. If the value of $Cov(X,Y)$ is negative, then as one feature increases, the other one decreases. In case where there is no correlation between the two features X and Y the covariance becomes zero, indicating that the two features are independent of each other. In the problem of face feature selection then one may find that the features have positive covariance, meaning that if X increases the other feature Y also increases. In case of a multi-dimensional feature vector, the covariance is measured between each pair of features. In practical pattern recognition problems, we compute a covariance matrix, where each element of the matrix gives a measure of the covariance between two features.

4.5.2 Eigenvectors and Eigenvalues

Before we discuss principal component analysis, we will briefly explain the concept of eigenvectors and eigenvalues of a matrix. Let us assume that we have a square matrix A of dimension $n \times n$, which when multiplied by a vector X of dimension $n \times 1$ yields another vector Y of dimension $n \times 1$, which is essentially the same as the original vector X that was chosen initially. Such a vector X is called an eigenvector which transforms a square matrix A into a vector, which is either the same vector X or a multiple of X (i.e., a scaled version of the vector X). The matrix A is called a transformation matrix, while the vector X is called an eigenvector. As is well known, any integer multiplication of the vector results in the same vector pointing to the same direction, with only its magnitude being scaled up (i.e., the vector is only elongated).

It is interesting to note here that eigenvectors can be determined only from the square matrices, while every square matrix does not necessarily yield an eigenvector. Also an $n \times n$ square transformation matrix may have only n number of eigenvectors. All these eigenvectors are perpendicular or orthogonal to each other. Every eigenvector is associated with a corresponding eigenvalue. The concept of an eigenvalue is that of a scale which when multiplied by the eigenvector yields the same scaled vector in the same direction.

4.5.3 Principal Component Analysis

While computing the principal component analysis, we represent an $N \times N$ image as a one-dimensional vector of $N * N$ elements, by placing the rows of the image one after another. Then we compute the covariance matrix of the entire data set. Next we compute the eigenvalues of this covariance matrix. The eigenvectors corresponding to the most significant eigenvalues will yield the principal components. To get the original data back we have to consider all the eigenvectors in our transformation. If we discard some of the less significant eigenvectors in the final transformation, then the retrieved data will lose some information. However, if we choose all the eigenvectors, we can retrieve the original data.

4.5.4 Singular Value Decomposition

The principal component analysis has also been developed based on the matrix theory for *Singular Value Decomposition* (SVD). According to singular value decomposition (SVD) theory, for any arbitrary $M \times N$ matrix F of rank L there exists an $M \times M$ unitary matrix U and an $N \times N$ unitary matrix V so that

$$U^T F V = \Lambda^{\frac{1}{2}}, \tag{4.32}$$

where

$$\Lambda^{\frac{1}{2}} = \begin{bmatrix} \lambda^{\frac{1}{2}}(1) & & & & & & \\ & \lambda^{\frac{1}{2}}(2) & & & & & \\ & & \ddots & & & & \\ & & & \lambda^{\frac{1}{2}}(L) & & & \\ & & & & 0 & & \\ & & & & & \ddots & \\ & & & & & & 0 \end{bmatrix}$$

is an $M \times N$ diagonal matrix and the first L diagonal elements $\lambda^{\frac{1}{2}}(i)$, for $i = 1, 2, \ldots, L$, are called the *singular* values of input matrix F. Since U and V are unitary matrices, we have

$$UU^T = I_M,$$

$$VV^T = I_N,$$

where I_M and I_N are the identity matrices of dimension M and N, respectively. As a result, the input matrix F can be decomposed as

$$F = U\Lambda^{\frac{1}{2}}V^T. \tag{4.33}$$

The columns of U are chosen as the eigenvectors u_m of the symmetric matrix FF^T so that

$$U^T(FF^T)U = \begin{bmatrix} \lambda(1) & & & & & & \\ & \lambda(2) & & & & & \\ & & \ddots & & & & \\ & & & \lambda(L) & & & \\ & & & & 0 & & \\ & & & & & \ddots & \\ & & & & & & 0 \end{bmatrix}, \tag{4.34}$$

where $\lambda(i)$, $i = 1, 2, \ldots, L$, are the nonzero eigenvalues of FF^T. Similarly, the columns of matrix V are eigenvectors v_n of the symmetric matrix F^TF as defined by

$$V^T(F^TF)V = \begin{bmatrix} \lambda(1) & & & & & & \\ & \lambda(2) & & & & & \\ & & \ddots & & & & \\ & & & \lambda(L) & & & \\ & & & & 0 & & \\ & & & & & \ddots & \\ & & & & & & 0 \end{bmatrix}, \tag{4.35}$$

where $\lambda(i)$, $i = 1, 2, \ldots, L$ are the corresponding nonzero eigenvalues of F^TF. The input matrix can be represented in series form by these eigenvalues and eigenvectors as

$$F = \sum_{i=1}^{L} \lambda^{\frac{1}{2}}(i)u_i v_i^T. \tag{4.36}$$

If the eigenvalues $\lambda(i)$, for $i = 1, 2, \ldots, L$ are sorted in decreasing order and only first K from the sorted list are significant ($K < L$), then we can approximate the input matrix F by a smaller-dimensional matrix $\bar{F}$ using these first K eigenvalues and corresponding eigenvectors only.

The eigenvector corresponding to the highest eigenvalue of $F^T F$ is called the *first principal component.* Likewise, the *second principal component* is the eigenvector corresponding to the next highest eigenvalue of $F^T F$, and so on. Hence the kth *principal component* is the eigenvector corresponding to the kth largest eigenvalue of $F^T F$.

4.6 SUMMARY

In this chapter, we have described the principles and mathematical formulations of different image transformation techniques. We have described the principles of one-dimensional and two-dimensional Fourier transform (both continuous and discrete) and their properties with some examples. We described the discrete Cosine transform (DCT) both in one and two dimensions and principle behind Walsh Hadamard transform (WHT) as well. We also discussed the principles behind Karhaunen-Loeve transform and its properties.

REFERENCES

1. W. K. Pratt, *Digital Image Processing*, 2nd Ed., Wiley, New York, 1991.

2. R. C. Gonzalez and R. E. Woods, *Digital Image Processing*, Addison-Wesley, Reading, MA, 1992.

3. K. R. Rao and P. Yip, *Discrete Cosine Transform: Algorithms, Advantages, Applications*, Acedemic Press, San Diego, 1990.

4. S. Mitra and T. Acharya, *Data Mining: Multimedia, Soft Computing, and Bioinformatics*, Wiley, Hoboken, NJ, 2003.

5

Discrete Wavelet Transform

5.1 INTRODUCTION

The Fourier transform is an analysis of global frequency content in the signal. There are applications in image processing that require the analysis to be localized in the spatial domain. This can be handled by introducing spatial frequency into Fourier analysis. The classical way of doing this is through what is called *Windowed Fourier Transform*. Central idea of windowing is reflected in *Short Time Fourier Transform* (STFT). The windowed transform of $f(x)$ is given as

$$F(\omega, \alpha) = \int_{-\infty}^{+\infty} f(x) g(x - \alpha) \exp^{-j\omega x} dx \tag{5.1}$$

where ω represents the frequency and α denotes the position of the window. It may be noted here that the Gaussian is well-localized around the time $x = \alpha$. Thus, Eq. 5.1 transforms the signal $f(x)$ in a small window around α. The STFT conveys the localized frequency component present in the signal during the short window of time. The same concept may be extended to a two-dimensional spatial image where the localized frequency components may be determined from the windowed transform. This is one of the basis of the conceptual understanding of wavelet transforms.

The concept of wavelet was hidden in the works of mathematicians even more than a century ago. In 1873, Karl Weierstrass mathematically described how a family of functions can be constructed by superimposing scaled versions of a given basis function [1]. Mathematically a "wave" is expressed as a sinusoidal (or oscillating) function of time or space. Fourier analysis expands an arbitrary signal in terms of an infinite number of sinusoidal functions of its harmonics and has been well studied by the signal processing community for decades. Fourier representation of signals is known to be very effective in analysis of time-invariant (stationary) periodic signals. In contrast to a sinusoidal function, a wavelet is a small wave whose energy is concentrated in time. The term *wavelet* was originally used in the field of seismology to describe the disturbances that emanate and proceed outward from a sharp seismic impulse [2]. In 1982, Morlet et al. first described how the seismic wavelets could be effectively modeled mathematically [3]. In 1984, Grossman and Morlet extended this work to show how an arbitrary signal can be analyzed in terms of scaling and translation of a single *mother wavelet* function (basis) [4]. Wavelets allow both time and frequency analysis of signals simultaneously because of the fact that the energy of wavelets is concentrated in time and still possesses the wave-like (periodic) characteristics. As a result, wavelet representation provides a versatile mathematical tool to analyze transient, time-variant (nonstationary) signals that are not statistically predictable especially at the region of discontinuities–a feature that is typical of images having discontinuities at the edges. For historical perspectives of wavelets, the reader is referred to the treatise by Ives Meyer [5].

5.2 WAVELET TRANSFORMS

Wavelets are functions generated from one single function (basis function) called the *prototype* or *mother wavelet* by *dilations* (scalings) and *translations* (shifts) in time (frequency) domain. If the mother wavelet is denoted by $\psi(t)$, the other wavelets $\psi_{a,b}(t)$ can be represented as

$$\psi_{a,b}(t) = \frac{1}{\sqrt{|a|}}\psi\left(\frac{t-b}{a}\right) \tag{5.2}$$

where a and b are two arbitrary real numbers. The variables a and b represent the parameters for *dilations* and *translations* respectively in the time axis. From Eq. 5.2, it is obvious that the mother wavelet can be essentially represented as

$$\psi(t) = \psi_{1,0}(t). \tag{5.3}$$

For any arbitrary $a \neq 1$ and $b = 0$, we can derive that

$$\psi_{a,0}(t) = \frac{1}{\sqrt{|a|}}\psi\left(\frac{t}{a}\right). \tag{5.4}$$

As shown in Eq. 5.4, $\psi_{a,0}(t)$ is nothing but a time-scaled (by a) and amplitude-scaled (by $\sqrt{|a|}$) version of the mother wavelet function $\psi(t)$ in Eq. 5.3. The parameter a causes contraction of $\psi(t)$ in the time axis when $a < 1$ and expansion or stretching when $a > 1$. That's why the parameter

a is called the *dilation* (scaling) parameter. For $a < 0$, the function $\psi_{a,b}(t)$ results in time reversal with dilation.

Mathematically, we can substitute t in Eq. 5.4 by $t-b$ to cause a translation or shift in the time axis resulting in the wavelet function $\psi_{a,b}(t)$ as shown in Eq. 5.2. The function $\psi_{a,b}(t)$ is a shift of $\psi_{a,0}(t)$ in right along the time axis by an amount b when $b > 0$ whereas it is a shift in left along the time axis by an amount b when $b < 0$. That's why the variable b represents the *translation* in time (*shift* in frequency) domain.

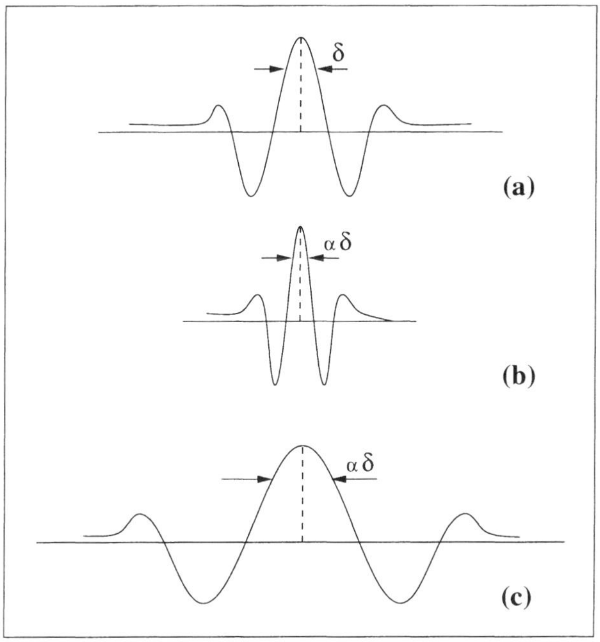

Fig. 5.1 (a) A mother wavelet $\psi(t)$, (b) $\psi(t/\alpha)$: $0 < \alpha < 1$, (c) $\psi(t/\alpha)$: $\alpha > 1$.

In Figure 5.1, we have shown an illustration of a mother wavelet and its dilations in the time domain with the dilation parameter $a = \alpha$. For the mother wavelet $\psi(t)$ shown in Figure 5.1(a), a contraction of the signal in the time axis when $\alpha < 1$ is shown in Figure 5.1(b) and expansion of the signal in the time axis when $\alpha > 1$ is shown in Figure 5.1(c). Based on this definition of wavelets, the *wavelet transform* (WT) of a function (signal) $f(t)$ is mathematically represented by

$$W(a, b) = \int_{-\infty}^{+\infty} \psi_{a,b}(t) f(t)\, dt. \tag{5.5}$$

The inverse transform to reconstruct $f(t)$ from $W(a,b)$ is mathematically represented by

$$f(t) = \frac{1}{C}\int_{a=-\infty}^{+\infty}\int_{b=-\infty}^{+\infty}\frac{1}{|a|^2}W(a,b)\psi_{a,b}(t)da\,db \tag{5.6}$$

where

$$C = \int_{-\infty}^{+\infty}\frac{|\Psi(\omega)|^2}{|\omega|}d\omega$$

and $\Psi(\omega)$ is the Fourier transform of the mother wavelet $\psi(t)$.

If a and b are two continuous (nondiscrete) variables and $f(t)$ is also a continuous function, $W(a,b)$ is called the *continuous wavelet transform* (CWT). Hence the CWT maps a one-dimensional function $f(t)$ to a function $W(a,b)$ of two continuous real variables a (dilation) and b (translation).

5.2.1 Discrete Wavelet Transforms

Since the input signal (e.g., a digital image) is processed by a digital computing machine, it is prudent to define the discrete version of the wavelet transform. Before we define the *discrete wavelet transform*, it is essential to define the wavelets in terms of discrete values of the *dilation* and *translation* parameters a and b instead of being continuous. There are many ways we can discretize a and b and then represent the *discrete wavelets* accordingly. The most popular approach of discretizing a and b is using Eq. 5.7,

$$a = a_0^m, \qquad b = nb_0a_0^m \tag{5.7}$$

where m and n are integers. Substituting a and b in Eq. 5.2 by Eq. 5.7, the discrete wavelets can be represented by Eq. 5.8.

$$\psi_{m,n}(t) = a_0^{-m/2}\psi\left(a_0^{-m}t - nb_0\right). \tag{5.8}$$

There are many choices to select the values of a_0 and b_0. We select the most common choice here: $a_0 = 2$ and $b_0 = 1$; hence, $a = 2^m$ and $b = n2^m$. This corresponds to sampling (discretization) of a and b in such a way that the consecutive discrete values of a and b as well as the sampling intervals differ by a factor of two. This way of sampling is popularly known as *dyadic sampling* and the corresponding decomposition of the signals is called the *dyadic decomposition.* Using these values, we can represent the discrete wavelets as in Eq. 5.9, which constitutes a family of orthonormal basis functions,

$$\psi_{m,n}(t) = 2^{-m/2}\psi\left(2^{-m}t - n\right). \tag{5.9}$$

In general, the wavelet coefficients for function $f(t)$ are given by

$$c_{m,n}(f) = a_0^{-m/2}\int f(t)\psi\left(a_0^{-m}t - nb_0\right)dt \tag{5.10}$$

and hence for dyadic decomposition, the wavelet coefficients can be derived accordingly as

$$c_{m,n}(f) = 2^{-m/2} \int f(t)\psi\left(2^{-m}t - n\right) dt. \tag{5.11}$$

This allows us to reconstruct the signal $f(t)$ in from the discrete wavelet coefficients as

$$f(t) = \sum_{m=-\infty}^{\infty} \sum_{n=-\infty}^{\infty} c_{m,n}(f)\psi_{m,n}(t). \tag{5.12}$$

The transform shown in Eq. 5.10 is called the *wavelet series*, which is analogous to the *Fourier series* because the input function $f(t)$ is still a continuous function whereas the transform coefficients are discrete. This is often called the *discrete time wavelet transform* (DTWT). For digital signal or image processing applications executed by a digital computer, the input signal $f(t)$ needs to be discrete in nature because of the digital sampling of the original data, which is represented by a finite number of bits. When the input function $f(t)$ as well as the wavelet parameters a and b are represented in discrete form, the transformation is commonly referred to as the *discrete wavelet transform* (DWT) of signal $f(t)$ [6, 7].

The *discrete wavelet transform* (DWT) became a very versatile signal processing tool after Mallat [6] proposed the multiresolution representation of signals based on wavelet decomposition. The advantage of the DWT over Fourier transformation is that it performs multiresolution analysis of signals with localization both in time and frequency, popularly known as time-frequency localization. As a result, the DWT decomposes a digital signal into different subbands so that the lower-frequency subbands have finer frequency resolution and coarser time resolution compared to the higher-frequency subbands. The DWT is being increasingly used for image compression due to the fact that the DWT supports features like progressive image transmission (by quality, by resolution), ease of compressed image manipulation, region of interest coding, etc. DWT is the basis of the new JPEG2000 image compression standard [8].

5.2.2 Gabor filtering

Gabor filter is an example of wavelet filters widely used in many image processing applications such as texture analysis, segmentation, classification, etc. [9]. In all such applications, it is necessary to analyze the spatial frequency components of an image in a localized fashion. For localized frequency analysis it is desirable to have a Gaussian envelope whose width adjusts with the frequency of the complex sinusoids. Gabor wavelets form class of self similar functions which yield better localization in space. The 2D Gabor filters optimally achieve joint resolution/localization in space and spatial frequency domains.

Gabor elementary functions are Gaussians modulated by complex sinusoids. The 2D Gabor functions are complex sinusoid (carrier) gratings modulated by 2D Gaussian functions in the space domain, and shifted Gaussians in the spatial frequency domain which means that they are complex valued functions. Regardless of the region of frequencies passed, the 2D Gabor functions uniquely minimize the 2D space-frequency uncertainty principle for complex-valued functionss. Hence Gabor functions can be interpreted as the product of a modulating amplitude envelope with a complex carrier function whose argument is a modulating phase envelope, both of which can be computed and analyzed separately.

In the spatial domain, the Gabor function is a complex exponential modulated by a Gaussian function. The Gabor function forms a complete and non-orthogonal basis set and its impulse response in the two-dimensional plane has the following general form :

$$G_x(x, y) = \frac{1}{2\pi\sigma_x\sigma_y} \exp\left[-\frac{1}{2}\left(\frac{x^2}{\sigma_x^2} + \frac{y^2}{\sigma_y^2}\right)\right] \exp\left(j2\pi u_0 x\right) \tag{5.13}$$

where u_0 denotes the radial frequency of the Gabor function. The space constants σ_x and σ_y define the Gaussian envelope along the X- and Y-axes. In a similar fashion, the Gabor function, obtained by modulation of complex exponential function in Y–direction by Gaussian function, is given as

$$G_x(x, y) = \frac{1}{2\pi\sigma_x\sigma_y} \exp\left[-\frac{1}{2}\left(\frac{x^2}{\sigma_x^2} + \frac{y^2}{\sigma_y^2}\right)\right] \exp\left(j2\pi v_0 y\right) \tag{5.14}$$

Figure 5.2 shows the perspective plot of a typical Gabor filter in the spatial domain. The real and imaginary components of the 2D Gabor function are shown in Figures 5.2(a) and 5.2(b) respectively.

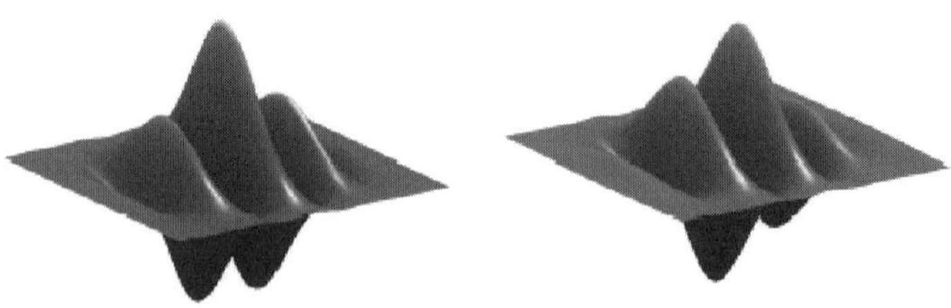

(a) (b)

Fig. 5.2 2D Gabor filter: (a) Real component, (b) Imaginary component.

Each of the complex Gabor filters has the real (even) and imaginary (odd) parts that are conveniently implemented as the spatial mask of $M \times M$ sizes. For a symmetric region of support, M is preferred to be an odd number.

A class of self-similar functions, referred to as Gabor wavelets, can be obtained by appropriate dilation and rotation of the mother Gabor function. This function can be dilated and rotated to get an array of filters. If we assume the number of dilations to be S and there are L number of orientations, the filter bank consists of $S \times L$ filters. The Gabor filter bank has the parameters (a) frequency half-peak bandwidth, (b) orientation half-peak bandwidth, (c) center frequency, and (d) orientation. The Gabor filters can be configured to have various shapes, bandwidths, center frequencies, and orientations by the adjustments of these parameters.

5.2.3 Concept of Multiresolution Analysis

There were a number of orthonormal wavelet basis functions of the form $\psi_{m,n}(t) = 2^{-m/2}\psi(2^{-m}t - n)$ discovered in the 1980s. The theory of *multiresolution analysis* presented a systematic approach to generate the wavelets [6, 10]. The idea of multiresolution analysis is to approximate a function $f(t)$ at different levels of resolution.

In multiresolution analysis, we consider two functions: the mother wavelet $\psi(t)$ and the *scaling function* $\phi(t)$. The dilated (scaled) and translated (shifted) version of the scaling function is given by $\phi_{m,n}(t) = 2^{-m/2}\phi(2^{-m}t - n)$. For fixed m, the set of scaling functions $\phi_{m,n}(t)$ are orthonormal. By the linear combinations of the scaling function and its translations we can generate a set of functions

$$f(t) = \sum_{n} \alpha_n \phi_{m,n}(t). \tag{5.15}$$

The set of all such functions generated by linear combination of the set $\{\phi_{m,n}(t)\}$ is called the span of the set $\{\phi_{m,n}(t)\}$, denoted by $\text{Span}\{\phi_{m,n}(t)\}$. Now consider V_m to be a vector space corresponding to $\text{Span}\{\phi_{m,n}(t)\}$. Assuming that the resolution increases with decreasing m, these vector spaces describe successive approximation vector spaces, $\cdots \subset V_2 \subset V_1 \subset V_0 \subset V_{-1} \subset V_{-2} \subset \cdots$, each with resolution 2^m (i.e., each space V_{j+1} is contained in the next resolution space V_j). In multiresolution analysis, the set of subspaces satisfies the following properties:

1. $V_{m+1} \subset V_m$, for all m: This property states that each subspace is contained in the next resolution subspace.

2. $\overline{\bigcup V_m} = L^2(\mathcal{R})$: This property indicates that the union of subspaces is dense in the space of square integrable functions $L^2(\mathcal{R})$; $\mathcal{R}$ indicates a set of real numbers (*upward completeness* property).

3. $\bigcap V_m = 0$ (an empty set): This property is called *downward completeness* property.

4. $f(t) \in V_0 \leftrightarrow f(2^{-m}t) \in V_m$: Dilating a function from resolution space V_0 by a factor of 2^m results in the lower resolution space V_m (*scale* or *dilation invariance* property).

5. $f(t) \in V_0 \leftrightarrow f(t-n) \in V_0$: Combining this with the scale invariance property above, this property states that translating a function in a resolution space does not change the resolution (*translation invariance property*).

6. There exists a set $\{\phi(t-n) \in V_0$: n is an integer$\}$ that forms an orthonormal basis of V_0.

The basic tenet of multiresolution analysis is that whenever the above properties are satisfied, there exists an orthonormal wavelet basis $\psi_{m,n}(t) = 2^{-m/2}\phi(2^{-m}t-n)$ such that

$$P_{m-1}(f) = P_m(f) + \sum c_{m,n}(f)\psi_{m,n}(t) \tag{5.16}$$

where P_j is the orthogonal projection of ψ onto V_j. For each m, consider the wavelet functions $\psi_{m,n}(t)$ span a vector space W_m. It is clear from Eq. 5.16 that the wavelet that generates the space W_m and the scaling function that generates the space V_m are not independent. W_m is exactly the orthogonal complement of V_m in V_{m-1}. Thus, any function in V_{m-1} can be expressed as the sum of a function in V_m and a function in the wavelet space W_m. Symbolically, we can express this as

$$V_{m-1} = V_m \oplus W_m. \tag{5.17}$$

Since m is arbitrary,

$$V_m = V_{m+1} \oplus W_{m+1}. \tag{5.18}$$

Thus,

$$V_{m-1} = V_{m+1} \oplus W_{m+1} \oplus W_m. \tag{5.19}$$

Continuing in this fashion, we can establish that

$$V_{m-1} = V_k \oplus W_k \oplus W_{k-1} \oplus W_{k-2} \oplus \cdots W_m \tag{5.20}$$

for any $k \geq m$.

Thus, if we have a function that belongs to the space V_{m-1} (i.e., the function can be exactly represented by the scaling function at resolution $m-1$), we can decompose it into a sum of functions starting with lower-resolution approximation followed by a sequence of functions generated by dilations of the wavelet that represent the loss of information in terms of details. Let us consider the representation of an image with fewer and fewer pixels at successive levels of approximation. The wavelet coefficients can then be considered as the additional detail information needed to go from a coarser to a finer approximation. Hence, in each level of decomposition the signal can be decomposed into two parts, one is the coarse approximation of the signal in the lower resolution and the other is the detail information that was lost because of the approximation. The wavelet coefficients derived by Eq. 5.10 or 5.11, therefore, describe the information (detail) lost when going from an approximation of the signal at resolution 2^{m-1} to the coarser approximation at resolution 2^m.

5.2.4 Implementation by Filters and the Pyramid Algorithm

It is clear from the theory of multiresolution analysis in the previous section that multiresolution analysis decomposes a signal into two parts–one approximation of the original signal from finer to coarser resolution and the other detail information that was lost due to the approximation. This can be mathematically represented as

$$f_m(t) = \sum_n a_{m+1,n}\phi_{m+1,n} + \sum_n c_{m+1,n}\psi_{m+1,n} \tag{5.21}$$

where $f_m(t)$ denotes the value of input function $f(t)$ at resolution 2^m, $c_{m+1,n}$ is the detail information, and $a_{m+1,n}$ is the coarser approximation of the signal at resolution 2^{m+1}. The functions, $\phi_{m+1,n}$ and $\psi_{m+1,n}$, are the dilation and wavelet basis functions (orthonormal).

In 1989, Mallat [6] proposed the multiresolution approach for wavelet decomposition of signals using a pyramidal filter structure of *quadrature mirror filter* (QMF) pairs. In multiresolution analysis, it can be proven that decomposition of signals using the discrete wavelet transform can be expressed in terms of FIR filters [6, 10] and all the discussions on multiresolution analysis boil down to the following algorithm (Eq. 5.22) for computation of the wavelet coefficients for the signal $f(t)$. For details see the original paper by Mallat [6].

$$\left.\begin{aligned} c_{m,n}(f) &= \textstyle\sum_k g_{2n-k}\, a_{m-1,k}(f) \\ a_{m,n}(f) &= \textstyle\sum_k h_{2n-k}\, a_{m-1,k}(f) \end{aligned}\right\} \tag{5.22}$$

where g and h are the high-pass and low-pass filters, $g_i = (-1)^i h_{-i+1}$ and $h_i = 2^{1/2}\int\phi(x-i)\phi(2x)\,dx$. Actually, $a_{m,n}(f)$ are the coefficients characterizing the projection of the function $f(t)$ in the vector subspace V_m (i.e., approximation of the function in resolution 2^m), whereas $c_{m,n}(f) \in W_m$ are the wavelet coefficients (detail information) at resolution 2^m. If the input signal $f(t)$ is in discrete sampled form, then we can consider these samples as the highest-order resolution approximation coefficients $a_{0,n}(f) \in V_0$ and Eq. 5.22 describes the multiresolution subband decomposition algorithm to construct $a_{m,n}(f)$ and $c_{m,n}(f)$ at level m with a low-pass filter h and high-pass filter g from $c_{m-1,n}(f)$, which were generated at level $m-1$. These filters are called the *analysis* filters. The recursive algorithm to compute DWT in different levels using Eq. 5.22 is popularly called Mallat's *Pyramid Algorithm* [6]. Since the synthesis filters h and g have been derived from the orthonormal basis functions ϕ and ψ, these filters give exact reconstruction

$$a_{m-1,i}(f) = \sum_n h_{2n-i}\, a_{m,n}(f) + \sum_n g_{2n-i}\, c_{m,n}(f) \tag{5.23}$$

Most of the orthonormal wavelet basis functions have infinitely supported ψ and accordingly the filters h and g could be with infinitely many taps.

However, for practical implementation of the DWT for image processing applications, it is desirable to have finite impulse response (FIR) filters with a small number of taps. It is possible to construct such filters by relaxing the orthonormality requirements and using biorthogonal basis functions. It should be noted that the wavelet filters are orthogonal when $(h', g') = (h, g)$, otherwise it is biorthogonal. In such a case the filters (h' and g', called the *synthesis* filters) for reconstruction of the signal can be different than the *analysis* filters (h and g) for decomposition of the signals. In order to achieve exact reconstruction, we can construct the filters such that it satisfies the relationship of the synthesis filter with the analysis filter as shown in Eq. 5.24:

$$\left.\begin{aligned} g'_n &= (-1)^n h_{-n+1} \\ g_n &= (-1)^n h'_{-n+1} \\ \textstyle\sum_n h_n h'_{n+2k} &= \delta_{k,0} \end{aligned}\right\}. \tag{5.24}$$

If $(h', g') = (h, g)$, the wavelet filters are called *orthogonal*, otherwise they are called *biorthogonal*. The popular (9, 7) wavelet filter adopted in JPEG2000 standard is one example of such a biorthogonal filter [8]. The signal is still decomposed using Eq. 5.22, but the reconstruction equation is now done using the synthesis filters h' and g' as shown in Eq. 5.25:

$$a_{m-1,i}(f) = \sum_n a_{m,n}(f) h'_{2n-i} + \sum_n c_{m,n}(f) g'_{2n-i}. \tag{5.25}$$

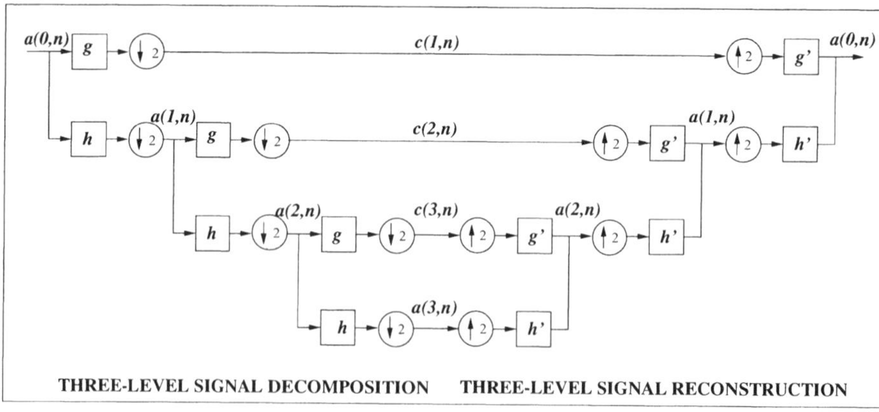

Fig. 5.3 Three-level multiresolution wavelet decomposition and reconstruction of signals using pyramidal filter structure.

Let's summarize the DWT computation here in terms of simple digital FIR filtering. Given the input discrete signal $x(n)$ (shown as $a(0, n)$ in Figure 5.3), it is filtered parallelly by a low-pass filter (h) and a high-pass filter (g) at each transform level. The two output streams are then subsampled by simply dropping the alternate output samples in each stream to produce the low-pass subband y_L (shown as $a(1, n)$ in Figure 5.3) and high-pass subband y_H (shown as $c(1, n)$ in Figure 5.3). The above arithmetic computation can be expressed as follows:

$$y_L(n) = \sum_{i=0}^{\tau_L - 1} h(i)x(2n - i), \quad y_H(n) = \sum_{i=0}^{\tau_H - 1} g(i)x(2n - i) \tag{5.26}$$

where τ_L and τ_H are the lengths of the low-pass (h) and high-pass (g) filters respectively. Since the low-pass subband $a(1, n)$ is an approximation of the input signal, we can apply the above computation again on $a(1, n)$ to produce the subbands $a(2, n)$ and $c(2, n)$ and so on. This multiresolution decomposition approach is shown in Figure 5.3 for three levels of decomposition. During the inverse transform to reconstruct the signal, both $a(3, n)$ and $c(3, n)$ are first upsampled by inserting zeros between two samples, and then they are filtered by low-pass (h') and high-pass (g') filters respectively. These two filtered output streams are added together to reconstruct $a(2, n)$ as shown in Figure 5.3. The same continues until we reconstruct the original signal $a(0, n)$.

5.3 EXTENSION TO TWO-DIMENSIONAL SIGNALS

The 2D extension of DWT is essential for transformation images. A two-dimensional signal (image) can be represented by a 2D array $X[M, N]$ with M rows and N columns, where M and N are nonnegative integers. The simple approach for 2D implementation of the DWT is to perform the one-dimensional DWT row-wise to produce an intermediate result and then perform the same one-dimensional DWT column-wise on this intermediate result to produce the final result. This is shown in Figure 5.4(a). This is possible because the two-dimensional scaling functions can be expressed as separable functions which are the product of two one-dimensional scaling functions such as $\phi_2(x, y) = \phi_1(x)\phi_1(y)$. The same is true for the wavelet function $\psi(x, y)$ as well. Applying the one-dimensional transform in each row, we produce two subbands in each row. When the low-frequency subbands of all the rows (L) are put together, it looks like a thin version (of size $M \times \frac{N}{2}$) of the input signal as shown in Figure 5.4(a). Similarly we put together the high-frequency subbands of all the rows to produce the H subband of size $M \times \frac{N}{2}$, which contains mainly the high-frequency information around discontinuities (edges in an image) in the input signal. Then applying a one-dimensional DWT column-wise on these L and H subbands (intermediate result), we produce four subbands LL, LH, HL, and HH of size $\frac{M}{2} \times \frac{N}{2}$ respectively, as shown in Figure 5.4(a). LL is a coarser version of the original input signal. LH, HL, and HH are the high-frequency subband containing the detail information. It should be noted that we could have applied the one-dimensional DWT column-wise first and then row-wise to achieve the same result.

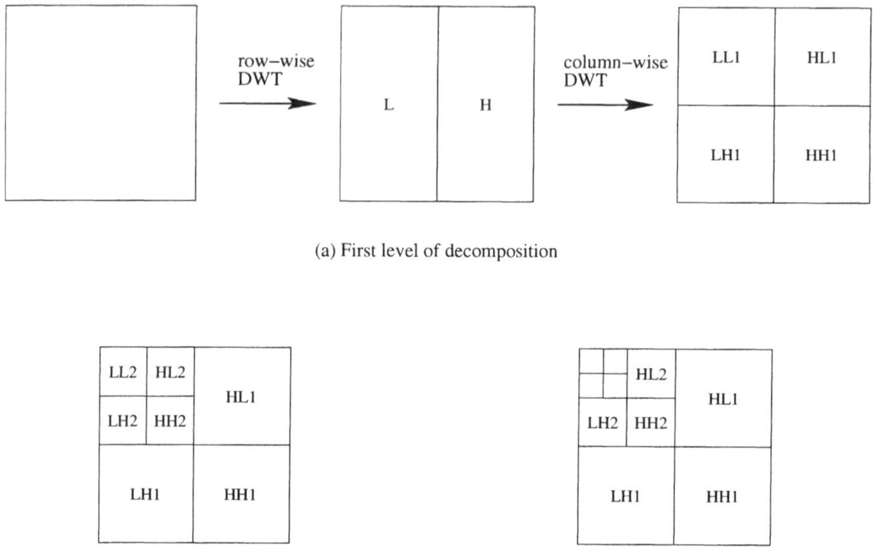

(a) First level of decomposition

(b) Second level decomposition

(c) Third level decomposition

Fig. 5.4 Row-column computation of two-dimensional DWT.

As shown in Figure 5.4(b), the LL1 subband can be further decomposed into four subbands LL2, HL2, LH2, and HH2 based on the principle of multiresolution analysis. The same computation can continue to further decompose LL2 into higher levels. In Figure 5.5, we show the result of the wavelet transform at different levels with a real-life image provided by the JPEG2000 standard committee. The subbands have been normalized to 8 bits for the purpose of display.

5.4 LIFTING IMPLEMENTATION OF THE DWT

The DWT implementation is basically frame-based as opposed to the DCT-type block-based implementation. Such an implementation requires both a large number of arithmetic computations and a large memory. Recently, a new mathematical formulation for wavelet transformation has been proposed by Swelden [11] based on spatial construction of the wavelets and a very versatile scheme for its factorization has been suggested in [12]. This new approach is called the lifting-based wavelet transform, or simply *lifting*. The main feature of the lifting-based DWT scheme is to break up the high-pass and low-pass wavelet filters into a sequence of smaller filters that in turn

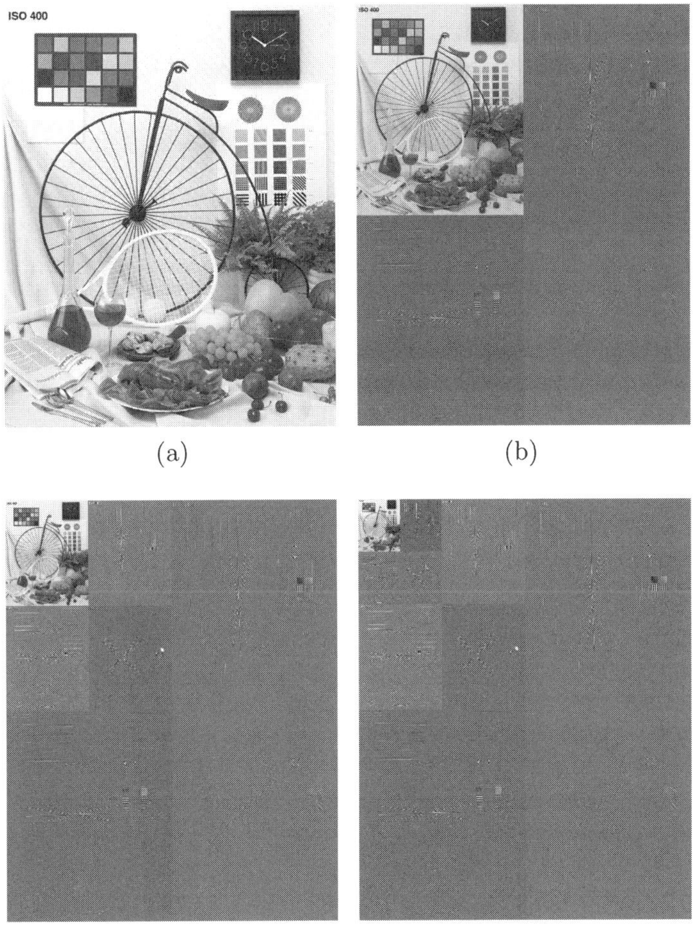

Fig. 5.5 (a) Original BIKE image and subbands; (b) after level 1, (c) after level 2, (d) after level 3 decomposition.

can be converted into a sequence of upper and lower triangular matrices, which will be discussed in the subsequent section. This scheme often requires far fewer computations compared to the convolution-based DWT [11, 12], and its computational complexity can be reduced up to 50%. It has several other advantages, including "in-place" computation of the DWT, integer-to-integer wavelet transform (IWT), symmetric forward and inverse transform, requiring no signal boundary extension, etc. Lifting has been suggested for implementation of the DWT in the upcoming JPEG2000 standard [8].

In FIR-based DWT implementation, the input signal (x) is filtered separately by a low-pass filter ($\tilde{h}$) and a high-pass filter ($\tilde{g}$). The two output

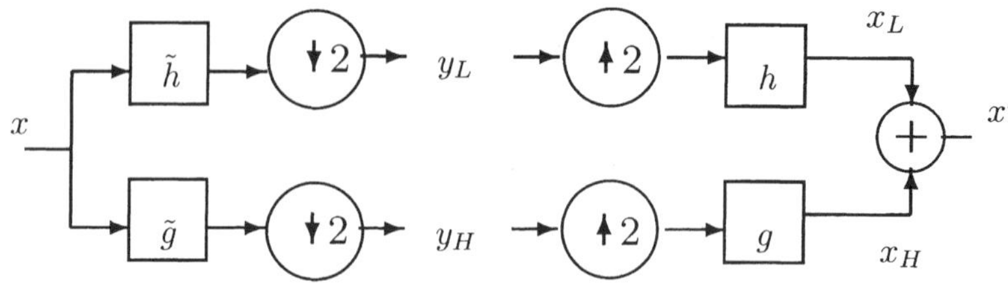

Fig. 5.6 Signal analysis and reconstruction in DWT.

streams are then subsampled by simply dropping the alternate samples to produce the low-pass (y_L) and high-pass (y_H) subbands as shown in Figure 5.6. These two filters ($\tilde{h}$,$\tilde{g}$) form the *analysis* filter bank. The original signal can be reconstructed by a *synthesis* filter bank (h,g) starting from y_L and y_H as shown in Figure 5.6. We have adopted the discussion on lifting from the celebrated paper by Daubechies and Sweldens [12]. It should also be noted that we adopted the notation ($\tilde{h}$, $\tilde{g}$) for the analysis filter and (h, g) as the synthesis filter in this section and onward in this chapter. Given a discrete signal $x(n)$, arithmetic computation of above can be expressed as follows:

$$y_L(n) = \sum_{i=0}^{\tau_L-1} \tilde{h}(i)x(2n-i), \quad y_H(n) = \sum_{i=0}^{\tau_H-1} \tilde{g}(i)x(2n-i) \tag{5.27}$$

where τ_L and τ_H are the lengths of $\tilde{h}$ and $\tilde{g}$ respectively. During the inverse transform, both y_L and y_H are first upsampled by inserting zeros between two samples and then they are filtered by low-pass (h) and high-pass (g) synthesis filters respectively. The two output streams are added to obtain the reconstructed signal (x') as shown in Figure 5.6.

5.4.1 Finite Impulse Response Filter and Z-transform

A digital *filter* $h = \{\cdots, h_{k-1}, h_k, h_{k+1}, \ldots\}$ is a linear time-invariant operator which can be completely defined by its *impulse* response $\{h_k \in R \,|\, k \in Z\}$. These *impulse* responses (h_k) are popularly called filter *coefficients*. The Z-transform of an FIR filter h is expressed as a Laurent polynomial $h(z)$ as shown in Eq. 5.28,

$$h(z) = \sum_{i=m}^{n} h_i z^{-i} \tag{5.28}$$

where m and n are positive integers. The degree of the above Laurent polynomial is defined as $|h(z)| = n - m$. Thus the length of the FIR filter h is $n-m+1$ (i.e., the degree of the associated Laurent polynomial plus one). The sum or difference of two Laurent polynomials is a Laurent polynomial. The product of two Laurent polynomials $a(z)$ and $b(z)$ is a Laurent polynomial of degree $|a(z)| + |b(z)|$. Let us assume that $b(z) \neq 0$ and $|a(z)| \geq |b(z)|$. In general, exact division of $a(z)$ by $b(z)$ is not possible. However, division with remainder is possible although this division is not unique. There always exists a quotient $q(z)$ and a remainder $r(z)$ (not necessarily unique) with $|q(z)| = |a(z)| - |b(z)|$ and $|r(z)| < |b(z)|$ so that

$$a(z) = b(z)q(z) + r(z). \tag{5.29}$$

5.4.2 Euclidean Algorithm for Laurent Polynomials

The Euclidean algorithm can be used to find the *greatest common divisor* (gcd) of two Laurent polynomials $a(z)$ and $b(s)$. If $b(z) \neq 0$ and $|a(z)| \geq |b(z)|$, we can state the algorithm as follows. By operations '/' and '%' in the algorithm, we mean to find the quotient and remainder of the division.

begin
- $k = 0$;
- $a_k(z) = a(z)$;
- $b_k(z) = b(z)$;
- **while** $b_k(z) \neq 0$ **do**
- {
  - $a_{k+1} = b_k(z)$;
  - $b_{k+1} = a_k(z) \% b_k(z)$;
  - $q_{k+1} = a_k(z) / b_k(z)$;
  - $k = k + 1$
- }
- $gcd = a_k(z)$;

end.

From the above algorithm, it is clear that the greatest common divisor (gcd) of $a(z)$ and $b(z)$ is a_n, where n is the smallest integer for which $b_n(z) = 0$. The number of iterations by the **while** loop in the above algorithms is bounded by $n \leq |n(z)| + 1$. From the above algorithm, we can establish that

$$\begin{bmatrix} a_{i+1}(z) \\ b_{i+1}(z) \end{bmatrix} = \begin{bmatrix} 0 & 1 \\ 1 & -q_i(z) \end{bmatrix} \begin{bmatrix} a_i(z) \\ b_i(z) \end{bmatrix} \tag{5.30}$$

which can be rewritten as

$$\begin{bmatrix} a_i(z) \\ b_i(z) \end{bmatrix} = \begin{bmatrix} q_i(z) & 1 \\ 1 & 0 \end{bmatrix} \begin{bmatrix} a_{i+1}(z) \\ b_{i+1}(z) \end{bmatrix}. \tag{5.31}$$

Thus,

$$\begin{bmatrix} a_0(z) \\ b_0(z) \end{bmatrix} = \begin{bmatrix} q_0(z) & 1 \\ 1 & 0 \end{bmatrix} \begin{bmatrix} a_1(z) \\ b_1(z) \end{bmatrix} = \begin{bmatrix} q_0(z) & 1 \\ 1 & 0 \end{bmatrix} \begin{bmatrix} q_1(z) & 1 \\ 1 & 0 \end{bmatrix} \begin{bmatrix} a_2(z) \\ b_2(z) \end{bmatrix}. \tag{5.32}$$

Since $a_0(z) = a(z)$ and $b_0(z) = b(z)$, we obtain the following factorization after iterating the above equation:

$$\begin{bmatrix} a(z) \\ b(z) \end{bmatrix} = \prod_{k=1}^{n} \begin{bmatrix} q_i(z) & 1 \\ 1 & 0 \end{bmatrix} \begin{bmatrix} a_n(z) \\ 0 \end{bmatrix}. \tag{5.33}$$

The above factorization algorithm will be used in Section 5.4.4.3 to show how the polyphase matrix for a filter pair can be factorized into lifting sequences.

5.4.3 Perfect Reconstruction and Polyphase Representation of Filters

For any practical signal transformation technique from one domain to another, the transformation should be reversible. For example, the Fourier transform converts a signal from the time domain to the frequency domain. When inverse Fourier transform is applied on the signal in frequency domain, the signal is converted back to the time domain. Ideally, if there is no additional processing or manipulation done in the frequency domain data after the transformation (i.e., if there is no loss of data or information at any form), the reconstructed signal after inverse Fourier transform should be an exact replica of the original one. The same principle applies for the DWT as well. Hence we need to choose the filter bank for DWT in such a way that perfect reconstruction is achieved. For the filter bank in Figure 5.6, the conditions for perfect reconstruction of a signal [12] are given by

$$\left. \begin{aligned} h(z)\tilde{h}(z^{-1}) + g(z)\tilde{g}(z^{-1}) &= 2 \\ h(z)\tilde{h}(-z^{-1}) + g(z)\tilde{g}(-z^{-1}) &= 0 \end{aligned} \right\} \tag{5.34}$$

where $h(z)$ is the Z-transform of the FIR filter h.

The *polyphase* representation of a filter h is expressed as

$$h(z) = h_e(z^2) + z^{-1}h_o(z^2) \tag{5.35}$$

where h_e contains the even filter coefficients and h_o contains the odd filter coefficients of the FIR filter h. In general by *polyphase* representation, we mean to split a sequence into several subsequences for possible parallel processing of the subsequences. From Eq. 5.35, we can intuitively split the filter into two smaller filters–one (h_e) with the even filter coefficients and the other (h_o)

with the odd filter coefficients delayed by a clock cycle, whose Z-transform can be expressed as

$$h_e(z) = \sum_k h_{2k} z^{-k}, \quad h_o(z) = \sum_k h_{2k+1} z^{-k} \tag{5.36}$$

and we can define a *polyphase matrix* for the filter h as

$$P(z) = \begin{bmatrix} h_e(z) & g_e(z) \\ h_o(z) & g_o(z) \end{bmatrix}. \tag{5.37}$$

Based on the discussion above, the polyphase representation of the filters $g(z)$, $\tilde{h}(z)$, and $\tilde{g}(z)$ is expressed as follows:

$$\left.\begin{aligned} g(z) &= g_e(z^2) + z^{-1} g_o(z^2) \\ \tilde{h}(z) &= \tilde{h}_e(z^2) + z^{-1} \tilde{h}_o(z^2) \\ \tilde{g}(z) &= \tilde{g}_e(z^2) + z^{-1} \tilde{g}_o(z^2) \end{aligned}\right\}. \tag{5.38}$$

Based on the above formulation, we can define two *polyphase matrices* as follows:

$$\tilde{P}(z) = \begin{bmatrix} \tilde{h}_e(z) & \tilde{h}_o(z) \\ \tilde{g}_e(z) & \tilde{g}_o(z) \end{bmatrix}, \quad P(z) = \begin{bmatrix} h_e(z) & g_e(z) \\ h_o(z) & g_o(z) \end{bmatrix}. \tag{5.39}$$

Often the polyphase matrix $P(z)$ is called the *dual* of the polyphase matrix $\tilde{P}(z)$. For perfect reconstruction, these two polyphase matrices $P(z)$ and $\tilde{P}(z)$ satisfy the following relation in Eq. 5.40,

$$P(z)\tilde{P}(z^{-1})^T = I \tag{5.40}$$

where I is the 2×2 *identity* matrix. Now based on the above formulation, the wavelet transform in terms of the polyphase matrix can be expressed as

$$\begin{bmatrix} y_L(z) \\ y_H(z) \end{bmatrix} = \tilde{P}(z) \begin{bmatrix} x_e(z) \\ z^{-1} x_o(z) \end{bmatrix} \tag{5.41}$$

for the forward DWT and

$$\begin{bmatrix} x_e(z) \\ z^{-1} x_o(z) \end{bmatrix} = P(z) \begin{bmatrix} y_L(z) \\ y_H(z) \end{bmatrix} \tag{5.42}$$

for the inverse DWT.

If the determinant of the polyphase matrix $P(z)$ is unity (i.e., $|P(z)| = h_e(z)g_o(z) - g_e(z)h_o(z) = 1$), then the matrix $P(z)$ is invertible. Hence we can apply Cramer's rule [12] in Eq. 5.40 as follows:

$$\tilde{P}(z^{-1}) = P(z)^{-1} = \frac{\begin{bmatrix} g_o(z) & -g_e(z) \\ -h_o(z) & h_e(z) \end{bmatrix}}{|P(z)|} = \begin{bmatrix} g_o(z) & -g_e(z) \\ -h_o(z) & h_e(z) \end{bmatrix}. \tag{5.43}$$

From Eq. 5.39, we find that

$$\tilde{P}(z^{-1}) = \begin{bmatrix} \tilde{h}_e(z^{-1}) & \tilde{h}_o(z^{-1}) \\ \tilde{g}_e(z^{-1}) & \tilde{g}_o(z^{-1}) \end{bmatrix}. \tag{5.44}$$

From Eqs. 5.43 and 5.44, we can conclude that

$$\left.\begin{aligned} \tilde{h}_e(z) &= g_o(z^{-1}), \\ \tilde{h}_o(z) &= -g_e(z^{-1}), \\ \tilde{g}_e(z) &= -h_o(z^{-1}), \\ \tilde{g}_O(z) &= h_e(z^{-1}) \end{aligned}\right\} \tag{5.45}$$

which implies that

$$\tilde{h}(z) = -z^{-1}g(-z^{-1}), \qquad \tilde{g}(z) = z^{-1}h(-z^{-1}) \tag{5.46}$$

and hence

$$h(z) = -z^{-1}\tilde{g}(-z^{-1}), \qquad g(z) = z^{-1}\tilde{h}(-z^{-1}). \tag{5.47}$$

When the determinant of $P(z)$ is unity, the synthesis filter pair (h,g) is called *complementary* and so is the analysis filter pair $(\tilde{h},\tilde{g})$. When $(h,g)=(\tilde{h},\tilde{g})$, the wavelet transformation is called orthogonal; otherwise it is biorthogonal.

When $h(z) = \tilde{h}(z) = g(z) = \tilde{g}(z) = 1$, the DWT simply splits an input signal $(x = \{x_k \mid k \in Z\})$ into two subsequences, one with all the odd samples (x_{2i+1}) and the other with all the even sequences (x_{2i}). This is called the *lazy* wavelet transform [11].

5.4.4 Lifting

There are two types of lifting. One is called *primal lifting* and the other is called *dual lifting*. We define these two types of lifting based on the mathematical formulations shown in the previous section.

5.4.4.1 Primal Lifting According to the lifting theorem [12], if the wavelet filter pair $(h,\ g)$ is complementary then any other FIR filter g^{new} that is complementary to h is of the form

$$g^{new}(z) = g(z) + h(z)s(z^2) \tag{5.48}$$

where $s(z^2)$ is a Laurent polynomial.

Proof: Expanding $g^{new}(z)$ in polyphase representation, we get

$$\begin{aligned} g^{new}(z) &= g(z) + h(z)s(z^2) \\ &= \{g_e(z^2) + z^{-1}g_o(z^2)\} + \{h_e(z^2) + z^{-1}h_o(z^2)\}s(z^2) \\ &= \{g_e(z^2) + h_e(z^2)s(z^2)\} + z^{-1}\{g_o(z^2) + h_o(z^2)s(z^2)\}. \end{aligned} \tag{5.49}$$

We know that

$$h(z) = h_e(z^2) + z^{-1}h_o(z^2). \tag{5.50}$$

Hence the new polyphase matrix can be defined as

$$\begin{aligned} P^{new}(z) &= \begin{bmatrix} h_e(z) & g_e(z) + h_e(z)s(z) \\ h_o(z) & g_o(z) + h_o(z)s(z) \end{bmatrix} \\ &= \begin{bmatrix} h_e(z) & g_e(z) \\ h_o(z) & g_o(z) \end{bmatrix} \begin{bmatrix} 1 & s(z) \\ 0 & 1 \end{bmatrix} \\ &= P(z) \begin{bmatrix} 1 & s(z) \\ 0 & 1 \end{bmatrix}. \end{aligned} \tag{5.51}$$

It can be easily verified that the determinant of $P^{new}(z)$ is 1 and hence it proves Eq. 5.48. From Eq. 5.40 we know that

$$P^{new}(z)\tilde{P}^{new}(z^{-1})^T = I. \tag{5.52}$$

Thus, we can derive that

$$\tilde{P}^{new}(z^{-1}) = [P(z)]^{-1} \begin{bmatrix} 1 & s(z) \\ 0 & 1 \end{bmatrix}^{-1} = \tilde{P}(z^{-1}) \begin{bmatrix} 1 & 0 \\ -s(z) & 1 \end{bmatrix}. \tag{5.53}$$

Consequently

$$\tilde{P}^{new}(z) = \tilde{P}(z) \begin{bmatrix} 1 & 0 \\ -s(z^{-1}) & 1 \end{bmatrix}. \tag{5.54}$$

Hence, the lifting created a new low-pass filter

$$\tilde{h}^{new}(z) = \tilde{h}(z) - \tilde{g}(z)s(z^{-2}). \tag{5.55}$$

As a result, we have lifted the low-pass subband with the help of the high-pass subband. This is called the *primal lifting.*

5.4.4.2 Dual Lifting By *dual lifting* we mean lifting the high-pass subband with the help the low-pass subband. If $(h,\ g)$ is complementary, then any other new FIR filter h^{new} complementary to g is of the form

$$h^{new}(z) = h(z) + g(z)t(z^2) \tag{5.56}$$

where $t(z^2)$ is a Laurent polynomial. Following the similar deduction as presented in the primal lifting section, the dual lifting creates a new high-pass filter

$$\tilde{g}^{new}(z) = \tilde{g}(z) - \tilde{h}(z)t(z^{-2}). \tag{5.57}$$

5.4.4.3 Lifting Factorization In this section, we show how a complementary filter pair for wavelet transformation can be factorized into lifting steps.

We can compute the *greatest common divisor* (*gcd*) of $\tilde{h}_e(z)$ and $\tilde{h}_o(z)$ by applying the Euclidean algorithm as shown in Section 5.4.2. If K is the gcd of $\tilde{h}_e(z)$ and $\tilde{h}_o(z)$, we can express $\tilde{h}_e(z)$ and $\tilde{h}_o(z)$ as follows:

$$\begin{bmatrix} \tilde{h}_e(z) \\ \tilde{h}_o(z) \end{bmatrix} = \prod_{k=1}^{n} \begin{bmatrix} q_i(z) & 1 \\ 1 & 0 \end{bmatrix} \begin{bmatrix} K \\ 0 \end{bmatrix}. \tag{5.58}$$

According to the theory of lifting discussed in Section 5.4.4, if $(\tilde{h}, \tilde{g})$ is a complementary filter pair, then we can always find another complementary filter $\tilde{g}^{new}$ so that the polyphase matrix can be represented as

$$\tilde{P}^{new}(z) = \begin{bmatrix} \tilde{h}_e(z) & \tilde{g}_e^{new} \\ \tilde{h}_o(z) & \tilde{g}_o^{new} \end{bmatrix} = \prod_{k=1}^{n} \begin{bmatrix} q_i(z) & 1 \\ 1 & 0 \end{bmatrix} \begin{bmatrix} K & 0 \\ 0 & 1/K \end{bmatrix}. \tag{5.59}$$

We can again rewrite Eq. 5.59 as

$$\tilde{P}^{new}(z) = \prod_{k=1}^{n/2} \begin{bmatrix} q_{2i-1}(z) & 1 \\ 1 & 0 \end{bmatrix} \begin{bmatrix} q_{2i}(z) & 1 \\ 1 & 0 \end{bmatrix} \begin{bmatrix} K & 0 \\ 0 & 1/K \end{bmatrix}. \tag{5.60}$$

It should be noted that

$$\begin{bmatrix} q_{2i-1}(z) & 1 \\ 1 & 0 \end{bmatrix} = \begin{bmatrix} 1 & q_{2i-1}(z) \\ 0 & 1 \end{bmatrix} \begin{bmatrix} 0 & 1 \\ 1 & 0 \end{bmatrix} \tag{5.61}$$

and

$$\begin{bmatrix} q_{2i}(z) & 1 \\ 1 & 0 \end{bmatrix} = \begin{bmatrix} 0 & 1 \\ 1 & 0 \end{bmatrix} \begin{bmatrix} 1 & 0 \\ q_{2i}(z) & 1 \end{bmatrix}. \tag{5.62}$$

Applying the rules in Eqs. 5.61 and 5.62 into Eq. 5.60, we can rewrite it as

$$\tilde{P}^{new}(z) = \prod_{k=1}^{n/2} \begin{bmatrix} 1 & q_{2i-1}(z) \\ 0 & 1 \end{bmatrix} \begin{bmatrix} 1 & 0 \\ q_{2i}(z) & 0 \end{bmatrix} \begin{bmatrix} K & 0 \\ 0 & 1/K \end{bmatrix}. \tag{5.63}$$

We also know from the lifting formulation that we can always construct filter $\tilde{g}$ by lifting $\tilde{g}^{new}$ as

$$\tilde{P}(z) = \tilde{P}^{new}(z) \begin{bmatrix} 1 & \tilde{s}(z) \\ 0 & 1 \end{bmatrix}. \tag{5.64}$$

By combining all of the above formulations, we can conclude that given a complementary filter pair $(\tilde{h}, \tilde{g})$, there always exist Laurent polynomials $\tilde{s}_i(z)$ and $\tilde{t}_i(z)$ for $1 \leq i \leq n$ and we can factorize the polyphase matrix $\tilde{P}(z)$ into a finite sequence of alternating upper and lower triangular matrices as follows,

$$\tilde{P}(z) = \left\{ \prod_{i=1}^{m} \begin{bmatrix} 1 & \tilde{s}_i(z) \\ 0 & 1 \end{bmatrix} \begin{bmatrix} 1 & 0 \\ \tilde{t}_i(z) & 1 \end{bmatrix} \right\} \begin{bmatrix} K & 0 \\ 0 & 1/K \end{bmatrix} \tag{5.65}$$

where K is a constant and acts as a scaling factor (so is $\frac{1}{K}$). In practice, $\tilde{s}_i(z)$ and $\tilde{t}_i(z)$ are usually of second- or lower-order polynomials, which correspond to usually one- to three-tap FIR filters. Computing the upper triangular matrix is known as *primal lifting*, and this is emphasized in the literature as lifting the low-pass subband with the help of the high-pass subband. Similarly, computation of the lower triangular matrix is called *dual lifting*, which is lifting of the high-pass subband with the help of the low-pass subband [11, 12]. Often these two basic lifting steps are called *update* and *predict* as well. The above factorization can also be formulated in the following way:

$$\tilde{P}(z) = \left\{ \prod_{i=1}^{m} \begin{bmatrix} 1 & 0 \\ \tilde{t}_i(z) & 1 \end{bmatrix} \begin{bmatrix} 1 & \tilde{s}_i(z) \\ 0 & 1 \end{bmatrix} \right\} \begin{bmatrix} K & 0 \\ 0 & 1/K \end{bmatrix}. \tag{5.66}$$

5.4.4.4 Lifting Algorithm Hence the lifting-based forward wavelet transform essentially means first applying the *lazy* wavelet transform on the input stream (split into even and odd samples), then alternately executing *primal* and *dual* lifting steps, and finally *scaling* the two output streams by $\frac{1}{K}$ and K respectively to produce low-pass and high-pass subbands, as shown in Figure 5.7(a). The inverse DWT using lifting can be derived by traversing the above steps in the reverse direction, first scaling the low-pass and high-pass subband inputs by K and $1/K$ respectively, and then applying the dual and primal lifting steps after reversing the signs of the coefficients in $\tilde{t}(z)$ and $\tilde{s}(z)$, and finally the inverse lazy transform by upscaling the output before merging them into a single reconstructed stream as shown in Figure 5.7(b).

Due to the linearity of the lifting scheme, if the input data are in integer format, it is possible to maintain data in integer format throughout the transform by introducing a rounding function in the filtering operation. Due to this property, the transform is reversible (i.e., lossless) and is called *integer wavelet transform* (IWT) [13]. It should be noted that filter coefficients need not be integers for IWT. However, if a scaling step is present in the factorization, IWT cannot be achieved. It has been proposed in [13] to split the scaling step into additional lifting steps to achieve IWT.

5.4.4.5 Example Consider the Le Gall (5,3) spline filter, with $\tilde{h} = (-\frac{1}{8}, \frac{1}{4}, \frac{3}{4}, \frac{1}{4}, -\frac{1}{8})$ and $\tilde{g} = (-\frac{1}{2}, 1, -\frac{1}{2})$. Hence,

$$\begin{aligned} \tilde{h}(z) &= -\tfrac{1}{8}z^{-2} + \tfrac{1}{4}z^{-1} + \tfrac{3}{4}z^{0} + \tfrac{1}{4}z - \tfrac{1}{8}z^{2}, \\ \tilde{g}(z) &= -\tfrac{1}{2}z^{-2} + z^{-1} - \tfrac{1}{2}z^{0}. \end{aligned}$$

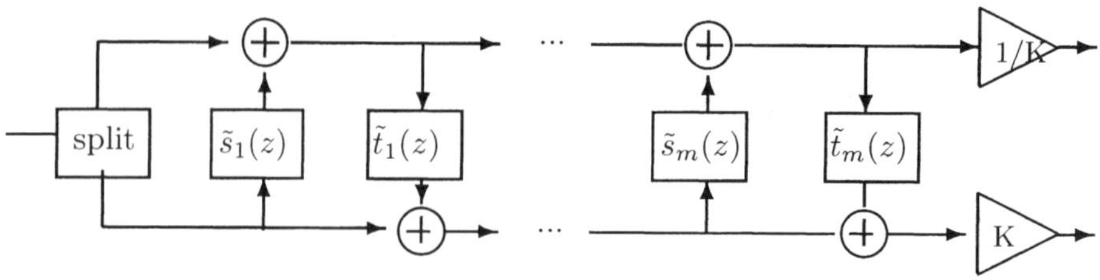

(a) Forward transform

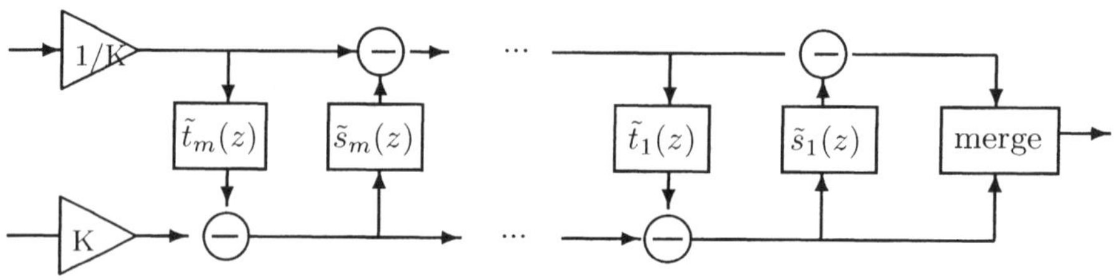

(b) Inverse transform

Fig. 5.7 Lifting-based forward and inverse DWT.

From the above equations, we can easily derive that

$$\tilde{h}_e(z^2) = -\tfrac{1}{8}z^{-2} + \tfrac{3}{4} - \tfrac{1}{8}z^2, \quad \tilde{h}_o(z^2) = \tfrac{1}{4} + \tfrac{1}{4}z^2,$$
$$\tilde{g}_e(z^2) = -\tfrac{1}{2}z^{-2} - \tfrac{1}{2}, \quad \tilde{g}_o(z^2) = 1.$$

As a result, the polyphase matrix of this filter bank is

$$\tilde{P}(z) = \begin{bmatrix} \tilde{h}_e(z) & \tilde{h}_o(z) \\ \tilde{g}_e(z) & \tilde{g}_o(z) \end{bmatrix} = \begin{bmatrix} -\tfrac{1}{8}z^{-1} + \tfrac{3}{4} - \tfrac{1}{8}z & \tfrac{1}{4} + \tfrac{1}{4}z \\ -\tfrac{1}{2}z^{-1} - \tfrac{1}{2} & 1 \end{bmatrix}.$$

Also based on conditions of perfect reconstructions of the complementary filters as described in Eq. 5.34, we can derive the corresponding synthesis filters as follows:

$$h(z) = -z^{-1}\tilde{g}(-z^{-1}) = \tfrac{1}{2}z^{-1} + 1 + \tfrac{1}{2}z,$$
$$g(z) = z^{-1}\tilde{h}(-z^{-1}) = -\tfrac{1}{8}z^{-3} - \tfrac{1}{4}z^{-2} + \tfrac{3}{4}z^{-1} - \tfrac{1}{4} - \tfrac{1}{8}z$$

and hence $h = (\frac{1}{2}, 1, \frac{1}{2})$ and $g = (-\frac{1}{8}, -\frac{1}{4}, \frac{3}{4}, -\frac{1}{4}, -\frac{1}{8})$.

Now based on the lifting factorization of the polyphase matrix, the possible factorization of $\tilde{P}(z)$ that leads to a band matrix multiplication is

$$\tilde{P}(z) = \begin{bmatrix} 1 & \tfrac{1}{4}(1+z) \\ 0 & 1 \end{bmatrix} \begin{bmatrix} 1 & 0 \\ -\tfrac{1}{2}(1+z^{-1}) & 1 \end{bmatrix}.$$

If the samples are numbered starting from 0, we consider the even terms of the output stream as the samples of low-pass subband and similarly the odd terms as the samples of high-pass subband. Accordingly, we can interpret the above matrices in the time domain as $y_{2i+1} = -\frac{1}{2}(x_{2i} + x_{2i+2}) + x_{2i+1}$ and $y_{2i} = \frac{1}{4}(y_{2i+1} + y_{2i+3}) + x_{2i}$ where $0 \leq i \leq \frac{N}{2}$ for an input stream x of length N and y's are the transformed signal values. Note that the odd samples are calculated from even samples and even samples are calculated from the updated odd samples. The corresponding matrices M_1 and M_2 are shown below, where $a = -\frac{1}{2}$ and $b = \frac{1}{4}$. The transform of the signal X is $Y = XM_1M_2$ while the inverse is $X = YM_2M_1$. The matrices M_1 and M_2 are as follows:

$$M_1 = \begin{bmatrix} 1 & a & 0 & \cdot & \cdot & \cdot & \cdot & \cdot & \cdot \\ 0 & 1 & 0 & 0 & \cdot & \cdot & \cdot & \cdot & \cdot \\ 0 & a & 1 & a & 0 & \cdot & \cdot & \cdot & \cdot \\ \cdot & 0 & 0 & 1 & 0 & 0 & \cdot & \cdot & \cdot \\ \cdot & \cdot & 0 & a & 1 & a & 0 & \cdot & \cdot \\ \cdot & \cdot & \cdot & 0 & 0 & 1 & 0 & 0 & \cdot \\ \cdot & \cdot & \cdot & \cdot & 0 & a & 1 & a & 0 \\ \cdot & \cdot & \cdot & \cdot & \cdot & 0 & 0 & 1 & 0 \\ 0 & \cdot & \cdot & \cdot & \cdot & \cdot & 0 & a & 1 \end{bmatrix}, \quad M_2 = \begin{bmatrix} 1 & 0 & 0 & \cdot & \cdot & \cdot & \cdot & \cdot & \cdot \\ 0 & 1 & b & 0 & \cdot & \cdot & \cdot & \cdot & \cdot \\ 0 & 0 & 1 & 0 & 0 & \cdot & \cdot & \cdot & \cdot \\ \cdot & 0 & b & 1 & b & 0 & \cdot & \cdot & \cdot \\ \cdot & \cdot & 0 & 0 & 1 & 0 & 0 & \cdot & \cdot \\ \cdot & \cdot & \cdot & 0 & b & 1 & b & 0 & \cdot \\ \cdot & \cdot & \cdot & \cdot & 0 & 0 & 1 & 0 & 0 \\ \cdot & \cdot & \cdot & \cdot & \cdot & 0 & b & 1 & 0 \\ 0 & \cdot & \cdot & \cdot & \cdot & \cdot & 0 & 0 & 1 \end{bmatrix}.$$

The other wavelet filter bank that has been proposed in JPEG2000 Part 1 is the (9, 7) filter. The most efficient factorization of the polyphase matrix for (9, 7) filter is as follows [12];

$$\tilde{P}(z) = \begin{bmatrix} 1 & a(1+z^{-1}) \\ 0 & 1 \end{bmatrix} \begin{bmatrix} 1 & 0 \\ b(1+z) & 1 \end{bmatrix} \begin{bmatrix} 1 & c(1+z^{-1}) \\ 0 & 1 \end{bmatrix} \begin{bmatrix} 1 & 0 \\ d(1+z) & 1 \end{bmatrix} \begin{bmatrix} K & 0 \\ 0 & \frac{1}{K} \end{bmatrix}$$

where a=−1.586134342, b=−0.0529801185, c=0.882911076, d=−0.443506852, K=1.149604398. In terms of banded matrix operation, the transform can be represented as $Y = XM_1M_2M_3M_4$, while the inverse transform is represented as $X = YM_4M_3M_2M_1$. The matrices M_1, M_2, M_3, and M_4 are as follows:

$$M_1 = \begin{bmatrix} 1 & a & 0 & \cdot & \cdot & \cdot & \cdot & \cdot & \cdot \\ 0 & 1 & 0 & 0 & \cdot & \cdot & \cdot & \cdot & \cdot \\ 0 & a & 1 & a & 0 & \cdot & \cdot & \cdot & \cdot \\ \cdot & 0 & 0 & 1 & 0 & 0 & \cdot & \cdot & \cdot \\ \cdot & \cdot & 0 & a & 1 & a & 0 & \cdot & \cdot \\ \cdot & \cdot & \cdot & 0 & 0 & 1 & 0 & 0 & \cdot \\ \cdot & \cdot & \cdot & \cdot & 0 & a & 1 & a & 0 \\ \cdot & \cdot & \cdot & \cdot & \cdot & 0 & 0 & 1 & 0 \\ 0 & \cdot & \cdot & \cdot & \cdot & \cdot & 0 & a & 1 \end{bmatrix}, \quad M_2 = \begin{bmatrix} 1 & 0 & 0 & \cdot & \cdot & \cdot & \cdot & \cdot & \cdot \\ 0 & 1 & b & 0 & \cdot & \cdot & \cdot & \cdot & \cdot \\ 0 & 0 & 1 & 0 & 0 & \cdot & \cdot & \cdot & \cdot \\ \cdot & 0 & b & 1 & b & 0 & \cdot & \cdot & \cdot \\ \cdot & \cdot & 0 & 0 & 1 & 0 & 0 & \cdot & \cdot \\ \cdot & \cdot & \cdot & 0 & b & 1 & b & 0 & \cdot \\ \cdot & \cdot & \cdot & \cdot & 0 & 0 & 1 & 0 & 0 \\ \cdot & \cdot & \cdot & \cdot & \cdot & 0 & b & 1 & 0 \\ 0 & \cdot & \cdot & \cdot & \cdot & \cdot & 0 & 0 & 1 \end{bmatrix}$$

$$M_3 = \begin{bmatrix} 1 & c & 0 & \cdot & \cdot & \cdot & \cdot & \cdot & \cdot \\ 0 & 1 & 0 & 0 & \cdot & \cdot & \cdot & \cdot & \cdot \\ 0 & c & 1 & c & 0 & \cdot & \cdot & \cdot & \cdot \\ \cdot & 0 & 0 & 1 & 0 & 0 & \cdot & \cdot & \cdot \\ \cdot & \cdot & 0 & c & 1 & c & 0 & \cdot & \cdot \\ \cdot & \cdot & \cdot & 0 & 0 & 1 & 0 & 0 & \cdot \\ \cdot & \cdot & \cdot & \cdot & 0 & c & 1 & c & 0 \\ \cdot & \cdot & \cdot & \cdot & \cdot & 0 & 0 & 1 & 0 \\ 0 & \cdot & \cdot & \cdot & \cdot & \cdot & 0 & c & 1 \end{bmatrix}, \quad M_4 = \begin{bmatrix} 1 & 0 & 0 & \cdot & \cdot & \cdot & \cdot & \cdot & \cdot \\ 0 & 1 & d & 0 & \cdot & \cdot & \cdot & \cdot & \cdot \\ 0 & 0 & 1 & 0 & 0 & \cdot & \cdot & \cdot & \cdot \\ \cdot & 0 & d & 1 & d & 0 & \cdot & \cdot & \cdot \\ \cdot & \cdot & 0 & 0 & 1 & 0 & 0 & \cdot & \cdot \\ \cdot & \cdot & \cdot & 0 & d & 1 & d & 0 & \cdot \\ \cdot & \cdot & \cdot & \cdot & 0 & 0 & 1 & 0 & 0 \\ \cdot & \cdot & \cdot & \cdot & \cdot & 0 & d & 1 & 0 \\ 0 & \cdot & \cdot & \cdot & \cdot & \cdot & 0 & 0 & 1 \end{bmatrix}$$

Most of the practical wavelet filters are decomposed either into 2 or 4 matrices (primal and dual). For example, each of the filter banks C(13, 7), S(13, 7), (2,6), (2, 10) can be decomposed into two matrices and (6, 10) can be decomposed into four matrices, as has been described in detail in [14].

5.4.5 Data Dependency Diagram for Lifting Computation

Computation of the lifting-based discrete wavelet transform can be explained via a data dependency diagram as shown by a block diagram in Figure 5.8. For the DWT requiring four lifting factors, such as the (9, 7) filter, the computation is done in four stages as shown in Figure 5.8. For the DWT filters requiring only two lifting factors, such as the (5, 3) filter, the intermediate two stages can simply be bypassed.

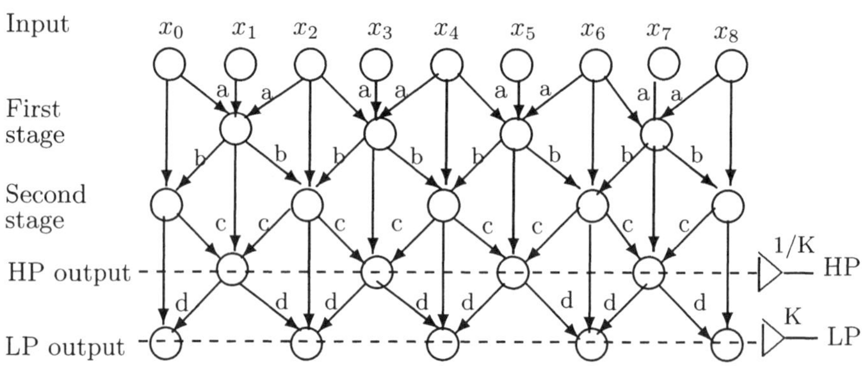

Fig. 5.8 Data dependency diagram with four lifting factors.

The results produced in the first stage of computation in the data dependency diagram can be stored immediately in the registers containing the odd samples of the input data because these odd samples are not used in later stages of computation in the data dependency diagram. Similarly the results produced in the second stage can be stored back to the registers allocated to the even samples of input data. Continuing in the same way, the high-pass (low-pass) output samples are stored into the registers where the odd (even) samples of the input data were originally stored at the beginning of the computation. As a result no extra memory is required at any stage. This property of lifting is popularly called "in-place" computation in the literature.

5.5 ADVANTAGES OF LIFTING-BASED DWT

The idea of lifting-based implementation of discrete wavelet transform is a relatively recent development and it is still an active area of research in mathematics and signal processing. The lifting-based DWT has many advantages over the convolution-based approach. Some of them are as follows.

- *Computational efficiency*: Usually the lifting-based DWT requires less computation (up to 50%) compared to the convolution-based approach. However, the savings depend on the length of the filters.
- *Memory savings*: During the lifting implementation, no extra memory buffer is required because of the in-place computation feature of lifting. This is particularly suitable for hardware implementation with limited available on-chip memory.
- *Integer-to-integer transform*: The lifting-based approach offers integer-to-integer transformation suitable for lossless image compression.
- *No boundary extension*: In lossless transformation mode, we can avoid the boundary extension (discussed in Section 17.6.1.3 of Chapter 17) of the input data because the original input can be exactly reconstructed by integer-to-integer lifting transformation.
- *Parallel processing*: From Figure 5.8, it is obvious that multiple MAC (*multiply* and *accumulate*) processors can produce the output samples in each stage in parallel. The computation of each MAC processor is of the form $a(x_i + x_{i+2}) + x_{i+1}$. The only sequential part is the order of the lifting operations.

5.6 SUMMARY

In this chapter, we have discussed the theoretical foundation of the discrete wavelet transform (DWT) both for convolution and lifting-based approaches. We have examined the multiresolution analysis feature of the wavelet transform, which makes it suitable for its application in image compression. We have discussed the pyramid algorithm for implementation of the DWT using the multiresolution approach. The properties of Gabor filter useful in image analysis has been discussed. Lifting-based implementation of discrete wavelet transform is new and became very popular for a number of efficient features in it. We have described the underlying theory behind the lifting algorithm for DWT and showed how it is implemented via banded matrix multiplication. We have given examples of the lifting factorization for the two default wavelet filter kernels (9, 7) and (5, 3) in the JPEG2000 standard. We have discussed the advantages of lifting-based DWT over the traditional convolution-based approach.

REFERENCES

1. K. Weierstrass, *Mathematische Werke*, Vol. II, Mayer & Muller, Berlin, 1895.
2. N. Ricker, "The Form and Nature of Seismic Waves and the Structure of Seismograms," *Geophysics*, 5(4), October 1940, 348–366.
3. J. Morlet, G. Arens, E. Fourgeau, and D. Giard, "Wave Propagation and Sampling Theory," *Geophysics*, 47(2), February 1982, 203–236.
4. A. Grossman and J. Morlet, "Decompositions of Functions into Wavelets of Constant Shape, and Related Transforms," in L. Streit, ed., *Mathematics and Physics: Lectures on Recent Results*, World Scientific, Singapore, 1985.
5. Y. Meyers, *Wavelet: Algorithms and Applications*. SIAM, Philadelphia, 1993 (Translated by Robert D. Ryan).
6. S. Mallat, "A Theory for Multiresolution Signal Decomposition: The Wavelet Representation," *IEEE Trans. on Pattern Analysis and Machine Intelligence*, Vol. 11, No. 7, pp. 674–693, July 1989.
7. R. M. Rao and A. S. Bopardikar, *Wavelet Transforms: Introduction to Theory and Applications*. Addison-Wesley, MA, 1998.
8. T. Acharya and P. Tsai, *JPEG2000 Standard for Image Compression: Concepts, Algorithms, and VLSI Architectures*, Wiley, Hoboken, NJ, 2004.
9. M. R. Turner, "Texture discrimination by Gabor functions," *Biolog. Cybern*, 55, 1986, 71–82.
10. I. Daubechies, *Ten Lectures on Wavelets*, SIAM, CBMS series, Philadelphia, 1992.
11. W. Sweldens, "The Lifting Scheme: A Custom-Design Construction of Biorthogonal Wavelets," *Applied and Computational Harmonic Analysis*, 3(15), 1996, 186–200.
12. I. Daubechies and W. Sweldens, "Factoring Wavelet Transforms into Lifting Schemes," *The J. of Fourier Analysis and Applications*, Vol. 4, 1998, 247–269.
13. M. D. Adams and F. Kossentini, "Reversible Integer-to-Integer Wavelet Transforms for Image Compression: Performance Evaluation and Analysis," *IEEE Trans. on Image Processing*, Vol. 9, June 2000, 1010–1024.
14. K. Andra, C. Chakrabarti, and T. Acharya,"A VLSI Architecture for Lifting-Based Forward and Inverse Wavelet Transform," *IEEE Trans. of Signal Processing*, 50(4), April 2002, 966–977.

6

Image Enhancement and Restoration

6.1 INTRODUCTION

Millions of pictures ranging from biomedical images to the images of natural surroundings and activities around us enrich our daily visual experience. All these images create elegant perception in our sensory organs. They also contain a lot of important information and convey specific meanings in diverse domains of application.

When such pictures are converted from one form to another by processes such as imaging, scanning, or transmitting, the quality of the output image may be inferior to that of the original input picture. There is thus a need to improve the quality of such images, so that the output image is visually more pleasing to human observers from a subjective point of view. To perform this task, it is important to increase the dynamic range of the chosen features in the image, which is essentially the process of image enhancement.

Enhancement has another purpose as well, that is to undo the degradation effects which might have been caused by the imaging system or the channel. The growing need to develop automated systems for image interpretation necessitates that the quality of the picture to be interpreted should be free from noise and other aberrations. Thus it is important to perform preprocessing operations on the image so that the resultant preprocessed image is better suited for machine interpretation. Image enhancement thus has both a subjective and an objective role and may be viewed as a set of techniques for improving the subjective quality of an image and also for enhancing the accuracy rate in automated object detection and picture interpretation.

Enhancement refers to accentuation or sharpening of image features, such as contrast, boundaries, edges, etc. The process of image enhancement, however, in no way increases the information content of the image data. It increases the dynamic range of the chosen features with the final aim of improving the image quality. Modeling of the degradation process, in general, is not required for enhancement. However, knowledge of the degradation process may help in the choice of the enhancement technique. The realm of image enhancement covers contrast and edge enhancement, noise filtering, feature sharpening, and so on. These methods find applications in visual information display, feature extraction, object recognition, and so on [1]–[3]. These algorithms are generally interactive, application dependent, and employ linear or nonlinear local or global filters.

Image enhancement techniques, such as contrast stretching, map each gray level into another gray level using a predetermined transformation function. One example of it is histogram equalization method, where the input gray levels are mapped so that the output gray level distribution is uniform. This is a powerful method for the enhancement of low-contrast images. Other enhancement techniques may perform local neighborhood operations as in convolution; transform operations as in discrete Fourier transform; and other operations as in pseudo-coloring where a gray level image is mapped into a color image by assigning different colors to different features. An important issue in image enhancement is quantifying the criterion for enhancement.

Many image enhancement techniques are empirical and require interactive procedures to obtain satisfactory results. Enhancement techniques are based on combinations of methods from spatial and frequency domains.

6.2 DISTINCTION BETWEEN IMAGE ENHANCEMENT AND RESTORATION

Sometimes we receive noisy images which are degraded by some degrading mechanism. One common source of degradation is the optical lens system in a digital camera which acquires the visual information. If the camera is not appropriately focused then we get blurred images. Here the cause of blur is the defocused camera. Very often one may come across images of outdoor scenes that were procured in a foggy environment. An outdoor scene captured on a foggy winter morning could result in a blurred image. In this case the degradation is due to the fog and mist in the atmosphere, and this type of degradation is known as atmospheric degradation. In some other cases there may be a relative motion between the object and the camera. Thus if the camera is given an impulsive displacement during the image capturing interval while the object is static, the resulting image will invariably be blurred and noisy.

Thus it may be observed that in all the above cases the resultant image is a degraded version of the original object. Here the sources of degradations are either the defocused camera as in the first case, or the atmospheric degradation caused by fog as in the second case, or a relative accelerated motion between the object and the focal plane of the lens of the camera during the capturing of the image as in the third case. The conventional enhancement techniques would not be suitable to get the original objects from the image. In such cases, if we can appropriately model the degrading mechanism then by some kind of deconvolution we may be able to reconstruct the original scene back. A number of strategies have been suggested by researchers for restoring the original object from the degraded scene. If we can mathematically model the cause of degradation, then it becomes easy to reconstruct or restore the original scene. Needless to say, these reconstruction or restoration techniques are different from the enhancement techniques which are employed essentially to get a better-quality picture and not necessarily the original object from the scene.

If the mathematical model of the source which causes degradation is known, then the standard techniques of inverse filtering or deconvolution work quite well. However, in many situations it may be difficult to model the degrading source appropriately. In such cases blind deconvolution strategies are employed to restore the original scene from the degraded image.

6.3 SPATIAL IMAGE ENHANCEMENT TECHNIQUES

The spatial filtering techniques used for noise reduction (or smoothing) are as follows:

- Spatial low-pass, high-pass and band-Pass filtering
- Unsharp masking and crisping
- Directional smoothing
- Median filtering.

6.3.1 Spatial Low-Pass and High-Pass Filtering

From our knowledge in signal processing theory we know that low-pass filtering attenuates the high-frequency components in the signal and is essentially equivalent to integrating the signal. Integration in turn implies summation and averaging the signal. Low-pass filtering of an image is a spatial averaging operation. It produces an output image, which is a smooth version of the original image, devoid of the high spatial frequency components that may be present in the image. In particular, this operation is useful in removing visual noise, which generally appears as sharp bright points in the image. Such high spatial frequencies associated with these spikes are attenuated by the low-pass filter.

High-pass filtering of an image, on the other hand, produces an output image in which the low spatial frequency components are attenuated. The cut off frequency at which lower frequencies are attenuated is varied by the selection of filter coefficients. High-pass filtering is used for edge enhancement. Since the sharpness of an image is related to the content of high-frequency components, low-pass filtering leads to blurring, while high-pass filtering is used for deblurring.

Such a filter can easily be implemented by subtracting the low-pass output from its input. Typically, the low-pass filter would perform a relatively long-term spatial average, on a 3×3 or 5×5 or larger window.

6.3.2 Averaging and Spatial Low-Pass Filtering

When each pixel is replaced by a weighted average of its neighborhood pixels, the resulting image a low pass filtered image. The output image in this case is expressed as

$$g(m,n) = \sum\sum a(k,1) f(m-k, n-1), \qquad (k,1) \in W \tag{6.1}$$

where $f(m,n)$ and $g(m,n)$ are the input and output images respectively, W is a suitably chosen neighborhood around the pixel at location (m,n), and $a(k,1)$ are the filter weights.

In general in spatial averaging filters all the weights are assigned equal values. Hence the mathematical representation of the filtering becomes:

$$g(m,n) = \frac{1}{N}\sum\sum f(m-k, n-1), \quad (k,1) \in W \tag{6.2}$$

where N is the number of pixels in the neighborhood W. Quite often each pixel is replaced by the average of its nearest four neighboring pixels only as given by

$$g(m,n) = 0.5[f(m,n)+0.25\{f(m-1,n)+f(m+1,n)+f(m,n-1)+f(m,n+1)\}] \tag{6.3}$$

The spatial averaging operation on an image may be used to smooth the noise. If the observed image is given as:

$$g(m,n) = f(m,n) + \eta(m,n) \tag{6.4}$$

then the spatial average yields:

$$g(m,n) = \frac{1}{N}\sum\sum f(m-k, n-1) + \bar{\eta}(m,n), \qquad (k,1) \in W \tag{6.5}$$

where $\bar{\eta}(m,n)$ is the spatial average of the noise component $\eta(m,n)$. If the noise has a variance σ^2, then it can be shown that $\bar{\eta}(m,n)$ is zero mean and has variance $\frac{\sigma^2}{N}$. This implies that by performing spatial averaging the image noise power is reduced by a factor equal to the number of pixels chosen in the neighborhood of the central pixel.

The conventional spatial filtering utilizes an averaging procedure to generate the smoothened image. The weights used to average are image data invariant. Thus all regions of the image which can be brought under an arbitrary neighborhood W are equally affected. Thus spatial filtering by averaging

(a) does not take into account the effect of the difference of gray levels between the central pixel and a neighboring pixel.

(b) does not always take into account the diminishing influence of the pixels that are situated in increasing distance from the central pixel.

6.3.3 Unsharp Masking and Crisping

As we have already observed, a sharp image can be obtained by high-pass filtering a blurred image. Alternatively, subtracting a blurred version of the image from the original image may also lead to the sharpening of the image. As the name suggests the unsharp masking technique is used for crisping the edges. Such a technique is used in the printing industries. A signal proportional to the unsharp or low-pass-filtered version of the original noisy image is subtracted from the image, such that the resulting image $v(m,n)$ is a crip high-contrast image. Here

$$v(m,n) = f(m,n) - g(m,n),$$

where $g(m,n)$ is a low-pass-filtered version of the original image $f(m,n)$.

From an alternative viewpoint, a gradient or a high-pass signal may be added to the original image, which may result in a better high-contrast image. From this view point, the unsharp masking operation can be represented by: $v(m,n) = f(m,n) + \gamma h(m,n)$ where $\gamma > 0$ and $h(m,n)$ is a suitably defined gradient at (m,n). This is also referred to as high emphasis filter, where the high frequency components are emphasized while retaining the low frequency components of the image. Any gradient function may be used, and these functions are discussed in chapter 7.

6.3.4 Directional Smoothing

Low-pass filters always result in blurring the image and quite often the crisp edges are blurred by averaging. To minimize this effect, directional averaging filter can be used. Spatial averages $g(m,n;\theta)$ are calculated in several directions θ as:

$$g(m,n;\theta) = \frac{1}{N_0} f(m-k, n-1), \quad (k,1) \in W_0 \tag{6.6}$$

where W_0 is the neighborhood selected in the direction θ.

The key to the implementation of effective directional smoothing is to identify a specific direction θ^* for which $|f(m,n) - g(m,n;\theta^*)|$ is minimum. Such a θ^* for which the above objective function is minimum yields the desired result. The directional smoothing operation often prevents the edges from getting blur resulting from the smoothing operation.

6.4 HISTROGRAM-BASED CONTRAST ENHANCEMENT

In a poorly contrasted image a large number of pixels occupy only a small portion of the available range of intensities. Through histogram modification we reassign each pixel with a new intensity value so that the dynamic range of gray levels is increased. The principle here is to stretch the dynamic range of the pixel values in such a way that the lighter pixels may turn still lighter, while the comparatively darker pixels may be still darkened. It is quite obvious that by suitably stretching the pixel values the overall contrast of the image will increase.

Linear contrast stretching and histogram equalization are two widely utilized methods for global image enhancement. The linear contrast stretching linearly adjusts the image's dynamic range and histogram equalization uses the input to output mapping relation obtained from the integral of the image histogram. Histogram equalization is a technique which consists of adjusting the gray scale of the image so that the gray level histogram of the input image is mapped onto a uniform histogram.

The basic assumption used here is that the information conveyed by an image is related to the probability of occurrence of gray levels in the form of histogram in the image. By uniformly distributing the probability of occurrence of gray levels in the image, it becomes easier to perceive the information content of the image. Thus through histogram modification we reassign each pixel with a new intensity value according to its original intensity. We first discuss the image histogram.

6.4.1 Image Histogram

Histogram of an image represents the relative frequency of occurrence of the various gray levels in the image. Mathematically speaking for a digital image with gray levels in the range $[0, L-1]$, the histogram is a discrete function $p(r_k) = \frac{n_k}{N}$, where r_k is the kth gray level, and n_k is the number of pixels in the image with that gray level. N is the total number of pixels in the image. It may be noted that $k = 0, 1, \cdots, L-1$.

The histogram gives primarily the global description of the image. For example, if the image histogram is narrow, then it means that the image is poorly visible because the difference in gray levels present in the image is generally low. Similarly a widely distributed histogram means that almost all the gray levels are present in the image and thus the overall contrast and visibility increases. The shape of the histogram of an image reveals important contrast information, which can be used for image enhancement. Histogram modeling techniques modify an image so that its histogram has some desired shape. This is useful in stretching the low-contrast levels of images with narrow histograms. The histogram modeling methods used for enhancement can be broadly classified into local and global methods. In local methods the operation is limited only for some limited number of pixels or more casually within some restricted region of the image. Global methods, on the other hand modify the entire image based on the overall image histogram information.

6.4.2 Histogram Equalization

Histogram equalization is a technique which consists of adjusting the gray scale of the image so that the gray level histogram of the input image is mapped onto a uniform histogram [1]–[3]. The histogram equalization technique is based on a transformation using the histogram of a complete image in histogram equalization, the goal is to obtain a uniform histogram for the output image. Let the variable r represents a random variable which indicates the gray level of an image. Initially we can assume that r is continuous and lies within the closed interval [0:1] with $r = 0$ representing black and $r = 1$ representing white. For any r in the specified interval let us consider a transformation of the form:

$$s = T(r).$$

The transformation produces a level s for every pixel value r in the original image. It is assumed that the transformation T satisfies the following criteria:

- $T(r)$ is a single valued function, monotonically increasing in the interval [0:1].
- $T(r)$ lies between 0 and 1.

The first condition preserves the order from black to white in the gray scale, and the second one guarantees that the function is consistent with the allowed range of pixel gray values. The inverse transform from s to r can be represented by

$$r = T^{-1}(s).$$

Let the original and transformed gray levels be characterized by their probability density functions $p_r(r)$ and $p_s(s)$ respectively. Then from elementary probability theory, if $p_r(r)$ and $p_s(s)$ are known and if $T^{-1}(s)$ satisfies the first condition stated above in (I) then the probability density function of the transformed gray level is given by:

$$P_s(s) = \left[P_r(r) \frac{dr}{ds} \right]_{r-T^{-1}(s)}. \tag{6.7}$$

If the transformation is given by:

$$s = T(r) = \int_0^r P_r(w)dw \tag{6.8}$$

then substituting $\frac{dr}{ds} = \frac{1}{p_r(r)}$ in Eq. 6.7 we obtain $P_s(s) = 1$. Thus it is possible to obtain a uniformly distributed histogram of an image by the transformation described by Eq. 6.8.

From the above discussions, it is clear that using a transformation function equal to the cumulative distribution of r (as given by Eq. 6.8) produces an image whose gray levels have a uniform density. This implies that such a transformation results in an increase in the dynamic range of the pixel gray values which produces a pronounced effect on the appearance of the image.

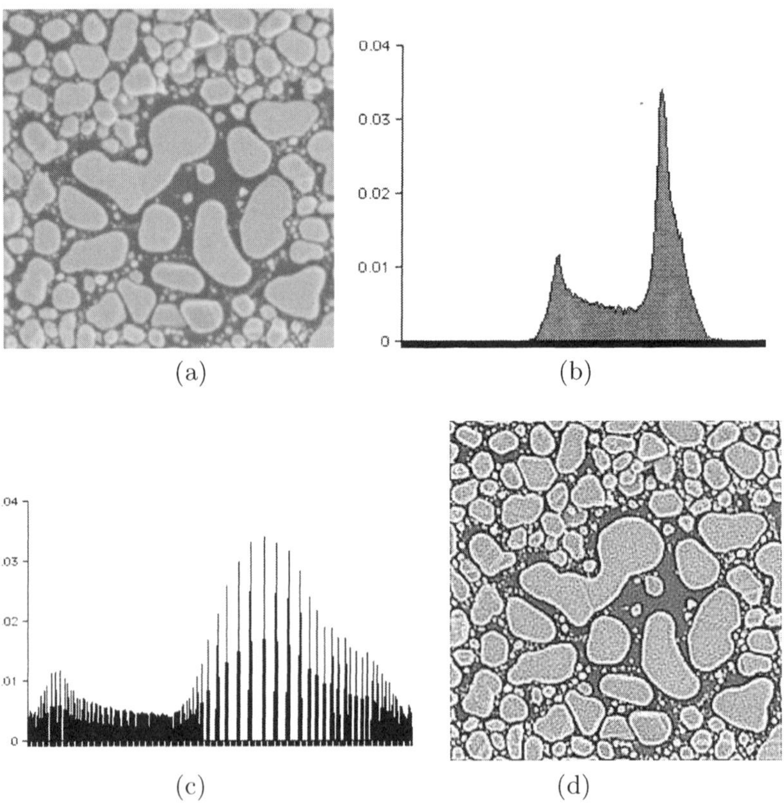

Fig. 6.1 Histogram equalization results: (a) original image, (b) histogram of the image, (c) equalized histogram (d) enhanced image.

A simple algorithm to implement histogram equalization for a gray level image is given below

- for every pixel in the image get gray value in variable i, $hist[i] = hist[i]+1$ when $i = 0$ to $L-1$ for a L level image.
- from the histogram array, get cumulative frequency of histogram $hist_{cf}[i] = hist_{cf}[i-1] + hist[i]$.
- generate the equalized histogram as

$$eqhist[i] = \left\lfloor \frac{[(L * hist_{cf}[i]) - N^2]}{N^2} \right\rfloor$$

  where, L is the number of gray levels present in the image, N^2 is number of pixel in NXN image, $\lfloor x \rfloor$ is truncation of x to the nearest integer.
- Replace the gray value i, by $eqhist[i]$ for each i. The $eqhist$ contains the new mapping of gray values.

It is shown in Figure 6.1 that while histogram of the original image is very nonuniform due to lot of undulations, the histogram of the equalized image has more or less a uniform density function. Often the unimodal histogram of images having dynamic gray intensity are confined in a very narrow range. Such images may be enhanced by equalizing the histogram which results from the extension of the dynamic range of the original image [4, 5].

6.4.3 Local Area Histogram Equalization

Although the above methods are simple they do not take into account local details and the global histogram equalization has the undesired effect of overemphasizing noise. Histogram equalization can also be performed locally, where the equalization is based on the histogram of the portion of the image under a two-dimensional sliding window that is centered over the pixel in the center of the rectangular region and its gray level is modified by equalization procedures. For the pixels that are not center pixels, bilinear interpolations of the four neighboring center pixel transformations are used to approximate the local area histogram transformation.

6.4.4 Histogram Specification

A generalization of the procedure given above can be used to modify the image so as to obtain the image histogram of some desired shape. The input gray level r is first transformed nonlinearly by some $f(r)$, and the output is uniformly quantized. As we have seen that for histogram equalization the function $f(r)$ can be given as:

$$f(r) = \sum_{x=0}^{u} P_r(x). \tag{6.9}$$

In general if we want to transform the input histogram $p_r(r)$ to the histogram given by the probability density function $p_s(s)$, then we first define a random variable w as:

$$w = \int_0^r P_r(x)dx = F_r(r) \tag{6.10}$$

which also satisfies:

$$w = \int_0^s P_s(x)dx = F_s(s) \tag{6.11}$$

Eliminating w we have

$$s = F_s^{-1}[F_r(r)]. \tag{6.12}$$

Histogram equalization like many other contrast enhancement methods, does not take into account contrast perception. This often results in a degradation of the subjective image quality.

6.4.5 Histogram Hyperbolization

To avoid the drawback, Frei has proposed a method called histogram hyperbolization [6]. His model is based on Weber's law, which expresses the logarithmic relationship between the perceived brightness and the luminosity of the object. Histogram hyperbolization is a technique which produces images with a uniform distribution of the perceived brightness levels. The human visual system has a logarithmic response to stimuli and thus the image information should be redistributed so as to produce images with uniform distribution of the perceived brightness levels. The method is based on Weber's law according to which $B = \log(I + C)$, where B the perceived brightness is a logarithmic function of the light intensity I incident on the eye. C is a constant. Weber's law follows from the experimentation on brightness discrimination that consists of exposing an observer to a uniform field of intensity I in which the intensity I of a disk is increased gradually with a quantity ΔI. The value of ΔI from which the observer perceives the disk is called the differential threshold.

To measure the performance of an image contrast enhancement method, a set of performance parameters like, image contrast value, local intensity variance, uniformity measure, white-black ratio and homogeneity measures may be used as tools for performance evaluation.

6.4.6 Median Filtering

In *median filtering* the input pixel is replaced by the median of the pixels contained in the neighborhood [7]. Symbolically this can be represented as:

$$v(m,n) = median\{y(m-k, n-1), \quad (k,1) \in W\} \tag{6.13}$$

where W is suitably chosen neighborhood. The algorithm for median filtering requires arranging the pixel gray values in the neighborhood in increasing or decreasing order and picking up the value at the center of the array. Generally the size of the neighborhood is chosen as odd number so that a well-defined center value exists. If, however, the size of the neighborhood is even the median is taken as the arithmetic mean of the two values at the center.

(a) (b) (c)

Fig. 6.2 Result of median filtering: (a) original image, (b) salt and pepper noisy image, (c) result of median filtering.

The result of median filtering is shown in Figure 6.2 as an example. Figure 6.2(a) shows the original image, Figure 6.2(a) is the original image corrupted with a salt-and-pepper noise, and Figure 6.2(c) shows the result of median filter applied on Figure 6.2(b). Some of the properties of median filter are:

- It is a nonlinear filter.
- It is useful in removing isolated lines or pixels while preserving spatial resolution. It is found that median filter works well on binary noise but not so well when the noise is Gaussian.
- Its performance is poor when the number of noise pixels is greater than or equal to half the number of pixels in the neighborhood.

6.5 FREQUENCY DOMAIN METHODS OF IMAGE ENHANCEMENT

When an image $f(x, y)$ is convolved with a linear operator $h(x, y)$, the resultant image $g(x, y)$ is given by

$$g(x, y) = h(x, y) * f(x, y).$$

The convolution theorem states that the convolution in spatial domain is equivalent to multiplication in frequency domain. This implies that

$$G(u, v) = H(u, v)F(u, v)$$

where $G(u,v)$, $H(u,v)$, and $F(u,v)$ are the Fourier transforms of $g(x,y)$, $h(x,y)$, and $f(x,y)$ respectively. Taking the inverse Fourier transform of $G(u,v)$, we get

$$g(x,y) = \Im^{-1}[H(u,v)F(u,v)].$$

It may be observed that by suitable selection of $h(x,y)$, we get a resultant image $g(x,y)$ which is an enhanced version of the original image $f(x,y)$. Assuming $f(x,y)$ to be point source of light, i.e., a unit impulse function whose Fourier transform is unity,

$$G(u,v) = H(u,v)$$

or

$$g(x,y) = \Im^{-1}[H(u,v)] = h(x,y)$$

which is known as point spread function (PSF). PSF is an inverse of optical transfer function. As discussed in Chapter 2, a point source of light (analogous to an impulse) is blurred or spread by the imaging device which is given by the PSF.

Enhancement in the frequency domain is achieved by high-pass, low-pass, and band-pass filtering of the original image. The task of enhancement in frequency domain involves computing the Fourier transform of the image $f(x,y)$ (i.e. $F(u,v)$)and the filter transfer function $H(u,v)$ and taking the inverse Fourier transform of the product $F(u,v)H(u,v)$. The variations in gray level in an image represents the frequency component present in the image. A uniformly homogeneous image with constant gray value has 0 frequency, while an image with adjacent black-and-while image has high spatial frequencies.

An ideal low-pass filter transfer function in two-dimension is given as

$$H(u,v) = \begin{cases} 1 & \text{when } D(u,v) \leq D_0 \\ 0 & \text{when } D(u,v) > D_0 \end{cases}$$

where $D(u,v) = \sqrt{u^2+v^2}$ represents the distance of (u,v) from origin. It may be observed from this definition that the transition from pass-band is very sharp at the cutoff frequency for an ideal 2D filter and all the frequencies inside a circle of radius D are passed without any attenuation while the rest of the frequencies are all attenuated. The transfer function of the Butterworth low-pass filter of order n is

$$H(u,v) = \frac{1}{1 + \left[\frac{D(u,v)}{D_0}\right]^{2n}}$$

The Butterworth filter transfer function has a smooth transition from pass-band to cutoff band and when $D(u,v) = D_0$, $H(u,v) = \frac{1}{2}$, which is 50% below its maximum value.

Similarly the Butterworth high-pass and band-pass filters can be defined by appropriate transfer functions.

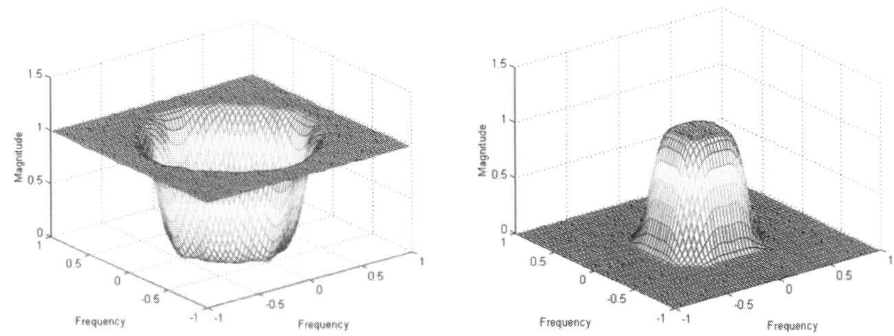

(a) (b)

Fig. 6.3 Two-dimensional filter response (a) high-pass, (b) low-pass.

The transfer function of the two-dimensional high-pass and low-pass filters are shown in Figure 6.3. The results of high-pass and low-pass filtered images are shown in Figure 6.4.

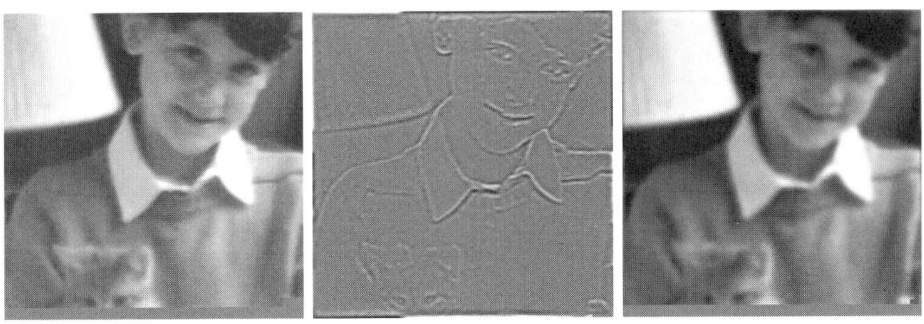

(a) (b) (c)

Fig. 6.4 (a) Input image, (b) High-passed filtered, (c) Low-pass filtered output images.

6.5.1 Homomorphic Filter

Homomorphic filters are widely used in image processing for compensating the effect of nonuniform illumination in an image. Pixel intensities in an image represents the light reflected from the corresponding points in the objects. As discussed in Chapter 2, image $f(x, y)$ may be characterized by two components: (1) the amount of source light incident on the scene being viewed, and (2) the amount of light reflected by the objects in the scene. These portions of light are called the illumination and reflectance components, and are denoted $i(x, y)$ and $r(x, y)$ respectively. The functions $i(x, y)$ and $r(x, y)$ combine multiplicatively to give the image function $f(x, y)$:

$$f(x, y) = i(x, y)r(x, y) \tag{6.14}$$

where $0 < i(x,y) < \alpha$ and $0 < r(x,y) < 1$. Homomorphic filters are used in such situations where the image is subjected to the multiplicative interference or noise as depicted in Eq. 6.14. We cannot easily use the above product to operate separately on the frequency components of illumination and reflection because the Fourier transform of $f(x,y)$ is not separable; that is

$$\Im(f(x,y)) \neq \Im\{i(x,y)\}.\Im\{r(x,y)\}. \tag{6.15}$$

We can separate the two components by taking the logarithm of the two sides

$$\ln\{f(x,y)\} = \ln\{i(x,y)\} + \ln\{r(x,y)\}. \tag{6.16}$$

Taking Fourier transforms on both sides we get,

$$\Im\{\ln\{f(x,y)\}\} = \Im\{ln\{i(x,y)\}\} + \Im\{\ln\{r(x,y)\}\} \tag{6.17}$$

that is, $F(u,v) = I(u,v) + R(u,v)$, where $F(u,v)$, $I(u,v)$ and $R(u,v)$ are the Fourier transforms of $\ln\{f(x,y)\}$, $\ln\{i(x,y)\}$, and $\ln\{r(x,y)\}$ respectively. The function F represents the Fourier transform of the sum of two images: a low-frequency illumination image and a high-frequency reflectance image.

If we now apply a filter with a transfer function that suppresses low-frequency components and enhances high-frequency components, then we can suppress the illumination component and enhance the reflectance component. Taking the inverse transform of $F(u,v)$ and then anti-logarithm, we get

$$f'(x,y) = i'(x,y) + r'(x,y) \tag{6.18}$$

The result of homomorphic filter is shown in Figure 6.5.

(a) (b) (c)

Fig. 6.5 Homomorphic filtering: (a) homomorphic filter response, (b) input image, (c) result of homomorphic filtering.

6.6 NOISE MODELING

Quite often an image gets corrupted by noise, which may arise in the process of acquiring the image, or during its transmission, or even during reproduction of the image. Removal of noise from an image is one of the most important tasks in image processing. Depending on the nature of the noise, such as additive or multiplicative type of noise, there are several approaches towards removing noise from an image. The nature of the noise is primarily inherent to the imaging procedure. The general model of such systems consists of a detector and a recorder. Mathematically the imaging procedure including corruption by noise can be represented by

$$v(x,y) = g[u(x,y)] + \eta(x,y) \tag{6.19}$$

$$g[u(x,y)] = \int\int h(x,y;x',y')u(x',y')dx'dy' \tag{6.20}$$

$$\eta(x,y) = f[g(u(x,y))]\eta_1(x,y) + \eta_2(x,y) \tag{6.21}$$

Here $u(x,y)$ represents the object (also called the original image), and $v(x,y)$ is the observed image. The image formation process can often be modeled by the linear system described in (6.20). Here $h(x,y;x',y')$ represents the impulse response of the image acquiring process. In case of a shift-invariant system $h(x,y;x',y') = h(x-x',y-y')$, functions $f()$ and $g()$ are generally nonlinear and represent the characteristics of detector or recording mechanisms. The term $\eta(x,y)$ represents the additive noise, which has an image dependent random component $f[g(w)]\eta_1$ and an image independent random component η_2.

In general the noise model given by (6.21) is applicable in many situations. For example, in photographic systems the noise in the electron beam current is often modeled as:

$$\eta(x,y) = \sqrt{g(x,y)}\eta_1(x,y) + \eta_2(x,y) \tag{6.22}$$

where $g()$ is the detector model and η_1 and η_2 are zero mean, mutually independent, *Gaussian white noise* fields. The detector function can, for example, return the illumination at the point (x,y).

The signal dependent term arises from the random deposition of silver grain on the film due to exposure and the term $\eta_2(x,y)$ represents additive wideband thermal noise, which is Gaussian in nature. In case of films $\eta_2(x,y)$ is not usually present.

The presence of signal dependent term in the noise modeling makes the restoration of noisy images more difficult.

A different type of noise in the coherent imaging of objects is called speckle noise. For low-resolution objects it is often multiplicative and occurs whenever the surface roughness of the object being imaged is of the order of the wavelength of the incident radiation. For example, when a photograph is digitized by using optical scanners, *speckle noise* can occur because the roughness of the paper surface is much of the order of the long wavelength of the light used for scanning. Speckle noise can be modeled as:

$$v(x,y) = u(x,y)s(x,y) + \eta(x,y) \tag{6.23}$$

where the speckle noise intensity is given by $s(x, y)$ and $\eta(x, y)$is a white Gaussian noise.

6.6.1 Types of Noise in An Image and Their Characteristics

The noise embedded in an image manifests in diverse varieties. The noise may be correlated or uncorrelated; it may be signal dependent or independent, and so on. The knowledge about the imaging system and the visual perception of the image helps in generating the noise model and estimating of the statistical characteristics of noise embedded in an image is important because it helps in seperating the noise from the useful image signal. For example, in case of Moire noise, which is an extremely narrow-band noise (i.e., energy is concentrated in a very narrow-bands), it is easy to detect the noise from the much wider image spectrum, since its spectrum contains few concentrated peaks in the background. We describe four important classes of noise here.

1. **Additive noise**: Sometimes the noises generated from sensors are thermal white Gaussian, which is essentially additive and signal independent, i.e., $g(x, y) = f(x, y) + n(x, y)$, where $g(x, y)$ is the result of the original image function $f(x, y)$ corrupted by the additive Gaussian noise $n(x, y)$.

2. **Multiplicative noise**: The graininess noise from photographic plates is essentially multiplicative in nature. The speckle noise from the imaging systems as in Coherent SAR, ultrasound imaging etc. are also multiplicative in nature, which may be modeled as $g(x, y) = f(x, y) * n(x, y)$, where $n(x, y)$ is the multiplicative noise.

3. **Impulse noise**: Quite often the noisy sensors generate impulse noise. Sometimes the noise generated from digital (or even analog) image transmission system is impulsive in nature, which can be modeled as

   $$g(x, y) = (1 - p)f(x, y) + p.i(x, y) \tag{6.24}$$

   where $i(x, y)$ is the impulsive noise and p is a binary parameter that assumes the values of either 0 or 1. The impulse noise may be easily detected from the noisy image because of the contrast anomalies. Once the noise impulses are detected, these are replaced by the signal samples. In the next section we will present a technique of noise removal when the image is embedded with impulse noise.

4. **Quantization noise**: The quantization noise is essentially a signal dependent noise. This noise is characterized by the size of signal quantization interval, which has been discussed in Chapter 2. Such noise produces image-like artifacts and may produce false contours around the objects. The quantization noise also removes the image details which are of low-contrast.

6.7 IMAGE RESTORATION

Improving the quality of images acquired by optical, electro-optical or electronic means is one of the basic tasks in digital image processing. Two-dimensional images may be degraded due to several reasons, e.g.,

- Imperfection of the imaging system
- Imperfection in the transmission channel
- Degradation due to atmospheric condition
- Degradation due to relative motion between the object and the camera

Thus the original scene is usually blurred due to convolution of the degrading mechanism as above. In many real-life situations a random noise is usually added to the degraded data. The random noise may originate from the image formation process, transmission medium, recording process, etc.

A typical cause of degradation may be defocusing of the camera lens system, sudden jerking motion of the imaging system, etc. The additive Gaussian noise results in the degraded image

$$g(x, y) = H[f(x, y)] + \eta(x, y)$$

where $f(x, y)$ is the pixel value of the original image at the point (x, y), H is the degradation model, and $g(x, y)$ is the pixel value of the degraded image at the point (x,y); $\eta(x, y)$ is the additive noise.

A simple degradation model may be achieved by convolving a 3×3 window with the original image. Thus $g(x, y)$ becomes

$$g(x, y) = \sum_{k,l \in w} H(k, l) f(x - k, y - l) + \eta(x, y).$$

Here w represents the convolution window. A simple 3x3 blurring functions may look like this:

$$\frac{1}{16} \begin{bmatrix} 1 & 1 & 1 \\ 1 & 8 & 1 \\ 1 & 1 & 1 \end{bmatrix}.$$

There are conventional methods like inverse filtering, Wiener filtering, Kalman filtering, Algebraic approach, etc., to restore the original object. In all these cases we assume that the blurring function H is known. If H is known, then obviously we can get back or restore the original image $f(x, y)$, simply by convolving the degraded image with the inverse of the blurring function. To remove the noise, it is important that the noise statistics should be known apriori.

The *algebraic* approach is based on the principle of estimating $\Im$ from f (i.e., the original image) which minimizes a predefined performance criterion.

If there is no additive noise, then a meaningful criterion function will be to find $\Im$ such that $H\Im$ is an approximation of g in a least square sense. These techniques will be explained in detail in the next section.

6.7.1 Image Restoration of Impulse Noise Embedded Images

The general strategy to remove noise from an image corrupted with impulse noise is a two-step process:

Step 1: Identify whether the pixel under consideration is noisy. If noisy, then go to step 2, otherwise do not change the pixel value.

Step 2: Replace the noisy pixel by another value to generate noise-free image.

To implement the above steps, we choose a window of size $(2M+1)\times(2M+1)$ around each pixel of the image. To detect if a pixel is noise corrupted, find the difference of the pixel from the median of the pixel values in the chosen window of size $(2M+1)\times(2M+1)$ around the pixel under test. If the difference is higher than a threshold, the pixel is detected as noisy; else it is a considered a noise-free pixel. The algorithm as presented above, however, cannot remove the impulse noise if the image is too much corrupted with noise. This is because of following two reasons:

1. the choice of a local window alone is unable to reflect the global details of the image

2. the choice of a small local neighborhood does not even consider the local region details.

To take into consideration the above two factors, several strategies of noise detection and cleaning may be employed. Wang and Zhang [8] have proposed a scheme, where they have chosen, in addition to a local neighborhood of size $(2M+1)\times(2M+1)$ around each noisy pixel, another window of the same size, located at a different place of the image in the vicinity of the pixel under test at position (i,j). This second window is selected with its center positioned at (k,l), while the first window is selected around the pixel location (i,j).

The second window should be chosen at (k,l) in such a way that it is covered by a larger search window of size $(2N+1)\times(2N+1)$; $N>M$ centered around the noisy pixel at (i,j) location and the pixel at (k,l) is a nonnoisy pixel. Thus there exists many such candidate second windows and the one that best matches with the first local window at (i,j) should be selected. The best match is identified based on the mean square error

$$MSE = \frac{\sum\sum(1-f_{i+m,j+n})(1-f_{k+m,l+n})(x_{i+m,j+n}-x_{k+m,l+n})^2}{\sum\sum(1-f_{i+m,j+n})(1-f_{k+m,l+n})}.$$

While computing the best match, only these corresponding pixel pairs should be used from the first and second windows which are both good. The one with minimum MSE becomes the winner. The corrupted pixel is replaced by the center pixel of the wining second window.

6.7.2 Restoration of Blurred Image

The basic principle of image restoration is as follows: Given an image $f(x, y)$, convolved with a blurring function $h(x, y)$, with additive noise $n(x, y)$, the task is to reconstruct the original image from the noisy image. In the frequency domain, the process of degradation mechanism can be written as

$$G(u, v) = F(u, v) \cdot H(u, v) + N(u, v) \tag{6.25}$$

where $G(u, v)$ is the Fourier transform of the input image $g(x, y)$, $F(u, v)$ is the Fourier transform of the original image $f(x, y)$, $H(u, v)$ is the Fourier transform of the blur mechanism h and $N(u, v)$ is the Fourier transform of the noise added to the image. From Eq. 6.25, we get

$$\hat{F}(u, v) = \frac{G(u, v)}{H(u, v)} = F(u, v) + \frac{N(u, v)}{H(u, v)}. \tag{6.26}$$

From the equation we observe that the original image cannot be recovered because the random noise function $N(u, v)$ is unknown. Moreover, if the degradation function $H(u, v)$ has very small or zero values, the ratio $\frac{N(u,v)}{H(u,v)}$ would be very large or infinite and the estimate $\hat{F}(u, v)$ would be dominated by this factor. This problem can be overcome by limiting the analysis to the frequencies near the origin $H(0, 0)$ as it represents the average value of $h(x, y)$ and it is normally the highest value of $H(u, v)$.

The blurring mechanism is essentially a low-pass filter, and thus $H(u, v)$ is large at low frequencies and it decreases at high frequencies. The power spectrum of noise is more or less uniform. The power spectrum of the image will generally be large at low frequencies and decreases at high-frequency. Thus at low frequencies $\frac{N(u,v)}{H(u,v)}$ is low and $F(u, v)$ is much larger than $\frac{N(u,v)}{H(u,v)}$.

6.7.3 Inverse Filtering

Inverse filtering approach toward image restoration assumes that the degradation of the image is caused by a linear function $h(i, j)$. The additive noise is assumed to be independent of the image signal [3, 9]. Thus the degradation can be expressed in the Fourier domain as

$$G(u, v) = F(u, v)H(u, v) + N(u, v) \tag{6.27}$$

where $F(u, v)$ is the Fourier transform of the uncorrupted image, $G(u, v)$ is the Fourier transform of the degraded image, $H(u, v)$ is the Fourier transform of the degradation process $h(i, j)$. Therefore from the above equation we can write

$$F(u, v) = G(u, v)H^{-1}(u, v) - N(u, v)H^{-1}(u, v) \tag{6.28}$$

From the above equation it is evident that the original image cannot be restored in case the nature of the random noise function is unknown.

In case there is no additive noise $N(u,v)$, the inverse filtering performs very well. If, however, the noise $N(u,v)$ is present, then its influence becomes significant at higher frequency (u,v), where the magnitude of $H(u,v)$ becomes quite low. The effect of noise in such cases influences the result of restoration. If however, we restrict the restoration to a small neighborhood region around the origin of (u,v), the $H(u,v)$ is usually large in this neighborhood and thus the restoration results are moderately acceptable.

6.7.4 Wiener Filter

Wiener filter is a discrete time linear FIR filter [9]. This has been widely used in reconstruction of one-dimensional signals and two-dimensional images. Although Wiener filter is sensitive to noise, yet it can be used for good restoration of the original image. The elegance of Wiener filter lies in the fact that it incorporates the prior knowledge about the noise embedded in the signal and also the spectral density of the object being imaged. As a result, Wiener filter provides a better and improved restoration of original signal since it takes care of the noise process involved in the filtering. The discrete version of the Wiener filter is a straightforward extension to the continuous Wiener filter.

In this section we discuss a method that restores an image in the presence of blur as well as noise. Here, we consider that both the image $f(x,y)$ and the noise $\eta(x,y)$ are the zero mean random processes and the objective is to obtain an estimate $\hat{f}(x,y)$ of the original image $f(x,y)$ such that the mean square error is

$$e^2 = E\left\{\left[f(x,y) - \hat{f}(x,y)\right]^2\right\} \tag{6.29}$$

is minimized, where $E\{.\}$ is the expected value of the argument. The resulting estimate is also called MMSE estimate. Assuming linear shift invariance

$$g(x,y) = h(x,y) * f(x,y) + n(x,y) \tag{6.30}$$

$$\hat{f}(x,y) = w(x,y) * g(x,y) \tag{6.31}$$

where $w(x,y)$ is the Wiener filter.

$$\begin{aligned} E&\left[\left\{f(x,y) - \hat{f}(x,y)\right\}^2\right] \\ &= E\left[f(x,y)^2\right] + E\left[\hat{f}(x,y)^2\right] - 2.E\left[f(x,y)\hat{f}(x,y)\right] \\ &= R_f(x,y) + E\left[\{w(x,y) * g(x,y)\}^2\right] - 2E\left[f(x,y)\{w(x,y) * g(x,y)\}\right] \end{aligned}$$

where

$$R_f(x,y) = E\left[f(x,y)^2\right]$$

From the above, we can express the error as

$$\begin{aligned} e^2 = {} & R_f(x,y) + \\ & w(x,y) * h(x,y) * R_f(x,y) * h(x,y) * w(x,y) w(x,y) * R_f(x,y) * w(x,y) \\ & -2w(x,y) * h(x,y) * R_f(x,y) \end{aligned}$$

The error can thus be written in matrix form as

$$e^2 = Tr\left[[R_f] + [W][H][R_f][H]^T[W]^T + [W][R_n][W]^T - 2[W][H][R_f]\right]$$

where $[W]$, $[H]$, etc. are stacked notations. For example, if the image matrix is represented as a $N^2 \times 1$ vector, then $[W]$ and $[H]$ are matrices of size $N^2 \times N^2$. Differentiating the above with respect to $[W]$ and equating to zero we get

$$[W]_{opt} = [R_f][H]^T\left[[H][R_f][H]^T + [R_n]\right]^{-1} \tag{6.32}$$

Since we have assumed that H is linear shift invariant and f is stationary, applying fourier transform leads to

$$W(u,v) = \frac{H^*(u,v)S_f(u,v)}{|H(u,v)|^2 S_f(u,v) + S_n(u,v)} \tag{6.33}$$

dividing by $S_f(u,v)$ both in the numerator and denominator we get

$$W(u,v) = \frac{H^*(u,v)}{|H(u,v)|^2 + S_n(u,v)/S_f(u,v)} \tag{6.34}$$

Now

$$\hat{F}(u,v) = W(u,v)G(u,v) \tag{6.35}$$

and assuming that the image and the noise are uncorrelated, and one of them has zero mean, the estimate of the original image is given in frequency domain from the above equation as

$$\hat{F}(u,v) = \frac{G(u,v)|H(u,v)| + N(u,v)|H^*(u,v)|}{|H(u,v)|^2 + S_n(u,v)/S_f(u,v)} \tag{6.36}$$

where $H(u,v)$ is the Fourier transform of the function representing the degrading mechanism and $G(u,v)$ is the spectrum of the degraded image. Power spectrum of the noise is

$$S_n(u,v) = |N(u,v)|^2$$

and the power specturm of the original image is

$$S_f(u,v) = |F(u,v)|^2$$

This is known as the *Wiener* filter or the *minimum mean square error* filter or the least square error filter. Here we can see that the filter does not have the same problem as the inverse filter even if the degraded function has zero values. If the noise is zero, the second term in the denominator vanishes and the Wiener filter reduces to the inverse filter. The results of restoration by the inverse filter and the Wiener filter and their comparison are shown in Figure 6.6.

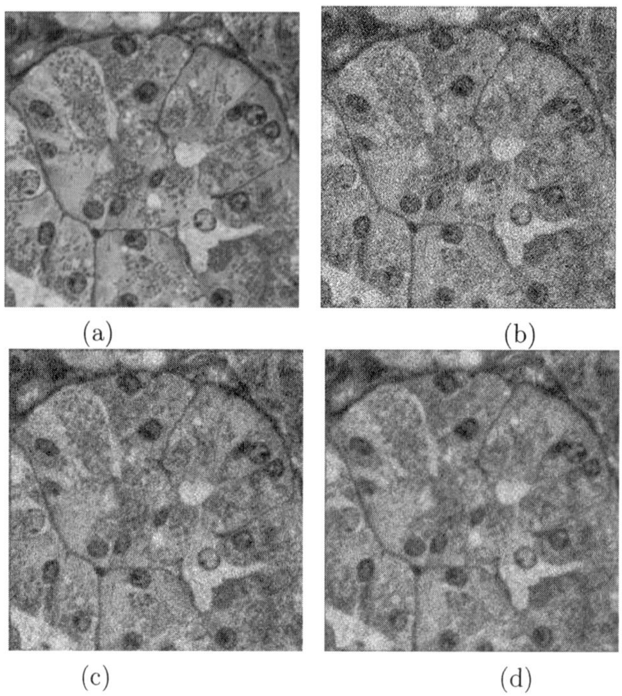

Fig. 6.6 The results of restoration: (a) original image, (b) degraded image, (c) inverse filtered result, (d) Wiener filtered result

The above analysis attempts to restore the object from a single degraded measurement and it is called single frame wiener restoration. If on the other hand we have multiple number of blurred and noisy images of the same objects, we can apply wiener restoration to each corrupted observation and average the outputs of Wiener filter [10]. Alternatively we can also average the measurements and then apply the Wiener filter to this averaged data. It may be noted that in the second case Wiener filtering will be different as the noise power gets reduced by averaging multiple observations. Now if we have multiple blurred and noisy image sequences as in a video sequence there will be additional correlation factors which may exist between the frames of the original image sequences. In such cases we need to make use of multi-frame

Wiener filter to restore the images [11]. This will help in making use of the information content in other frames in restoring a particular frame.

6.8 IMAGE RECONSTRUCTION BY OTHER METHODS

6.8.1 Image Restoration by Bispectrum

In many practical situations, the available data is a set of noisy frames, in which the objects are not known and are shifted from frame to frame. Due to the misalignment of objects in each frame, a simple ensemble averaging cannot be used to improve the signal to noise ratio. In such cases, phase information may be more useful than the corresponding magnitude information [12]–[15]. Thus, while reconstructing the object in an image, it is necessary to obtain a phase estimate as clear as possible of the original object. Bispectrum [13] retains the phase and magnitude information of the signal while suppressing the additive Gaussian noise. As a matter of fact, bispectrum suppresses the linear phase component, which represents the linear shift and thus averaging of bispectrum can be used for object reconstruction [13]–[15].

The bispectrum $B(\omega_1, \omega_2)$ of a real discrete time sequence $f(i)$ may be represented as

$$B(\omega_1, \omega_2) = \sum_{m=-\infty}^{+\infty} \sum_{n=-\infty}^{+\infty} R_f(m,n) e^{-j(m\omega_1 + n\omega_2)} \tag{6.37}$$

where $R_f(m,n)$ is the third order moment of the sequence $f(i)$ represented as

$$R_f(m,n) = \sum_{i=-\infty}^{+\infty} f(i) f(i+m) f(i+n) \tag{6.38}$$

The bispectrum defined in Eq. 6.37, may be expressed as

$$B_f(\omega_1, \omega_2) = F(\omega_1) F(\omega_2) F^*(\omega_1 + \omega_2) \tag{6.39}$$

where $F(w)$ is the Fourier transform of the sequence $f(i)$, corresponding to the object.

The bispectrum of the object signal $f(i)$ needs to be estimated when the degraded signal $g(i)$ corrupted with additive Gaussian zero-mean, stationary noise $n(i)$ is available to us. The bispectrum of $f(i)$ may be estimated from the set of noisy observations $g(i)$ as

$$\hat{B_f}(\omega_1, \omega_2) = \frac{1}{N} \sum_{k=1}^{N} G_k(\omega_1) G_k(\omega_2) G_k^*(\omega_1 + \omega_2) \tag{6.40}$$

where $G_k(\omega)$ is the Fourier transform of the kth record $g_k(i)$ and $k = 1, \cdots, N$, where N is the number of noisy frames. The problem of extracting phase of

$f(i)$ has been addressed in [13]–[15]. Once the phase estimate is available, the object can be reconstructed.

6.8.2 Tomographic Reconstruction

Tomographic reconstruction finds lot of application in biomedical imaging, geological exploration, etc. Seismic tomography is an efficient exploration tool where localization of subsurface resources, detection of hazardous subsurface region, etc. are carried out using electromagnetic and seismic waves. The presence of anomalies in subsurface geological formations can be detected and their location, size, shape, and other parameters may be estimated using backprojection techniques, algebraic reconstruction technique, simultaneous iterative reconstruction technique and evolutionary programming based methods [16, 17].

6.9 SUMMARY

In this chapter, we have discussed the basic principles involved in the process of image enhancement and restoration. The enhancement strategies have been classified as spatial domain and frequency domain techniques. We have presented a couple of techniques in both spatial domain and frequency domain. The distinction between enhancement and restoration has been explained. Two-dimensional image restoration using Wiener filter, restoration of impulse noise corrupted image signals, and restoration of blurred noise corrupted signals have been presented.

REFERENCES

1. R. C. Gonzalez and R. E. Woods, *Digital Image Processing*, Addison-Wesley, Reading, MA, 1992.

2. W. K. Pratt, *Digital Image Processing*, 2nd ed., Wiley, New York, 1991.

3. A. Rosenfeld and A. C. Kak, *Digital Picture Processing*, 2nd ed., Vol. 1, Academic Press, 1982.

4. R. A. Hummel, "Image Enhancement by Histogram Transformation," *Computer Graphics and Image Processing*, 6(2), 1977, 184–195.

5. S. M. Pizer et al., "Adaptive Histogram Equalization and its variations," *Computer Vision, graphics and Image Processing*, 39(3), September 1987, 355–368.

6. W. Frei, "Image Enhancement by Histogram Hyperbolisation," *Computer Graphics and Image Processing*, 6(3), June 1977, 286–294.

7. T. S. Huang, G. I.Yang, and G. Y. Tang, "A Fast Two-Dimensional Median Filtering Algorithm," *IEEE Trans. Acoustics, Speech, and Signal processing*, ASSP-27, 1, February 1979, 13–18.

8. Z. Wang and D. Zhang, "Restoration of impulse noise corrupted images using Long-Range correlation," *IEEE Signal Processing Letters*, 5(1), January 1998, 4–7.

9. H. C. Andrews and B. R. Hunt, *Digital Image Restoration*, Prentice Hall, Englewood Cliffs, NJ, 1977.

10. Galatsanos et el, "Least Square Restoration of Multi Channel images," *IEEE Trans., on Signal Processing*, 39(10), October 1991.

11. Ozken et el., "Efficient Multiframe Wiener Restoration of Blurred and Noisy Image Sequences", *IEEE Trans., Image Processing*, 1(4), October 1992.

12. A. V. Openheim and J. S. Lim, "The importance of phase in signals," *Proc. IEEE*, 69, May 1981, 521–541.

13. A. P. Petropulu and C. L. Nikias, "Signal reconstruction from the phase of the Bispectrum," *IEEE Trans. Signal Processing*, 40(3), 1992, 601–610.

14. R. S. Holambe, A. K. Ray, and T. K. Basu, "Phase Only Blind Deconvolution Using Bicepstrum Iterative Reconstruction Algorithm," *IEEE Trans. Signal Processing*, 44(9), 1996, 2356–2359.

15. R. S. Holambe, A. K. Ray, and T. K. Basu, "Signal Phase Recovery Using the Bispectrum," *Signal Processing*, 55, 321–337, 1996.

16. R. S. Bichkar and A. K. Ray, "Tomographic reconstruction of Circular and Elliptical Objects Using Genetic Algorithms," *IEEE Signal Processing Letter*, 5(10), 1998.

17. R. S. Bichkar and A. K. Ray, "Genetic Algorithm approach to the detection of subsurface voids in cross-hole seismic tomography," *Pattern Recognition Letters*, 19, 527–536, 1998.

7

Image Segmentation

7.1 PRELIMINARIES

The growing need for automated image analysis and interpretation in a wide range of applications necessitates the development of *segmentation* algorithms. Segmentation involves partitioning an image into a set of homogeneous and meaningful regions, such that the pixels in each partitioned region possess an identical set of properties or attributes [1]–[3]. These sets of properties of the image may include gray levels, contrast, spectral values, or textural properties. The result of segmentation is a number of homogeneous regions, each having an unique label. An image is thus defined by a set of regions that are connected and nonoverlapping, so that each pixel in the image acquires a unique region label that indicates the region it belongs to. The set of objects of interest in an image, which are segmented, undergoes subsequent processing, such as object classification and scene description.

A complete segmentation of an image R involves identification of a finite set of regions $\langle R_1, R_2, R_3, \cdots, R_N \rangle$ such that

1. $R = R_1 \cup R_2 \cup \cdots R_N$.
2. $R_i \cap R_j = \Phi, \forall i \neq j$.
3. $P(R_i) = \text{True}, \forall\ i$
4. $P(R_i \cup R_j) = \text{False},\ i \neq j$

There may exist a number of possible partitions, but the selection of an appropriate set of regions depends on the choice of the property P association in the region [4]. In addition, connectivity among the pixels in a region need to be considered. Segmentation algorithms are based on one of the two basic properties of gray-level values–*discontinuity* and *similarity* among the pixels.

In the first category of algorithms, we partition an image based on abrupt changes in gray level. The principal areas of interest within this category are the detection of lines and edges in an image. Thus if we can extract the edges in an image and link them, then the region is described by the edge contour that contains it.

We may view the process of segmentation from another perspective. From this point of view, the connected set of pixels having more or less the same homogeneous intensity form the regions. Thus the pixels inside the regions describe the region and the process of segmentation involves partitioning the entire scene in a finite number of regions.

The principal approaches in the second category are based on the *similarity* among the pixels within a region. While segmenting an image, various local properties of the pixels are utilized. The well-established segmentation techniques are:

- Histogram-based thresholding
- Region growing
- region splitting and merging
- Clustering/Classification
- Graph theoretic approach
- Rule-based or knowledge-driven approach

This chapter discusses the issues associated with segmentation of an image. We present some of the basic algorithms of image segmentation, starting with the discussions on edge detection process.

7.2 EDGE, LINE, AND POINT DETECTION

Edge detection: Edges, lines, and points carry a lot of information about the various regions in the image. These features are usually termed as local features, since they are extracted from the local property alone. Though the edges and lines are both detected from the abrupt change in the gray level, yet there is an important difference between the two. An edge essentially demarcates between two distinctly different regions, which means that an edge is the border between two different regions. A line, on the other hand, may be embedded inside a single uniformly homogeneous region. For example, a thin line may run between two plots of agricultural land, bearing the same vegetation. A point is embedded inside a uniformly homogeneous region and its gray value is different from the average gray value of the region in which it is embedded. This is analogous to a spike. The changes in the gray levels in case of a perfect step edge, line, ramp edge are shown in the form of an edge profile in Figure 7.1. The diverse forms and nature of ideal edges and lines, such as step edge, ramp edge, line, step line, are shown in Figure 7.2.

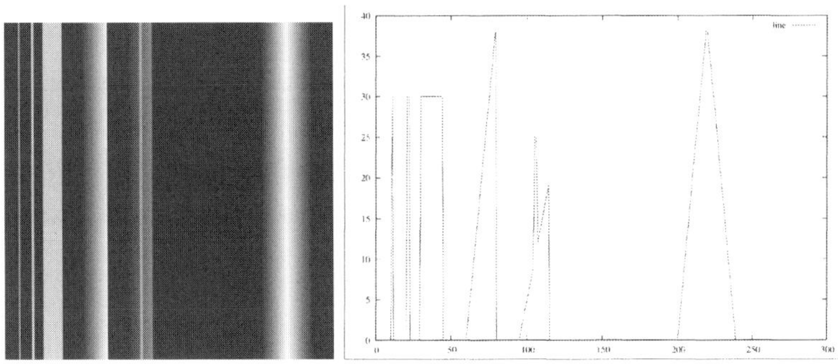

Fig. 7.1 Edge profile of one row of a synthetic image.

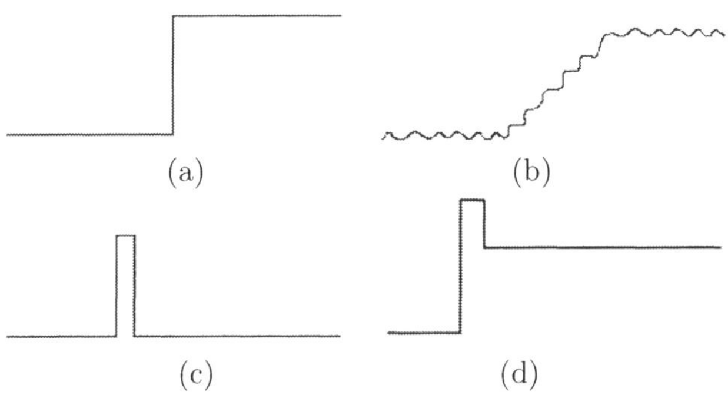

Fig. 7.2 Different types of edges: (a) step, (b) ramp, (c) line, (d) step-line.

The *edge detection* operation is essentially an operation to detect significant local changes in the intensity level in an image. The change in intensity level is measured by the gradient of the image. Since an image $f(x, y)$ is a two-dimensional function, its gradient is a vector

$$\begin{bmatrix} G_x \\ \\ G_y \end{bmatrix} = \begin{bmatrix} \frac{df}{dx} \\ \\ \frac{df}{dy} \end{bmatrix}.$$

The magnitude of the gradient may be computed in several ways

$$\left.\begin{aligned} G[f(x,y)] &= \sqrt{G_x^2 + G_y^2} \\ G[f(x,y)] &= |G_x| + |G_y| \\ G[f(x,y)] &= max\,\{|G_x|, |G_y|\} \end{aligned}\right\}. \tag{7.1}$$

The direction of the gradient is

$$\theta(x, y) = \tan^{-1}(G_y/G_x) \tag{7.2}$$

where the angle θ is measured with respect to the X-axis.

Gradient operators compute the change in gray level intensities and also the direction in which the change occurs. This is calculated by the difference in values of the neighboring pixels, i.e, the derivatives along the X-axis and Y-axis. In a two-dimensional image the gradients are approximated by

$$G_x = f(i+1, j) - f(i, j)$$

and

$$G_y = f(i, j+1) - f(i, j).$$

Gradient operators require two masks, one to obtain the X-direction gradient and the other to obtain the Y-direction gradient. These two gradients are combined to obtain a vector quantity whose magnitude represents the strength of the edge gradient at a point in the image and whose angle represents the gradient angle. Figure 7.3 shows the gradient images of a checker board image along horizontal, vertical directions, and also along both the directions.

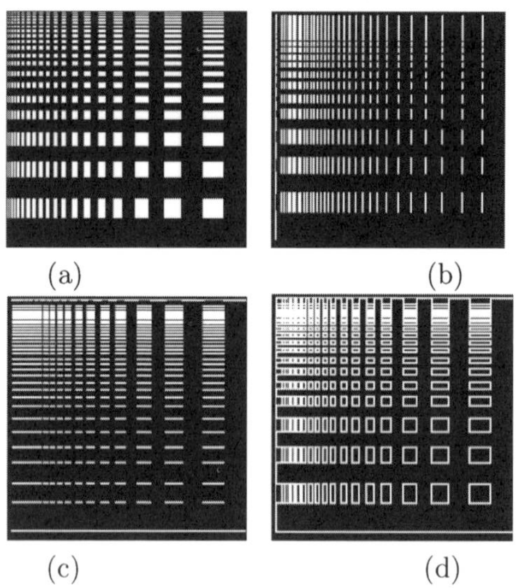

Fig. 7.3 (a) Input image, (b) vertical edges, (c) horizontal edges, (d) edge image along both directions.

Alternative approach to computation of edge gradients involves convolving the image with a set of edge templates, chosen in say eight equispaced directions–east, northeast, north, northwest, west, southwest, south, and southeast. The image is convolved with each of these gradient templates. Each template responds to the edges in a particular direction.

7.2.0.1 Ideal edge detection process An ideal edge detector is required to detect an edge point precisely in the sense that a true edge point in an image should not be missed, while a false, nonexistent edge point should not be erroneously detected. These two requirements often conflict with each other. The decision regarding the existence of an edge point is based on a threshold. Thus if the magnitude of the gradient is greater than a threshold, then we infer that an edge point exists at that point, else there is no edge point. If the selected threshold is large, then there is a possibility that true edge points may be undetected, while if the threshold is low, many noisy points may be falsely detected as edge points. The goal of an ideal edge detector is to choose the threshold appropriately.

Since gradient operation enhances image noise, the best way to reduce this problem is to filter high spatial frequencies containing noise prior to applying the gradient operators.

7.3 EDGE DETECTOR

A number of edge detectors based on a single derivative have been developed by various researchers. Amongst them most important operators are the Robert operator, Sobel operator, Prewitt operator, Canny operator, Krisch operator [1]–[5], etc. In each of these operator-based edge detection strategies, we compute the gradient magnitude in accordance with the formula given below. If the magnitude of the gradient is higher than a threshold, then we detect the presence of an edge. Below we discuss some of these operators.

7.3.1 Robert Operator-Based Edge Detector

The Robert Cross operator is a simple gradient operator based on a 2 × 2 gradient operator. This operator provides the simplest approximation of the gradient magnitude given as

$$G[f(i,j)] = [f(i,j) - f(i+1,j+1)] + [f(i+1,j) - f(i,j+1)]$$

The convolution mask for the Robert's operator is shown below. Since the Robert kernel is only a 2 × 2 mask, it is quite sensitive to noise.

1	1
−1	−1

7.3.2 Sobel Operator-Based Edge Detector

Sobel operator is a 3 × 3 neighborhood based gradient operator. The convolution masks for the Sobel operator are defined by the two kernels shown in Figure 7.4. The two masks are separately applied on the input image

1	2	1
0	0	0
–1	–2	–1

(a)

1	0	–1
2	0	–2
1	0	–1

(b)

Fig. 7.4 Sobel masks to compute (a) gradient G_x and (b) gradient G_y

to yield two gradient components G_x and G_y in the horizontal and vertical orientations respectively.

$$G_x = [f(i-1,j-1) + 2f(i-1,j) + f(i-1,j+1)] - \\ [f(i+1,j-1) + 2f(i+1,j) + f(i+1,j+1)]$$

and

$$G_y = [f(i-1,j-1) + 2f(i,j-1) + f(i+1,j-1)] - \\ [f(i-1,j+1) + 2f(i,j+1) + f(i+1,j+1)]$$

The gradient magnitude is usually computed as

$$G[f(x,y)] = \sqrt{G_x^2 + G_y^2}$$

However, the other two formulae in Eq. 7.1 can also be used to compute the gradient magnitude. The direction of gradient is computed as in Eq. 7.2.

The result of an edge image generated by the Sobel operator is shown in Figure 7.5. The edge images have been computed using threshold values 110, 90 and 70. It may be observed from the images that as the treshold value is decreased, more and more number of non edge points become edge points.

7.3.3 Prewitt Operator-Based Edge Detector

The Prewitt operator is defined by a set of eight masks, four of which are shown in Figure 7.6. Others can be generated by rotation of 90^0 successively. The mask that produces maximal response yields the direction of the gradient.

The magnitude and direction of the edge gradients can be computed in a similar fashion as in the Sobel operator. The result of an edge image generated by the Prewitt operator is shown in Figure 7.7.

7.3.4 Kirsch Operator

Similar to the Prewitt operators, Kirsch masks can be defined in eight directions, which yields the estimated gradients in these directions. In Figure 7.8, we show the Kirsch masks which detect gradients in four directions only.

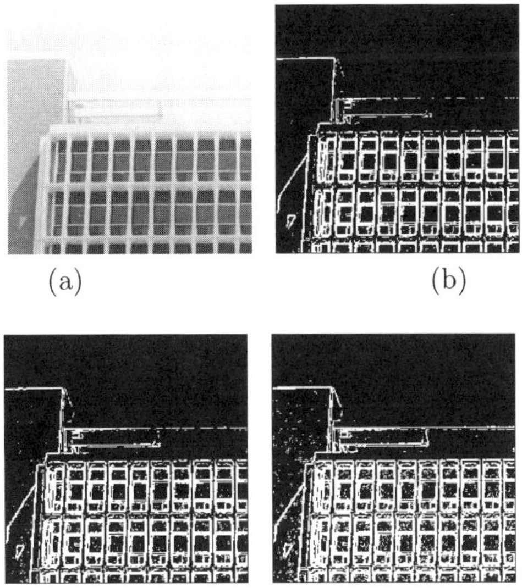

(a) (b) (c) (d)

Fig. 7.5 Edge Image using Sobel operator (a) Original image, edge detection using threshold value (b) 110, (c) 90, (d) 70.

1	1	1
0	0	0
−1	−1	−1

(a)

0	1	1
−1	0	1
−1	−1	0

(b)

−1	0	1
−1	0	1
−1	0	1

(c)

−1	−1	0
−1	0	1
0	1	1

(d)

Fig. 7.6 Prewitt masks in 90^0 successive rotations.

7.3.5 Canny's Edge Detector

Canny's edge detector ensures good noise immunity and at the same time detects true edge points with minimum error [5]. Canny has optimized the edge detection process by

1. Maximizing the signal-to-noise ratio of the gradient

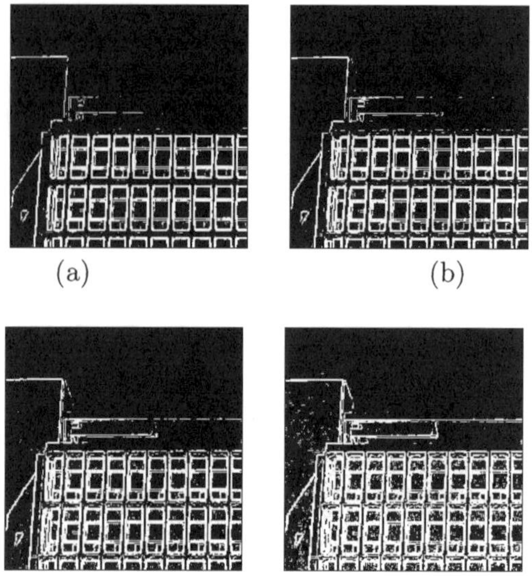

(a) (b) (c) (d)

Fig. 7.7 Prewitt results with thresholds (a) 110, (b) 90, (c) 70 (d) 50.

3	3	3
3	0	3
–5	–5	–5

(a)

–5	3	3
–5	0	3
–5	3	3

(b)

5	–3	–3
5	0	–3
5	–3	–3

(c)

–3	–3	–3
–3	0	–3
5	5	5

(d)

Fig. 7.8 Kirsch masks to compute gradients in four directions.

2. An edge localization factor, which ensures that the detected edge is localized as accurately as possible

3. Minimizing multiple responses to a single edge

The signal-to-noise ratio of the gradient is maximized when true edges are detected and false edges are avoided. Thus by discarding the false responses when there are multiple number of responses to a single edge, the noise-corrupted edges may be removed. In this method the image is first convolved with Gaussian smoothing filter with standard deviation σ. This operation is followed by gradient computation on the resultant smoothed image.

7.3.5.1 Non-Maxima Suppression The Canny's edge detector produces thick edges wider than a pixel. The operation of non maxima suppression thins down the broad ridges of gradient magnitude. There are several techniques for such a thinning operation. In one technique, the edge magnitudes of two neighboring edge pixels, perpendicular to the edge direction are considered and the one with lesser edge magnitude is discarded.

7.3.5.2 Double Thresholding The gradient image obtained after non-maxima suppression may still contain many false edge points. To remove false edge points, an appropriate threshold is selected such that all the edge points having magnitude greater than the threshold may be preserved as true edge points, while others are removed as false edge points. If the threshold is small, then a number of false edge points may be detected as true edge points, otherwise some true edge points may be missed. To avoid this problem, two thresholds T_1 and T_2 may be chosen to create two different edge images E_1 and E_2, where $T_2 \approx 1.5T_1$. E_1 will contain some false edge points, whereas E_2 will contain very few false edge points and miss a few true edge points. Threshold selection algorithm starts with the edge points in E_2, linking the adjacent edge points in E_2 forming a contour, and the process continues till no more adjacent edge points are available. At the boundary of the contour the algorithm searches for the next edge points from the edge image E_1 in its 8-neighborhood. The gaps between two edge contours may be filled by taking edge points from E_1 till the gap has been completely filled up. This process yields complete contour constituted by the true edges of the image.

7.3.5.3 Edge Threshold Selection The detection of edges is based on comparing the edge gradient with a threshold. This threshold value can be chosen low enough only when there is no noise in the image, so that all true edges can be detected without miss. In noisy images, however, the threshold selection becomes a problem of maximum likelihood ratio optimization based on Bayes decision theory, which has been discussed at length in Chapter 8.

Let the apriori probability of detection of true edges be $P(Edge)$ and that of no edge be $P(no\,edge)$ while $p(edge)$ and $p(no\,edge)$ are the conditional probability density functions of edge and no-edge classes. The probability of true edge detection P_{True} and the probability of false edge detection P_{False} can be computed using Bayes Decision Theory discussed in Chapter 8. The result of an edge image generated by the Canny's edge detector is shown in Figure 7.9.

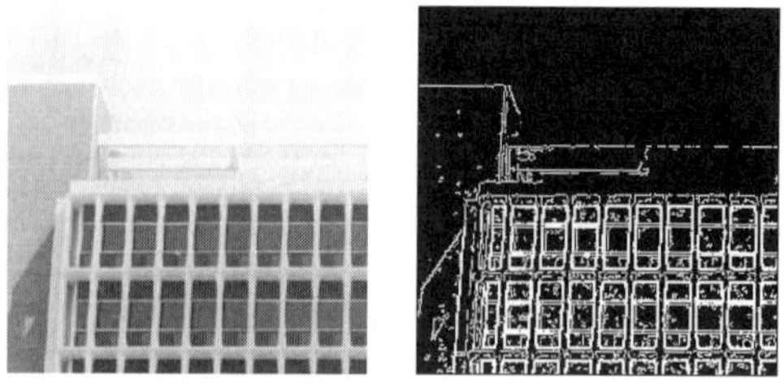

(a) (b)

Fig. 7.9 (a) Original image, (b) edge image by Canny's edge detector.

7.3.6 Operators-Based on Second Derivative

The principle of edge detection based on double derivative is to detect only those points as edge points which possess local maxima in the gradient values. In this case, we get a peak in the first derivative and a zero crossing at the second derivative at the edge points. Hence the points at which the second derivative has a zero crossing are treated as edge points. Laplacian operator is the most commonly used second derivative-based edge operator.

7.3.6.1 Laplacian Operator The gradient operators that we have discussed so far are anisotropic, i.e., they are not rotation invariant. If we apply the isotropic operators to an image and then rotate the resultant image, it will yield the same results as rotating the image first and then applying the operator. The utility of isotropic edge operator is that we can extract direction invariant enhanced edges. An isotropic gradient operator involves derivatives of even order. The *Laplacian* operator is one such isotropic rotation invariant operator. The Laplacian operator is expressed by

$$\frac{d^2 f}{dx^2} = \frac{dG_x}{dx} = \frac{d\left[f(i,j) - f(i,j-1)\right]}{dx} = \frac{df(i,j)}{dx} - \frac{df(i,j-1)}{dx}$$

$$= [f(i,j+1) - f(i,j)] - [f(i,j) - f(i,j-1)]$$

$$= f(i,j+1) - 2f(i,j) + f(i,j-1).$$

Similarly

$$\frac{d^2 f}{dy^2} = f(i+1,j) - 2f(i,j) + f(i-1,j).$$

The corresponding convolution masks along X and Y directions for a 4-neighbor Laplacian impulse response is shown in Figure 7.10.

Combining the two convolution masks in Figure 7.10, the gain normalized 4-neighbor Laplacian impulse response kernel may be represented as shown in Figure 7.11(a). In a like-wise fashion, the gain normalized 8-neighbor Laplacian impulse response kernel is shown in Figure 7.11(b).

0	0	0
1	−2	1
0	0	0

(a)

0	1	0
0	−2	0
0	1	0

(b)

Fig. 7.10 Laplacian masks (a)in X-direction, and (b) in Y-direction.

$$\frac{1}{4}\begin{bmatrix} 0 & 1 & 0 \\ 1 & -4 & 1 \\ 0 & 1 & 0 \end{bmatrix} \qquad \frac{1}{8}\begin{bmatrix} 1 & 1 & 1 \\ 1 & -8 & 1 \\ 1 & 1 & 1 \end{bmatrix}$$

(a) (b)

Fig. 7.11 Laplacian Impulse Response Kernels.

The Laplacian can also be computed as $f - \hat{f}$, where for a 4-neighbor case

$$\hat{f} = \frac{1}{5}\left[f(x, y) + f(x+1, y) + f(x-1, y) + f(x, y+1) + f(x, y-1)\right]$$

Laplacian, being the second derivative operator has zero response to linear ramp. In case of a ramp edge, it responds to the two sides, but not in the middle part of the ramp edge. On one side of the edge it gives positive response while on the other we get negative response.

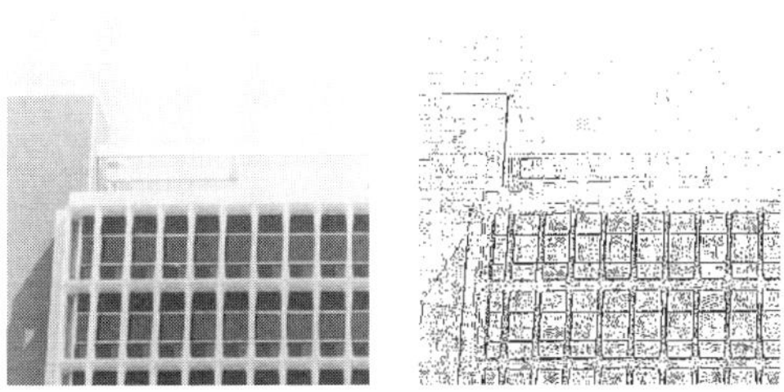

(a) (b)

Fig. 7.12 (a) Original image, (b) edge image by Laplacian edge detector.

The principle of edge detection based on a double derivative is to detect only those points as edge points which possess local maxima in the gradient values. Thus at edge points, we get a peak in the first derivative and a zero crossing at the second derivative. Hence the points at which the second derivative has a zero crossing are treated as edge points. The result of Laplacian edge detector is shown in Figure 7.12.

7.3.6.2 Laplacian of Gaussian (LOG) Operator Laplacian operator is susceptible to noise. To reduce the noise susceptibility, Laplacian of Gaussian (LOG) operator can be used. LOG first performs the Gaussian smoothing, which is followed by the Laplacian operation. It is less susceptible to noise because Gaussian function reduces the noise and the resultant Laplacian mask minimizes the probability of detection of false edges. The LOG function for convolution is defined as

$$LOG(x,y) = \frac{1}{\pi\sigma^4}\left[1 - \frac{x^2+y^2}{2\sigma^2}\right] e^{-\frac{x^2+y^2}{2\sigma^2}}. \tag{7.3}$$

The three-dimensional plot of the LOG function is shaped like a *Mexican hat* as shown in Figure 7.13(a). As σ increases, we need a wider convolution mask. The first zero-crossing of the LOG function takes place at $\sqrt{2}\sigma$ and the two-dimensional cross section of the zero-crossing of the *Mexican hat* is shown in Figure 7.13(b).

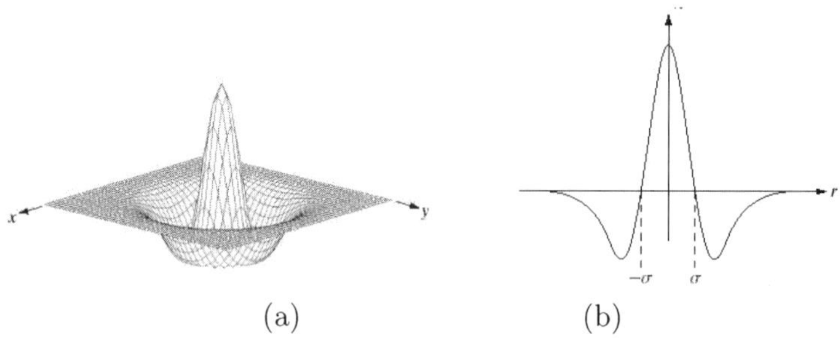

Fig. 7.13 (a) 3D plot of LOG function, (b) cross section showing zero crossing.

The 5×5 convolution mask to implement LOG edge detector is

$$\begin{bmatrix} 0 & 0 & -1 & 0 & 0 \\ 0 & -1 & -2 & -1 & 0 \\ -1 & -2 & 16 & -2 & -1 \\ 0 & -1 & -2 & -1 & 0 \\ 0 & 0 & -1 & 0 & 0 \end{bmatrix}.$$

7.3.6.3 Difference of Gaussian (DOG) operator It is possible to approximate the LOG filter by taking the difference of two differently sized Gaussians. Such a filter is known as a DOG filter (Difference of Gaussians) given as

$$DOG(x,y) = \frac{e^{-\frac{x^2+y^2}{2\sigma_1^2}}}{\pi\sigma_1^2} - \frac{e^{-\frac{x^2+y^2}{2\sigma_2^2}}}{\pi\sigma_2^2}. \tag{7.4}$$

The DOG operator is implemented by convolving an image with a mask which is obtained by subtracting two Gaussian masks with two different *sigma* values as shown in Eq. 7.4. The ratio $\frac{\sigma_1}{\sigma_2}$ between 1 to 2 yields good results for edge detection. The 7 × 7 convolution mask to implement the DOG edge detector is as shown below:

$$\begin{bmatrix} 0 & 0 & -1 & -1 & -1 & 0 & 0 \\ 0 & -2 & -3 & -3 & -3 & -2 & 0 \\ -1 & -3 & 5 & 5 & 5 & -3 & -1 \\ -1 & -3 & 5 & 16 & 5 & -3 & -1 \\ -1 & -3 & 5 & 5 & 5 & -3 & -1 \\ 0 & -2 & -3 & -3 & -3 & -2 & 0 \\ 0 & 0 & -1 & -1 & -1 & 0 & 0 \end{bmatrix}.$$

7.3.7 Limitations of Edge-Based Segmentation

The principal limitations of edge detection methods are:

(a) The edges extracted using the classical methods often do not necessarily correspond to boundary objects. In many low-quality images, captured using low quality imaging devices, some of the conventional methods produce spurious edges and gaps and their applicabilities are thus limited.

(b) The edge detection techniques depend on the information contained in the local neighborhood of the image. Most of the edge detection techniques do not consider model-based information embedded in an image.

(c) In most of the cases the edge detection strategies ignore the higher order organization which may be meaningfully present in the image.

(d) After the edge points are extracted from the image, these points are linked in order to determine boundaries. This is usually done by first associating edge elements into edge segments and then by associating segments into boundaries. The edge linking process sometimes lead to discontinuities and gaps in the image.

(e) The edge linking methods often resort to arbitrary interpolation in order to complete boundary gaps.

(f) It is often difficult to identify and classify spurious edges.

7.4 IMAGE THRESHOLDING TECHNIQUES

Gray level thresholding techniques are computationally inexpensive methods for partitioning a digital image into mutually exclusive and exhaustive regions. The thresholding operation involves identification of a set of optimal thresholds, based on which the image is partitioned into several meaningful regions. A survey of various thresholding schemes may be found in [6].

One of the earliest thresholding algorithms was suggested by Otsu [7], which is based on the principle that the gray-level for which the inter-class variance is maximum is selected as the threshold. For a gray level k all the gray-value $\leq k$ will form a class (C_0) and all the others will form a different class (C_1). Select that k as threshold for which the between class variance $V(k)$ is maximum. The criterion proposed by Otsu maximizes the between-class variance of pixel intensity. This method results in more computational complexity because of complexities involved in computation of the between-class variance.

7.4.1 Bi-level Thresholding

Bi-level thresholding is employed on images which have bimodal histograms. In bi-level thresholding, the object and background form two different groups with distinct gray levels. Examples are:

1. The alphanumeric characters in a book are generally darker than the background white paper.

2. The chromosomes in an image of mitotic cells, are darker than the background.

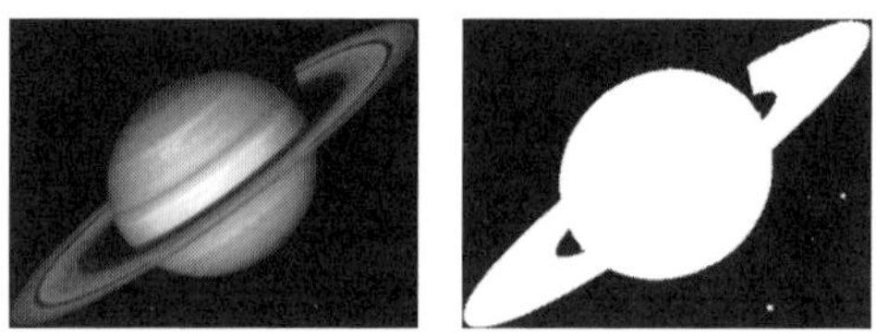

Fig. 7.14 Bimodal Image Thresholding.

In both these cases, the shapes of the histograms are bimodal with peaks corresponding to the object and background regions and a valley in between. The valley point is usually chosen as the threshold. In bimodal thresholding all gray values greater than threshold T are assigned the *object* label and all other gray values are assigned the *background* label, thus separating the object pixels from the background pixels. Thresholding thus is a transformation of an input image A into a segmented output image B as follows:

(a) $b_{ij} = 1$ for $a_{ij} \geq T$.

(b) $b_{ij} = 0$ for $a_{ij} < T$, where T is the threshold.

Here $b_{ij} = 1$, for the object pixels and $b_{ij} = 0$, for the background pixels. In Figure 7.14, we show the bi-level image thresholding of the saturn image. A simple iterative algorithm for threshold selection in a bimodal image is presented below.

Step 1: Choose an initial threshold $T \leftarrow T_0$.

Step 2: Partition the image using T in two regions–background and foreground (object).

Step 3: Compute mean gray values μ_1 and μ_2 of background and object regions respectively.

Step 4: Compute the new threshold $T \leftarrow \frac{\mu_1+\mu_2}{2}$;

Step 5: Repeat Steps 2 to 4 until there is no change of T.

7.4.2 Multilevel Thresholding

In *multilevel thresholding*, the image is partitioned into different segments using multiple threshold values. The histograms in such cases are multimodal, with valleys in between.

If objects are disjoint and their gray levels are clearly distinct from the background, then the histogram is multimodal with each peak distinctly separate from the other. While segmenting such an image, the valleys in between the peaks are chosen as the threshold values. Such a multiobject segmentation is shown in figure Figure 7.15.

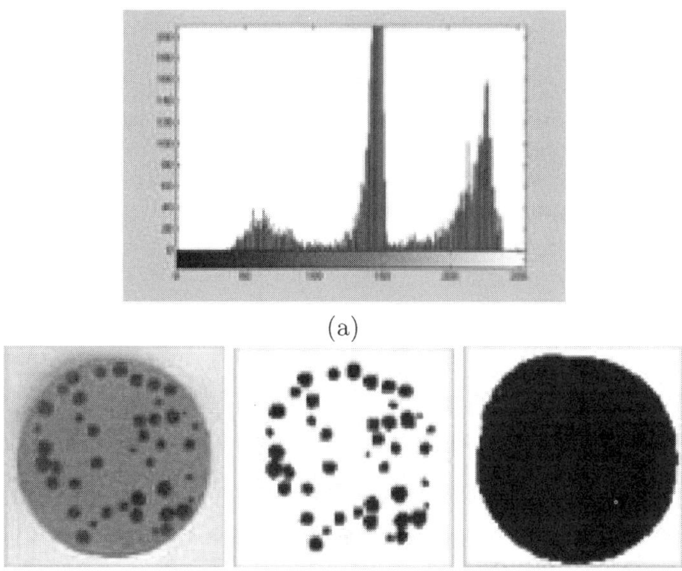

(a) (b) (c) (d)

Fig. 7.15 Multilevel Image thresholding: (a) Multimodal Histogram, (b) input image, (c) segmented foreground objects, (d) segmented background.

The threshold T may be global or local depending on the following cases.

1. When T depends only on gray value $g(x, y)$, the thresholding is global.
2. The thresholding is local when T depends on both $g(x, y)$ and $N(x, y)$, where $N(x, y)$ denotes a local image property at the point (x, y). The local image property may be computed from the local image statistics, e.g., the average gray level over a neighborhood around the point (x, y).

If additionally the threshold T depends on the time as well as on $g(x, y)$ and $N(x, y)$, then the thresholding scheme is called dynamic.

7.4.3 Entropy-Based Thresholding

Entropy based thresholding is widely used in bilevel thresholding. Entropy is a measure of information in an image defined by Shannon [8]. The variants of Shannon's entropy have been effectively used for estimation of thresholds in image segmentation [6]–[11]. In entropy-based thresholding, the entropy of foreground (object) and background regions are used for optimal selection of thresholds.

In Kapur's thresholding technique [9], the foreground and background region entropies are

$$H_f(T) = -\sum_{g=0}^{T} \frac{p(g)}{P_f(T)} \ln \frac{p(g)}{P_f(T)} \tag{7.5}$$

and

$$H_b(T) = -\sum_{g=T+1}^{L-1} \frac{p(g)}{P_b(T)} \ln \frac{p(g)}{P_b(T)} \tag{7.6}$$

where the foreground gray values range from 0 to T and background pixels lie in $[T+1, L-1]$ in an L-level gray image. In Eqs. 7.5 and 7.6, $p(g)$ is the probability mass function

$$p(g) = \frac{h(g)}{N}, \quad g = 0, 1, \cdots, L-1$$

where $h(g)$ is the histogram of gray value g and N is the total number of pixels. The foreground and background area probability values are

$$P_f(T) = \sum_{g=0}^{T} p(g) \text{ and } P_b(T) = \sum_{g=T+1}^{L-1} p(g)$$

The thresholding strategy maximizes the total entropy of both the foreground and the background regions. The desired threshold is 1 for which the total entropy assumes the maximum value.

Renyi's entropy has also been used for image thresholding [6]. The Renyi's entropy for foreground and background regions are

$$H_f(T) = \frac{1}{1-\rho} \ln \left[\sum_{g=0}^{T} \left(\frac{p(g)}{P(T)} \right)^{\rho} \right], \quad H_b(T) = \frac{1}{1-\rho} \ln \left[\sum_{g=T+1}^{L-1} \left(\frac{p(g)}{1-P(T)} \right)^{\rho} \right]$$

In this thresholding technique, the total entropy of foreground and background regions is computed for various ρ and appropriate ρ value is chosen that yields the best thresholding results.

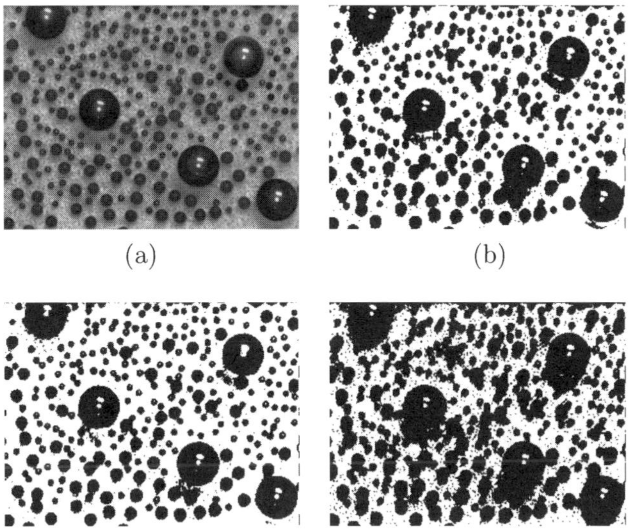

(a) (b) (c) (d)

Fig. 7.16 Comparison of thresholding techniques: (a) original image, (b) Otsu, (c) Kapoor, and (d) Renyi's Entropy.

Figure 7.16 shows the thresholded images by using Otsu, Kapoor, and Renyi's entropy based thresholding algorithms, as an example.

7.4.4 Problems Encountered and Possible Solutions

Quite often we encounter situations where two local maxima belong to the same global maximum in an image. Several techniques may be employed to avoid detection of such local maxima:

- A minimum distance of gray levels between these maxima is usually required for a good thresholding. Even if the histogram is bimodal, correct segmentation may not occur with objects located in a background having diverse gray levels. The histogram may be smoothed to ensure good thresholding.

- The influence of the pixels with a high image gradient may be reduced. This implies that the histogram will contain the gray levels that belong mostly either to the object or to the background, deleting the border pixels having intermediate gray values. This will produce a deeper valley and thus allow easier determination of the threshold.

- Another technique uses only gray level gradient pixels to form the gray level histogram in which the peak corresponds to the gray level borders between the object and the background. This histogram should be unimodal and the threshold is selected as the gray level of the peak.

7.5 REGION GROWING

Region growing refers to the procedure that groups pixels or subregions into larger regions. Starting with a set of seed points, the regions are grown from these points by including to each seed point those neighboring pixels that have similar attributes like intensity, gray level texture, color, etc. It is an iterative process where each seed pixel grows iteratively until every pixel is processed and thereby forms different regions whose boundaries are defined by closed polygons.

The important issues in the region growing are:

- Selection of initial seeds that represent regions and the selection of suitable properties for including the points in various regions during the growing process.
- Growing the pixels based on certain properties of the image may not ascertain good segmentation. Connectivity or adjacency information should also be used in the region-growing process.
- Similarity: The similarity denotes the minimum difference in the gray level observed between two spatially adjacent pixels or average gray level of a set of pixels, which will yield different regions. If this difference is less than the similarity threshold value, the pixels belong to the same region.
- Area of region: The minimum area threshold is associated with the smallest region size in pixels. In the segmented image, no region will be smaller than this threshold, defined by the user.

Region growing post-processing: Region growing often results in undergrowing or overgrowing as a result of nonoptimal parameter setting. A variety of postprocessors has been developed, which combine segmentation information obtained from region growing and edge-based segmentation. Simpler postprocessors are based on general heuristics and decrease the number of small regions in the segmented image that cannot be merged with any adjacent region according to the originally applied homogeneity criteria.

7.5.1 Region Adjacency Graph

The adjacency relation among the regions in a scene can be represented by a *region adjacency graph* (RAG). The regions in the scene are represented by a set of nodes $N = \{N_1, N_2, \cdots, N_m\}$ in the RAG, where node N_i represent the region R_i in the scene and properties of the region R_i is stored in the node data structure N_i. The edge $e_{i,j}$ between N_i and N_j represent the adjacency between the regions R_i and R_j. Two regions R_i and R_j are adjacent if there exist a pixel in region R_i and a pixel in region R_j which are adjacent to each other. The adjacency can be either 4-connected or 8-connected. The adjacency relation is *reflexive* and *symmetric*, but not necessarily *transitive*. In Figure 7.17, we show the adjacency graph of a scene with five distinct regions.

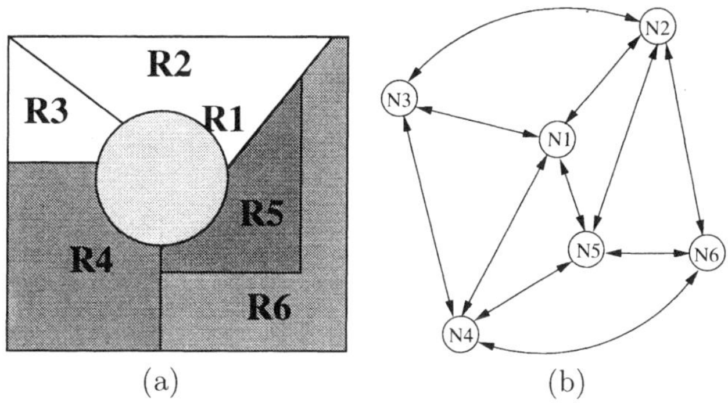

Fig. 7.17 (a) A scene with 6 distinct regions, (b) the adjacency graph of the scene.

A binary matrix A is called an *adjacency matrix* when it represents a region adjacency graph (RAG). When the nodes N_i and N_j in RAG are adjacent, $a_{i,j}$ in A is 1. Since adjacency relation is reflexive, the diagonal elements of the matrix are all 1. The adjacency matrix (relation) of a multiregion scene in Figure 7.17(a) is shown below.

$$A = \begin{bmatrix} & N1 & N2 & N3 & N4 & N5 & N6 \\ N1 & 1 & 1 & 1 & 1 & 1 & 0 \\ N2 & 1 & 1 & 1 & 0 & 1 & 1 \\ N3 & 1 & 1 & 1 & 1 & 0 & 0 \\ N4 & 1 & 0 & 1 & 1 & 1 & 1 \\ N5 & 1 & 1 & 0 & 1 & 1 & 1 \\ N6 & 0 & 1 & 0 & 1 & 1 & 1 \end{bmatrix}$$

7.5.2 Region Merging and Splitting

A segmentation algorithm can produce too many small regions because of fragmentation of a single large region in the scene. In such a situation, the smaller regions need to be merged based on similarity and compactness of the smaller regions. A simple region merging algorithm is presented below.

Step 1: Segment the image into $R_1, R_2, \cdots, R_m$ using a set of thresholds.

Step 2: Create a *region adjacency graph* (RAG) from the segmented description of the image.

Step 3: For every R_i, $i = 1, 2, \cdots, m$, identify all R_j, $j \neq i$ from the RAG such that R_j is adjacent to R_i.

Step 4: Compute an appropriate similarity measure S_{ij} between R_i and R_j, for all i and j.

Step 5: If $S_{ij} > T$, then merge R_i and R_j.

Step 6: Repeat steps 3 to 5 until there is no region to be merged according to the similarity criteria.

Another strategy for merging is based on the intensity of the edges between two regions. In this method, merging between adjacent regions is based on the edge intensity along the length of the demarcating boundary between two regions. If the edge intensity is small, i.e., the edge points are weak, then the two regions can be merged if the merger does not change the mean pixel intensity values considerably.

There are situations, when too little regions are generated because of inaccurate preliminary segmentation. This is due to wrong merger of different regions into a single one. In such situation, the variances of the gray values in a segmented region may be above a threshold (T) and hence the region needs to be split in smaller regions such that each of the smaller regions has uniform small variances.

Splitting and merging can be combined together for segmenting complex scenes where a rule-based may guide the applications of split and merge operations.

7.5.3 Clustering Based Segmentation

Data driven segmentation techniques can be *histogram-oriented* or *cluster-oriented*. Histogram-oriented segmentation produces an individual segmentation for each feature of the multifeature data, and then overlaps the segmentation results from each feature to produce more fragmented regions. Cluster-oriented segmentation uses the multidimensional data to partition the image pixels into clusters. Cluster-oriented techniques may be more appropriate than histogram-oriented ones in segmenting images, where each pixel has several attributes and is represented by a vector. Cluster analysis has attracted much attention since the 1960's and has been applied in many fields such as OCR (optical character Recognition) system, fingerprint identification, remote sensing, biological image segmentation, and so on.

In *cost minimization* clustering techniques, each clustering configuration is assigned a value or cost to measure its goodness. An appropriate cost function measures the *goodness* of a cluster. Usually, the cost for a cluster configuration is its squared error, i.e., the sum of squared Euclidean distances of each point to its cluster center. Thus low values of such a cost indicate better clustering results. It is found that this cost surface is complicated in nature with many poor local minima.

K-means is a popular cluster based segmentation method where each pixel is iteratively assigned to the nearest cluster and the cluster center position is recalculated. After each iteration the cost will decrease until the cluster configuration converges at a stable state, at which point the cost is at a local minimum.

Many image segmentation systems use an iterative moving method which attempts to search for a cluster configuration to determine whether the cost will decrease if a pixel is assigned from its current cluster to another. The pixel will be moved only if the cost decreases because of this assignment. Obviously the final result of this method depends on the initial configuration and the sequences of pixels which are tested.

7.6 WATERFALL ALGORITHM FOR SEGMENTATION

Waterfall algorithm is an interesting morphological segmentation technique. The watershed transformation is the paradigm of segmentation of morphology and has proven to be a powerful tool used in the solution of multiple applications. The principles and application of waterfall algorithm may be found in [12, 13]. In this technique a gray level image is considered as a topographic relief, where every pixel is assigned to a catchment basin of a regional minima. Thus every regional minima is considered to have a zone of influence. The watershed lines are the lines which separate these regions or zones of influence. Each smallest topographical region has sufficient information regarding its surrounding topographical map. The rainfall-pro regions are the only regions that get rainfall. Every region knows about the resultant flow of water caused due to the rain. This is because of the a-priori knowledge about the slope of the region and the surface tension of the water in that region.

(a) (b) (c)

Fig. 7.18 (a) An original image, (b) interframe difference image, (c) segmented human object.

In video sequence of images, consecutive image frames change only in their edges and the object is detected by linking the edges. It is the inherent ability of human beings to correlate an object with its outline or edge information. Thus starting with the edge information as seed points (rainfall-pro regions), region growing is performed recursively as long as the neighboring pixels (surrounding regions) are within a certain global limit (surface tension). For different cases the surface tension will change with global and local criteria. This is quite similar to the waterfall model in the plateau area where the water spreads in the surrounding regions, based on the porosity of the soil and surface tension of the flowing water.

Results of waterfall algorithm of images are shown in Figure 7.18. The main drawback of this method is the oversegmentation produced. Two approaches are proposed in the literature to overcome this drawback:

- the selection of markers, which supposes that the characteristics of the interesting objects are known.
- hierarchical approaches, that are able to rank the importance of each region.

7.7 CONNECTED COMPONENT LABELING

Connected component labeling is defined on a set of objects. In an image, connected component is said to be a separate and independent object. Independency comes with the connectivity. A connected component may be 4-connected or 8-connected. In 4-connected case, any element having coordinate (x, y) in that connected component has at least one element having coordinate in the following set:

$$\{(x, y-1),\ (x, y+1),\ (x-1, y),\ (x+1, y)\}$$

and all the elements having coordinate in that set should belong to the same connected component. For 8-connected the neighboring set of coordinate will be

$$\{(x-1, y-1),\ (x-1, y),\ (x-1, y+1),\ (x, y-1),\ (x, y+1),\ (x+1, y-1),\ (x+1, y),\ (x+1, y+1)\}.$$

To perform the connected component labeling there are many methods. Further detail in connected component labeling and a two pass algorithm are presented in Section 14.4 in Chapter 14.

7.8 DOCUMENT IMAGE SEGMENTATION

Document image segmentation is a part of *document image analysis* (DIA), which is concerned with automatic interpretation of printed and handwritten documents, including, text, drawings, maps, etc [14]–[16]. Document analysis is used to extract geometric structures of a document. The document is broken down into several blocks which represent different constituent parts of a document, such as text lines, headings, graphics, and half-tones with or without the knowledge regarding the specific format. Due to its immense potential for commercial applications research in this field supports a rapidly growing industry including OCR and vision systems for DIA. General OCR systems show high accuracy in interpreting text portions but fail to properly process other components like graphics, half-tones (general picture), mathematical formulas, and equations. To provide a complete support to OCR it is required to segment the document in block level. Major independent parts of commonly used documents are as follows

1. Half-tone

2. Text

3. Graphics

Half-tones are the pictures having gray value distribution as observed in newspaper, magazine, etc. Texts are the set of alphabets, numeric, and symbols, while graphics are the pictorial interpretation of graphs, geometrical sketches and line arts. Based on the spatial orientation, the text may be (a) Tabular text, (b) Non-tabular text.

In the domain of tabular text the different sub-parts are (a) Table, (b) TOC (Table of Content) page, and (c) Index page. Tables are the horizontally and vertically separated data set, having different rows and columns. TOC pages are the table of content pages that indicates the relation of orderly arranged subjects and the related page numbers. Index pages are alphabetically ordered pages with indexing terms.

Non-tabular texts are the normal pages found in any document page. It is divided into (a) Normal text, (b) Heading, (c) Math equations. Normal text is self-explanatory and observed mainly in all the paper documents whereas heading is text with bigger font, while math equations are enriched with special mathematical symbols, which are difficult for OCR to retrieve.

The first step in document image segmentation system involves preprocessing followed by segmentation. Preprocessing includes binarization, noise reduction, skew detection, and character thinning [14]. It is important to obtain good quality document page before layout analysis is attempted. Various filtering strategies can be employed to filter the noise and reconstruct the broken lines which may result from skeletonization process. Pixel classification followed by region classification are attempted for segmenting the document in a set of homogeneous regions. In view of the fact that different regions in a document have different textures, texture based segmentation is well suited for document page segmentation.

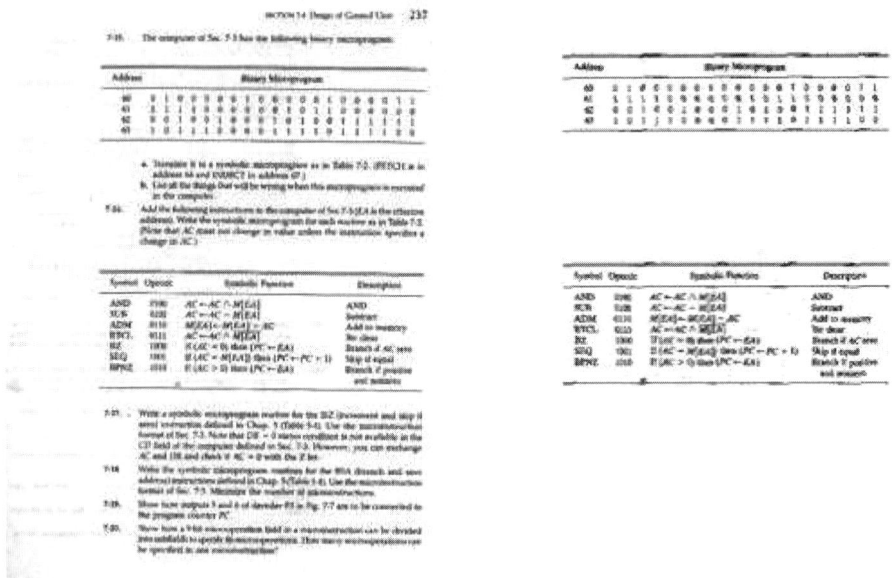

(a) (b)

Fig. 7.19 (a) A document image, (b) text portion segmented

Figure 7.19(a) shows a noisy document page having text and table from which the text portion has been segmented. The resulting tabular structure is shown in Figure 7.19(b).

Document processing leads to the form processing system whose objective is to extract the user-filled data. The layout structure of the form must be known in advance to identify as well as to exclude pre-printed data such as text, logo, boxes and lines. A blank form without any user-filled data may be used to extract the layout structure as reference prototype. These reference prototypes may be stored in a Prototype database and can be compared with the instances of a filled-in input form.

7.9 SUMMARY

The problem of image segmentation has been viewed as a problem of detection of edges and also as a problem of partitioning an image into several partitions. In this chapter we have presented both the approaches towards image segmentation. The objective of segmentation is to minimize the classification errors and reduce the statistical uncertainty. This has lead to the development of several algorithms based on local and global properties of the pixels in an image. Thresholding techniques are simple and elegant. A number of histogram thresholding schemes have been presented here. Other segmentation strategies, mainly region growing, region split and merge, have been discussed. Waterfall algorithms find applications in many image processing tasks where edge linking is avoided after edge detection. This algorithm has been deeply discussed here. Finally, an application of segmentation in the area of document image analysis has been briefly reviewed.

REFERENCES

1. A. Rosenfeld and A. C. Kak, *Digital Picture Processing*, 2nd ed., Vol. 1, Academic Press, 1982.

2. R. C. Gonzalez and R. E. Woods, *Digital Image Processing*, Addison-Wesley, Reading, MA, 1992.

3. W. K. Pratt, *Digital Image Processing*, 2nd ed., Wiley, New York, 1991.

4. T. Pavlidas, *Structural Pattern Recognition*, Springer Verlag, New York, 1977.

5. J. Canny, "A Computational approach to edge detection," *IEEE Transaction on Pattern Analysis and machine Intelligence*, 8(6), 1986, 679–698.

6. P. K. Sahoo, S. Soltani, A. K. C. Wong, and Y. C. Chen, "A Survey of Thresholding Techniques," *Computer Vision, Graphics & Image Processing*, 1988, 233–260.

7. N. Otsu, "A Threshold Selection Method From Gray Level Histograms," *IEEE Transaction on System, Man, and Cyber*, 9, 1979, 62–66.

8. C. E. Shannon and W. Weaver, *The Mathematical Theory of Communication*, University of Illinois Press, Urbana, IL, 1949.

9. J. N. Kapur, P. K. Sahoo, and A. K. C Wong, "A New Method of Gray Level Picture Thresholding Using the Entropy of the Histogram," *Computer Vision, Graphics & Image Processing*, 29, 1985, 273–285.

10. A. D. Brink and N. E. Pendcock, "Minimum Cross-Entropy Threshold Selection," *Pattern Recognition*, 29, 1996, 179–188.

11. A. S. Abutaleb, "Automatic Thresholding of Gray Level Pictures Using Two-Dimensional Entropy," *Computer Vision Graphics and Image Processing*, 47, 1989, 22–32.

12. F. Meyer, "An Overview of Morphological Segmentation," *International Journal of Pattern Recognition an Artificial Intelligence*, 15(7), 2001, 1089–1118.

13. M. Grimaud, "New Measure of Contrast: Dynamics," *Prof. SPIE, Image Algebra and Morphological Processing III*, San Diego CA, 1992.

14. L. O'Gorman and R. Kasturi, "Document Image Analysis," *IEEE Press*, Los Alamitos, CA, 1995.

15. G. Nagy, "Twenty Years of Document Image Analysis", *IEEE Tran. Pattern Analysis and Machine Intelligence*, 22(1), 2000, 38–62.

16. S. P. Chowdhury, S. Mandal, A. K. Das, and B. Chanda, "Automated segmentation of math-zones from document images", *7th International Conference on Document Analysis and Recognition*, Edinburgh, U.K., 2003, Vol. II, 755–759.

8

Recognition of Image Patterns

8.1 INTRODUCTION

Once an image is segmented, the next task is to recognize the segmented objects or regions in the scene. Hence, the objective in pattern recognition is to recognize objects in the scene from a set of measurements of the objects. Each object is a pattern and the measured values are the features of the pattern. A set of similar objects possessing more or less identical features are said to belong to a certain pattern class.

Pattern recognition is an integral part of machine vision and image processing [1]–[6]and finds its applications in biometric and biomedical image diagnostics to document classification, remote sensing, and so on.

There are many types of features and each feature has a specific technique for measurement. In addition, higher-order features are formed by combinations or distributions of the simpler set of features. As an example, each letter in the English alphabet is composed of a set of features like horizontal, vertical, slant straight lines, as well as some curvilinear line segments. While the letter 'A' is described by two slant lines and one horizontal line, letter 'B' has a vertical line with two curvilinear segments, joined in a specific structural format. Some of the features of a two- or three-dimensional object pattern are the area, volume, perimeter, surface, etc. which can be measured by counting pixels. Similarly the shape of an object may be characterized by its border. Some of the attributes to characterize the shape of an object pattern are Fourier descriptors, invariant moments, medial axis of the object, and so on.

The color of an object is an extremely important feature, which can be described in various color spaces. Also various types of textural attributes characterize the surface of an object. The techniques to measure the features are known as *feature extraction* techniques. Patterns may be described by a set of features, all of which may not have enough discriminatory power to discriminate one class of patterns from another. The selection and extraction of appropriate features from patterns poses the first major problem in pattern recognition.

8.2 DECISION THEORETIC PATTERN CLASSIFICATION

A number of pattern classification techniques have been used for the recognition of patterns. Some of these techniques are known as decision theoretic techniques, in which the classification of an unknown pattern is decided based on some deterministic or statistical or even fuzzy set theoretic principles. The block diagram of a decision theoretic pattern classifier is shown in Figure 8.1.

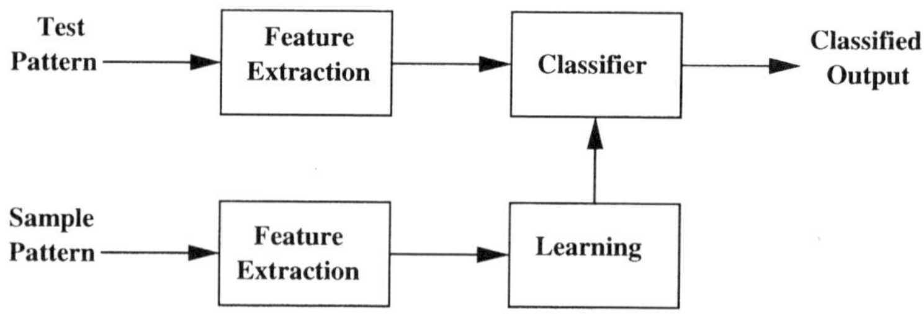

Fig. 8.1 Block diagram of a decision theoretic pattern classifier.

The decision theoretic pattern recognition techniques are mainly of two types

1. Classification methods based on supervised learning
2. Classification methods using unsupervised techniques.

The supervised classification algorithms can further be classified as

- Parametric classifiers
- Nonparametric classifiers

In parametric supervised classification, the classifier is trained with a large set of labeled training pattern samples in order to estimate the statistical parameters of each class of patterns such as mean, variance, etc. By the term '*labeled pattern samples*', we mean the set of patterns whose class memberships are known in advance. The input feature vectors obtained during the training phase of the supervised classification are assumed to be Gaussian in nature.

The *minimum distance classifier* and the *maximum likelihood classifier* are some of the frequently used supervised algorithms.

On the other hand, the parameters are not taken into consideration in the nonparametric supervised classification techniques. Some of the nonparametric techniques are K-nearest neighbor, Parzen window technique, etc.

In unsupervised case, the machine partitions the entire data set based on some similarity criteria. This results in a set of clusters, where each cluster of patterns belong to a specific class.

8.3 BAYESIAN DECISION THEORY

Bayesian decision theory is a very good tool for pattern classification. Assume that there are N classes of patterns $C_1, C_2, \cdots, C_N$, and an unknown pattern x in a d-dimensional feature space $x = [x_1, x_2, x_3, \cdots, x_d]$. Hence the pattern is characterized by d number of features. The problem of pattern classification is to compute the probability of belongingness of the pattern x to each class C_i, $i = 1, 2, \cdots, N$. The pattern is classified to the class C_k if probability of its belongingness to C_k is maximum.

While classifying a pattern based on Bayesian decision theory, we distinguish two kinds of probabilities: (1) Apriori probability, and (2) Aposteriori probaility. The apriori probability indicates the probability that the pattern should belong to a class, say C_k, based on the prior belief or evidence or knowledge. This probability is chosen even before making any measurements, i.e., even before selection or extraction of a feature. Sometimes this probability may be modeled using Gaussian distribution, if the previous evidence suggests it. In cases where there exists no prior knowledge about the class membership of the pattern, usually a uniform distribution is used to model it. For example, in a four class problem, we may choose the apriori probability as 0.25, assuming that the pattern is equally likely to belong to any of the four classes.

The aposteriori probability $P(C_i|x)$, on the other hand, indicates the final probability of belongingness of the pattern x to a class C_i. The aposteriori probability is computed based on the

- feature vector of the pattern,
- class conditional probability density functions $p(x|C_i)$ for each class C_i,
- Apriori probability $P(C_i)$ of each class C_i.

Bayesian decision theory states that the aposteriori probability of a pattern belonging to a pattern class C_k is given by

$$P(C_k|x) = \frac{p(x|C_k)P(C_k)}{\sum_{i=1}^{N}(p(x|C_i)P(C_i)}.$$

The denominator $\sum_{i=1}^{N} p(x|C_i)P(C_i)$ in the above expression is the scaling term which yields the normalized value of the aposteriori probability that the pattern x belongs to class C_i. Hence, x belongs to class C_p when

$$P(C_p|x) = \max\{P(C_1|x), P(C_2|x), \cdots, P(C_N|x)\}$$

8.3.1 Parameter Estimation

Given a large number of labeled training sample patterns, the next task is to estimate the parameters of the Gaussian distribution $p(x|C_i)$. Since in most of the cases we require more than one feature, we have to select multivariate Gaussian distribution, which is given as

$$p(x|C_i) = \frac{e^{\{-\frac{1}{2}(x-\mu_i)^T \Sigma_i^{-1}(x-\mu_i)\}}}{(2\Pi)^{d/2} |\Sigma_i|^{1/2}}.$$

The parameters that we need to estimate are (1) mean feature vector μ_i, and (2) covariance matrix Σ_i for each pattern class C_i. If the chosen features are statistically independent, then the off diagonal elements of the covariance matrix Σ_i will be zero. In such a situation, the covariance matrix Σ_i will contain only the diagonal elements which represent the variances for each feature dimension and hence computing the inverse of the diagonal covariance matrix becomes simpler. During the training phase, the parameters such as the elements of the mean feature vector μ and the covariance matrix Σ are estimated. The next step is to compute the aposteriori probability of the test pattern sample to one of the classes.

8.3.2 Minimum Distance Classification

Distance functions are used to measure the similarity or dissimilarity between two classes of patterns. The smaller the distance between two classes of patterns, the larger is the similarity between them. The *minimum distance classification* algorithm is computationally simple and commonly used. It can result in classification accuracies that are comparable to other more computationally intensive algorithms like maximum likelihood class. The classifier finds the distances from a test input data vectors to all the mean vectors representative of the target classes. The unknown pattern is assigned to that class from which its distance is smaller than its distances to all other classes.

Let us consider an N class problem. If the class C_i contains a single prototype pattern μ_i selected by appropriate measure and distance of the unknown pattern $X = [x_1, x_2, \cdots, x_n]$ from the pattern class C_i is $d(X, \mu_i)$, then the pattern X belongs to the class C_k if

$$D_k = \min_i\{d(X, \mu_i)\}$$

for all $i = 1, \cdots, N$. The distance function d is supposed to satisfy three properties as below.

1. **Reflexitivity:** The distance between a pattern $X = [x_1, x_2, \cdots, x_n]$ and itself is always less than or equal to its distance with any other pattern. This implies that $D(X, X) \leq D(X, Y)$, for all $Y \neq X$. In fact such a distance is in most of the cases zero, i.e., $D(X, X) = 0$.

2. **Symmetry:** The distance between pattern $X = [x_1, x_2, \cdots, x_n]$ and pattern $Y = [y_1, y_2, \cdots, y_n]$ is always equal to the distance between pattern Y and pattern X, i.e., $D(X, Y) = D(Y, X)$.

3. **Triangular law of inequality:** The sum of the distances between pattern $X = [x_1, x_2, \cdots, x_n]$ and pattern $Y = [y_1, y_2, \cdots, y_n]$ and between pattern Y and pattern $Z = [z_1, z_2, \cdots, z_n]$ is always greater than the distance between pattern X and pattern Z, i.e, $D(X, Y) + D(Y, Z) > D(X, Z)$.

8.3.2.1 Minkowski Distance In the numeric domain, a popular distance measure is the Minkowski distance. This is a generalized distance measure between two data objects or patterns A and B computed as

$$D_p(A, B) = \sqrt[p]{\sum_{i=1}^{n} |ai - bi|^p}.$$

- When $p = 1$, the distance is called the *city block* or *Manhattan* distance, which is computed as

$$D_1(A, B) = \sum_{i=1}^{n} |ai - bi|.$$

- When $p = 2$, the distance is called the *Euclidean* distance, which is computed as

$$D_2(A, B) = \sqrt{\sum_{i=1}^{n} |ai - bi|^2}.$$

8.3.2.2 Mahalanobish Distance The Mahalanobish distance essentially indicates the distance between a test pattern and a pattern class. If the parameters of the distribution of a specific pattern class are assumed to be Gaussian with mean feature vector μ and the covariance matrix Σ, then the Mahalanobish distance between the test pattern with the feature vector x and that pattern class C is given by

$$D_3(x, C) = (x - \mu)^T \Sigma^{-1} (x - \mu),$$

where μ is the mean feature vector of the pattern samples belonging to the pattern class C and Σ is the covariance matrix of C.

8.3.2.3 Bounded Distance In many pattern classification problems, it may be useful to work with a bounded distance function, which lies in the range $[0, 1]$. Any given distance function $D(X, Y)$ may be transformed into a bounded distance function $d(X, Y)$, where

$$d(X, Y) = \frac{D(X, Y)}{D(X, Y) + 1}.$$

It is obvious that when distance is measured with the function d, the maximum distance between any two patterns lies between 0 and 1, regardless of their location.

It is interesting to note that even when $D(X, Y)$ is not bounded, $d(X, Y)$ is. Thus both distances induce the same notion of nearness between the two patterns X and Y. There are many such functions, like $f(a) = \frac{a}{a+1}$, which produce new distance functions. These functions may be found to be useful in pattern classification.

8.4 NONPARAMETRIC CLASSIFICATION

The nonparametric classification strategies are not dependent on the estimation of parameters. There are several nonparametric classifiers, e.g., *K-nearest neighbor* classifier, *Parzen window*-based classifier, etc. [1, 7].

8.4.1 K-Nearest-Neighbor Classification

In many situations we may not have the complete statistical knowledge about the underlying joint distribution of the observation or feature vector x and the true class C_i to which the pattern belongs.

The K-nearest neighbor rule is a nonparametric pattern classifier, which is simple, yet yields good classification accuracy. Let us consider a two class problem. Assume that all the samples belonging to each of the two particular classes are independently and identically distributed according to the distribution $p(x)$. The samples which are similar most likely will have the same aposteriori probabilities. For an unknown test sample, K nearest rule suggests that it should be assigned to the class to which majority of its K-nearest neighbors belong.

There are, however, certain problems in classifying an unknown pattern using nearest neighbor rule. If there are N number of sample patterns, then to ascertain the nearest neighbor, we need to compute N distances from the test pattern to each of the sample points. Also it is important to store all these N sample points. This leads to increase of the computational as well as storage complexity of the K-nearest neighbor problem. As the number of features increases, we require more number of training data samples and hence it increases the storage and computational complexities as well.

To reduce these complexities various researchers have taken different measures.

- Remove the redundant data from the data set, which will reduce the storage complexity.
- The training samples need to be sorted to achieve better data structure for reducing the computational complexities.
- The distance measure to be used for computation should be simple.

8.4.1.1 A two-class pattern classification example A two-class pattern classification problem is exemplified in Figure 8.2. We consider $k = 5$ nearest neighbors from the unknown pattern X, marked by the dotted line within an appropriate neighborhood around the test pattern X. As shown in Figure 8.2, there are three patterns from class C_2 and two from class C_1. The majority decision enables assignment of the unknown pattern X to class C_2.

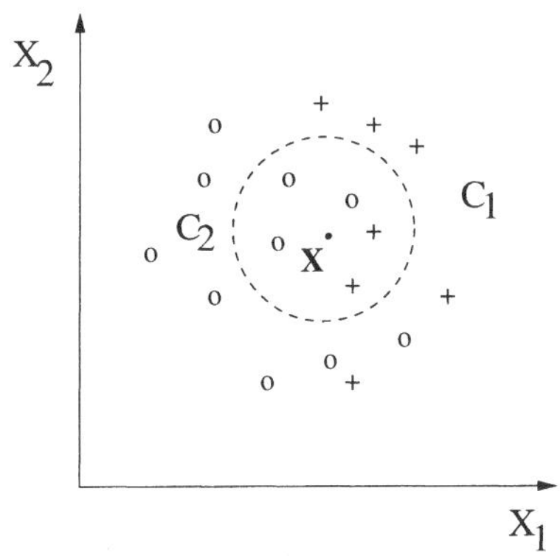

Fig. 8.2 A k-nearest neighbor example for $k = 5$.

To reduce the computational and storage complexity for implementation of K-nearest neighbor algorithm, a number of techniques have been suggested in [1]. The asymptotic error in case of K-nearest neighbor is less than twice that of Baysean classification error.

8.5 LINEAR DISCRIMINANT ANALYSIS

An image can be described by a set of local features, These features can be extracted at each pixel of the image. The features may be, Gabor Wavelet feature, shape feature, color feature or texture feature and so on. Let $f_k(p)$ denotes the kth feature at pixel p. The raster scanning of the entire image yields a vector f_k at each pixel of the image. This results in a large matrix, whose dimensionality is equal to the number of the pixels in the image. If each pixel in an image is associated with d number of features, we have a matrix $F = \{f_1, \cdots, f_d\}$ of dimension $n \times d$, where n is the total number of pixels in the image. It may be noted here that this matrix contains lot of local information of the entire image, much of which is redundant. The discriminant analysis is employed to find which variables discriminate between two classes and is essentially analogous to the analysis of variance. In discriminant analysis, we assume that the discriminant function is linear, i.e.,

$$g(x) = w^t x + x_0 = 0$$

is a linear decision surface, called as hyperplane, which partitions the feature space in two subspaces. In Fisher's linear discriminat approach, the d-dimensional patterns x are projected onto a line, such that the projection of data

$$y = w^t x$$

are well separated. The measure of this separation can be chosen as

$$J(\vec{w}^t) = \frac{(m_1 - m_2)^2}{\tilde{S}_1^2 + \tilde{S}_2^2},$$

where m_1 and m_2 are the projection means for classes C_1 and C_2 and $\tilde{S}_1^2$ and $\tilde{S}_2^2$ are the within class variances of the projected data [1]. Here

$$\tilde{S}_i^2 = \sum (y - m_i)^2$$

gives a measure of scatter of the projected set of data points y. The objective function $J(\vec{w}^t)$ is maximized for the weight $\vec{w}$ such that

$$\vec{w} = S_w^{-1} (\tilde{m}_1 - \tilde{m}_2) .$$

The discriminant function can be extended to a generalized polynomial

$$g(x) = w_0 + \sum_{i=1}^{d} w_i x_i + \sum_{i=1}^{d} \sum_{j=1}^{d} w_{ij} x_i x_j + \cdots$$

where the first three terms yield a quadratic discriminant function.

The Fisher linear discriminant function is widely used for identifying the linear separating vector between pattern classes. The procedure uses the maximization of between class scatter while minimizing the intra-class variances.

8.6 UNSUPERVISED CLASSIFICATION STRATEGIES - CLUSTERING

The ability to cluster data in a finite set of groups is an essential feature of human intelligence. The elements or patterns in a cluster have more similarities among them compared to their similarities from other clusters. Thus in a clustering problem, we have a set of patterns, that have to be partitioned in a set of clusters such that the patterns within a cluster are more similar to each other than the patterns from other clusters or partitions. Thus central to the goals of cluster analysis lies the notion of similarity. There are a couple of methods of clustering. We can divide these methods into the following three classes:

1. Hierarchical methods
2. K-means methods
3. Graph theoretic methods

In hierarchical algorithms, the data set is partitioned in a number of clusters in a hierarchical fashion. The hierarchical clustering methods may again be subdivided into the following two categories.

1. *Agglomerative clustering*: In agglomerative clustering, we start with a set of singleton clusters, which are merged in each step, depending on some similarity criterion, and finally we get the appropriate set of clusters.
2. *Divisive clustering*: In divisive clustering, as the name suggests, the whole set of patterns initially is assumed to belong to a single cluster, which subsequently is divided in several partitions in each step.

The hierarchical clustering may be represented by *dendograms*, a tree structure which demonstrates the merging (fusion) or division of points in each step of hierarchical partitioning. Agglomerative clustering is the bottom up clustering procedure where each singleton pattern (leaf nodes at the bottom of the dendogram) merges with other patterns, according to some similarity criterion. In divisive algorithm, on the other hand, starting with the root node S, we recursively partition the set of patterns until singleton patterns are reached at the bottom of the tree.

8.6.1 Single Linkage Clustering

The single linkage or nearest neighbor agglomerative clustering technique involves grouping of patterns based on a measure of intercluster (distance between two clusters). The distance between two clusters, each containing finite number of pattern samples is defined as the distance between the closest pair of patterns, where the pair is made up of one pattern belonging to each cluster. Assuming two clusters P_1 and P_2, each containing finite number of patterns, in single linkage method, the distance between P_1 and P_2 is given by

$$D_{min}(P_1, P_2) = \min_{i,j}\{d(p_i, p_j)\},$$

where pattern p_i is in cluster P_1 and pattern p_j is in cluster P_2.

8.6.2 Complete Linkage Clustering

In complete linkage clustering, distance between two clusters is defined as the distance between the most distant pair of patterns, each pattern belonging to one cluster. This method may thus be called the *farthest-neighbor* method. In complete linkage method, the distance between P_1 and P_2 is given by

$$D_{max}(P_1, P_2) = \max_{i,j}\{d(p_i, p_j)\},$$

where pattern p_i is in cluster P_1 and pattern p_j is in cluster P_2.

8.6.3 Average Linkage Clustering

In average linkage clustering, the distance between two clusters is given by the average of all distances between all pairs of patterns. In this method, the distance between P_1 and P_2 is

$$D_{Avg}(P_1, P_2) = Avg_{i,j}\{d(p_i, p_j)\},$$

where pattern p_i is in cluster P_1 and pattern p_j is in cluster P_2. If there are n_1 number of patterns in Cluster P_1 and n_2 patterns in cluster P_2, then the average distance between two clusters is given by

$$D_{Avg}(P_1, P_2) = \frac{1}{2}\frac{\sum_{i,j} d(p_i, p_j)}{n_1 n_2},$$

where pattern p_i is in cluster P_1 and pattern p_j is in cluster P_2.

8.7 K-MEANS CLUSTERING ALGORITHM

In K-means clustering approach, we partition the set of input patterns S into a set of K partitions, where K is known in advance. The method is based on the identification of the centroids of each of the K clusters. Thus, instead of computing the pairwise interpattern distances between all the patterns in all the clusters, here the distances may be computed only from the centroids. The method thus essentially boils down to searching for a best set of K centroids of the clusters as follows:

Step 1: Select K initial cluster centers $C_1, C_2, \cdots, C_K$.

Step 2: Assign each pattern $X \in S$ to a cluster C_i $(1 \le i \le K)$, whose centroid is nearest to pattern X.

Step 3: Recompute the centroids in each cluster C_j $(1 \le j \le K)$ in which there has been any addition or deletion of pattern points.

Step 4: Jump to step 2, until convergence is achieved.

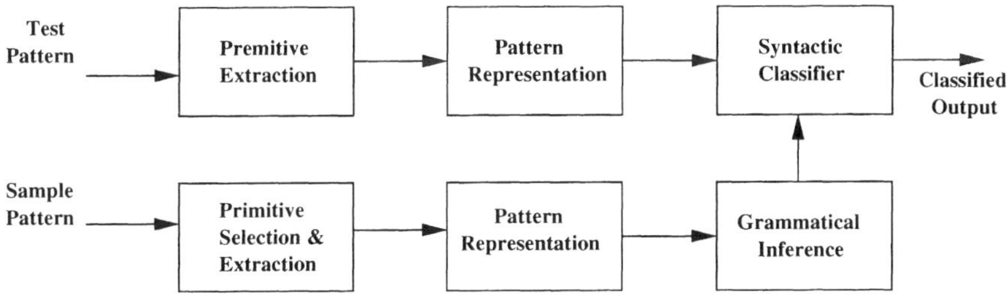

Fig. 8.3 Syntactic pattern recognition system.

The major problem is the selection of initial cluster configurations. It is possible either to select the first k samples as the initial cluster centers or to randomly select K samples from the pool of patterns as the cluster centers. A rough partition in K clusters may, however, yield a better set of initial cluster centers.

8.8 SYNTACTIC PATTERN CLASSIFICATION

It may be noted that there exists an inherent structure inside a pattern and there is a positive interrelationship among the primitive elements which form a pattern [2, 3, 5, 8, 9]. The limitations of decision theoretic pattern classification techniques lie in their incapability to articulate this interrelationship of the pattern substructures. This has led to a new era of structural or syntactic pattern recognition. The interrelationship between pattern elements called primitives and the articulated description of a pattern in terms of such relations provide a basis of structural or linguistic approach to pattern recognition.

In syntactic pattern recognition each pattern is characterized by a string of primitives and the classification of a pattern in this approach is based on analysis of the string with respect to the grammar defining that pattern class. The syntactic approach to pattern recognition involves a set of processes, viz.,

1. Selection and extraction of a set of primitives (segmentation problem)

2. Analysis of pattern description by identification of the interrelationship among the primitives

3. Recognition of the allowable structures defining the interrelationship between the pattern primitives

A typical block diagram of a syntactic pattern recognition system is shown in Fig 8.3.

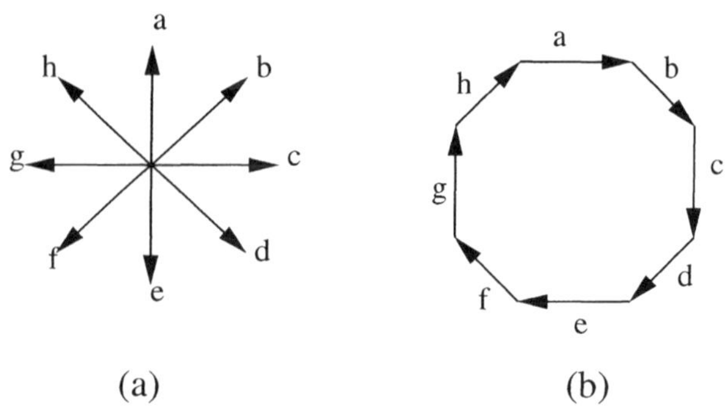

Fig. 8.4 (a) Example of a Freeman chain code (b) Freeman chain code of an octagon.

8.8.1 Primitive selection Strategies

As mentioned earlier, segmentation of patterns poses the first major problem in syntactic pattern recognition. A pattern may be described by a string of subpatterns or primitives, which may easily be identified. If each subpattern is complex in structure, each of them may again be described by simpler subpatterns which are easily identifiable.

The problem of primitive selection is a fundamental one and various approaches to primitive selection have been suggested in the literature [2, 4, 8]. One of the most frequently used schemes of boundary descriptions is the chain code method by Freeman [10]. Under this approach, a rectangular grid is overlaid on a two-dimensional pattern and straight line segments are used to connect the adjacent grid points covering the pattern.

Let us consider a sequence of n points $\{p_1, p_2, \cdots, p_n\}$ which describe a closed curve. Here the point p_i is a neighbor of p_{i-1} and p_{i+1} when $i < n$, whereas the point p_n is the neighbor of p_{n-1} and the point p_0 and also p_0 is the neighbor of p_1 and p_n. The Freeman chain code contains n vectors $p_i p_{i-1}$ and each of these vectors is represented by an integer $m = 0, 1, \cdots, 7$ as shown in Figure 8.4(a). Each line segment is assigned an octal digit according to its slope and the pattern is represented by a chain of octal digits. This type of representation yields patterns composed of a string of symbolic valued primitives. Figure 8.4(b) shows the Freeman chain code of a closed pattern. This method may be used for coding any arbitrary two-dimensional figures composed of straight line or curved segments and has been widely used in many shape recognition applications. The major limitation of this procedure is that the patterns need adequate preprocessing for ensuring proper representation.

Once a satisfactory solution to the primitive selection and extraction problem is available, the next step is the identification of structural interrelationship among the extracted pattern primitives. A pattern may be described as sets of strings or sentences belonging to specific pattern classes. First order logic may be used for describing the primitive interrelationship where a pattern is described by certain predicates and objects occurring in the pattern may be defined using the same predicates. When the patterns are represented as strings of primitives they may be considered as sentences of a *regular*, *context-free*, or *context-sensitive* languages. Thus suitable grammars may be defined for generating pattern languages by specifying a set of production rules which generate the sentences in the said pattern language. The corresponding computing machines known as *automata* have the capability of recognizing whether a string of primitives belongs to a specific pattern class [8, 11].

8.8.2 High-Dimensional Pattern Grammars

The string representation of patterns is quite adequate for structurally simpler forms of patterns. The classical string grammars are, however, weak in handling noisy and structurally complex pattern classes. This is because the only relationship supported by string grammars is the concatenation relationship between the pattern primitives. Here each primitive element is attached with only two other primitive elements–one to its right and the other to its left. Such a simple structure thus may not be sufficient to characterize more complex patterns, which may require better connectivity relationship for their description. An appropriate extension of string grammars has been suggested in the form of high-dimensional grammars. These grammars are more powerful as generators of language and are capable of generating complex patterns like chromosome patterns, nuclear bubble chamber photographs, etc. [2, 3, 4].

In a string grammar each primitive symbol is attached with only two other primitive elements, one to the right and the other to the left of the element. A class of grammars was suggested by Fedder [3], where a set of primitive elements may be used with multiple connectivity structure. These grammars are known as PLEX grammars. PLEX grammar involving primitive structures called n-attaching point entity (NAPE) and a set of identifiers associated with each NAPE has been used for pattern generation. The n-attaching point entities are primitive elements in which there are n number of specified points on the primitive elements where other attaching elements may be connected. Thus this class of grammars have more generating capabilities compared to the string grammars.

8.9 SYNTACTIC INFERENCE

A key problem in syntactic Pattern Recognition is inferring an appropriate grammar using a set of samples belonging to different pattern classes.

In syntactic pattern recognition, the problem of grammatical inference is one of central importance. This approach is based on the underlying assumption of the existence of at least one grammar characterizing each pattern class.

The identification and extraction of the grammar characterizing each pattern class forms the core problem in the design of a syntactic pattern classifier. The problem of grammatical inference involves development of algorithms to derive grammars using a set of sample patterns which are representatives of a pattern class under study. This may thus be viewed as a learning procedure using a finitely large and growing set of training patterns. In syntactic pattern classification, the strings belonging to a particular pattern class may be considered to form sentences belonging to the language corresponding to the pattern class. A machine is said to recognize a pattern class if for every string belonging to that pattern class, the machine decides that it is a member of the language and for any string not in the pattern class, it either rejects or loops forever. A number of interesting techniques have been suggested for the automated construction of automaton which accepts the strings belonging to a particular pattern class [11].

8.10 SYMBOLIC PROJECTION METHOD

Here we will present a scene interpretation scheme based on a work by Jungert [12]. The structure is called symbolic projections and its principles are relatively simple and straightforward. The basic idea is to project the positions of all objects in a scene or image along each coordinate axis and then generate a string corresponding to each one of the axes. Each string contains all the objects in their relative positions, that is, one object is either equal to or less than any of the others. Figure 8.5 shows how simple objects can be projected along the X- and Y-coordinate axis. The two operators used are *equal to* and *less then*. The strings are called U- and the V-strings, where the U-string corresponds to the projections of the objects along the X-axis, and the V-string to the Y-axis. The symbolic projections are best suited for describing relative positions of objects, which is important in spatial reasoning in images of our discussion.

One may use several spatial relational operators, such as *equal*, *less then*, *greater than*, etc., as follows:

- **Equal (=):** Two objects A and B are said to be equal in spatial dimension, i.e., $A = B$ if and only if centroid of A is same as the centroid of B.

- **Less than (<):** Two objects A and B separated by a distance may be spatially related by $A < B$ if and only if $\max(A_x) < \min(B_x)$, where $\max(A_x)$ (or $\min(B_x)$) indicates the maximum (or minimum) values of the projection of all the pixels in object A (or object B) along the X-direction. Similar relationships can be defined along the Y-axis.

- **Greater than (>):** Two objects A and B separated by a distance may be spatially related by $A > B$ if and only if $\min(A_x) > \max(B_x)$.

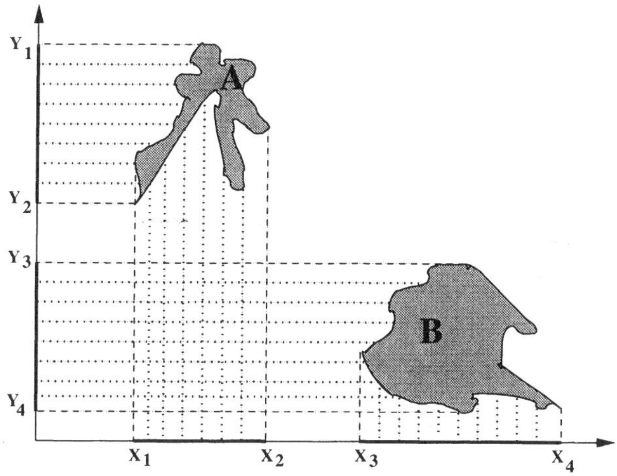

Fig. 8.5 Example of a symbolic projection.

- **Top** and **Bottom:** Two objects A and B separated by a distance may be spatially related by *A top of B* if and only if $\min(A_y) > \max(B_y)$ and hence *Bbottom of A*'.

The illustration of the *less-than* operator is shown as an example in Figure 8.5. From the figures, it is clear that the maximum value of the projection (x_2) of object A along X-direction is less than the minimum value of the projection (x_3) of object B. This implies that the object A is to the left of the object B. Similarly, the minimum value of the projection (y_2) of object A along Y-direction is greater than the maximum value of the projection (y_3) of object B. Hence the object A is top of object B.

To describe all possible formations of the object, it is required that we should be able to identify the spatial relationships among the objects in eight possible directions, namely, North, South, East, West, Northeast, Northwest, Southeast, and Southwest.

8.11 ARTIFICIAL NEURAL NETWORKS

Artificial Neural networks have been extensively used in image segmentation and object classification problems [13]–[17]. These networks are essentially learning networks, which are used for Classifying pixels or objects in a scene. They are large set of interconnected neurons, which execute in parallel to perform the task of learning. The neurons are modeled after the biological neurons and hence they are termed as neural networks. Based on the type of learning process, these networks may be supervised or unsupervised. In this chapter, we discuss a multilayered perceptron neural network based on one supervised learning and Kohonen's self organizing feature map, which is based on unsupervised learning.

8.11.1 Evolution of Neural Networks

Historically it was the work on *linear threshold units*, modeled after the interconnected set of human neurons, by McCulloch and Pitt in 1943, which was the first attempt towards building a neural network [18]. Their model consisted of a network of simple units, called linear threshold units, which could perform logical AND, OR operations. The subsequent attempt by Hebb on Hebbian learning in 1949 and his postulates are the founding ones based on which most of the later day learning rules have been derived [19]. It was, however, possibly the concepts of *perceptron* by Rosenblatt in 1957 which was the first attempt to design a classifier based on learning network [20]. The limitations of perceptron were soon brought out by Minsky and Papert, and it was found that perceptron was unable to discriminate between the Pattern classes which are not linearly separable [21]. The current wave of popularity of Neural networks is due to Hopfeld networks and work by Rumelhart et. al. [16].

Although there exists many models and representations of ANNs, each one of these networks possesses four tuple attributes $< N_C, W, \sigma, \delta >$, where N_C is a finite set of highly interconnected neurons with outputs $n_1, n_2, \cdots, N_k$; W denotes a finite set of weights which represents the strength w_{ij} of the interconnection between neurons n_i and n_j; σ is a propagation rule which shows how the input signals to a neuron n_i propagates through it. A typical propagation rule may be $\sigma(i) = \sum n_j w_{ij}$ and δ is an activation function which is usually a nonlinear function like sigmoid function or a hard limiter.

8.11.2 Multilayer Perceptron

The most popular neural network model is the *multilayer perceptron* (MLP), which is an extension of the single layer perceptron proposed by Rosenblatt [16, 20]. Multilayer perceptrons, in general, are feedforward network, having distinct input, output, and hidden layers. The architecture of multilayered perceptron with error backpropagation network is shown in Figure 8.6.

In an M-class problem where the patterns are N-dimensional, the input layer consists of N neurons and the output layer consists of M neurons. There can be one or more middle or hidden layer(s). We will consider here a single hidden layer case, which is extendable to any number of hidden layers. Let the hidden layer consists of p neurons. The output from each neuron in the input layer is fed to all the neurons in the hidden layer. No computations are performed at the input layer neurons. The hidden layer neurons sum up the inputs, passes them through the sigmoid non-linearity and fan-out multiple connections to the output layer neurons.

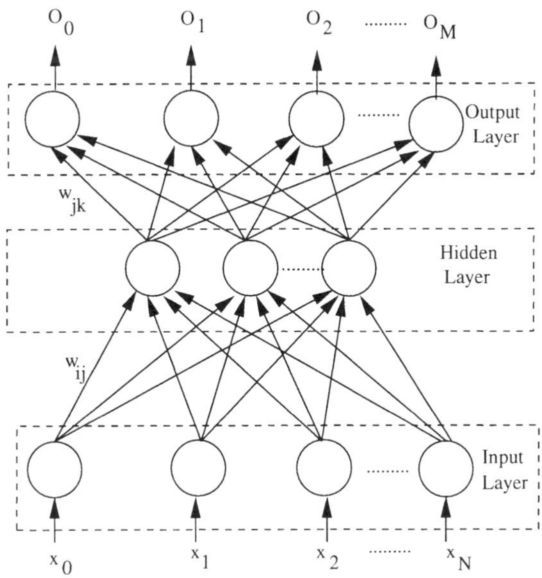

Fig. 8.6 Architecture of a Backpropagation Neural Network.

In feed forward activation, neurons of the first hidden layer compute their activation and output values and pass these on to the next layer as inputs to the neurons in the output layer, which produce the networks actual response to the input presented to neurons at the input layer. Once the activation proceeds forward from the input to the output neurons, the network's response is compared to the desired output corresponding to each set of labeled pattern samples belonging to each specific class, there is a desired output. The actual response of the neurons at the output layer will deviate from the desired output which may result in an error at the output layer. The error at the output layer is used to compute the error at the hidden layer immediately preceding the output layer and the process continues.

In view of the above, the net input to the jth hidden neuron may be expressed as

$$I_j^h = \sum_{n=1}^{N} x_n w_{ij}^h + \theta_j^h.$$

The output of the jth hidden layer neuron is

$$O_j = f_j^h(I_j^h) = \frac{1}{1 - \exp(-I_j^h)}$$

where $x_1, \cdots, x_n$ is the input pattern vector, weights w_{ij} represents the weight between the hidden layer and the input layer, and θ_j^h is the bias term associated with each neuron in the hidden layer. Identical equations with change of subscripts hold good for the output layer. These calculations are known as forward pass. In the output layer, the desired or target output is set as T_k and the actual output obtained from the network is O_k. The error $(T_k - O_k)$ between the desired signal and the actual output signal is propagated backward during the backward pass. The equations governing the backward pass are used to correct the weights. Thus the network learns the desired mapping function by back propagating the error and hence the name 'error backpropagation'. The generalized delta rule originates from minimizing the sum of squared error between the actual network output and desired output responses (T_k) over all the patterns. The average error E is a function of weight as shown:

$$E(w_{jk}) = \frac{1}{2}\sum_{k=1}^{M}(T_k - O_k)^2.$$

To minimize the error E, we have to find the root of the partial derivatives

$$\sum_{k=1}^{M}\frac{\partial E}{\partial w_{jk}} = 0.$$

E can be minimized bý taking incremental steps to correct the weights by using the Delta rule.

$$\nabla w_{jk}(t) = -\eta\frac{\partial E}{\partial w_{jk}} + \alpha\nabla w_{jk}(t-1);$$

$\frac{\partial E}{\partial w_{jk}}$ for the output layer can be obtained by chain rule:

$$\frac{\partial E}{\partial w_{jk}} = O_k(1 - O_k)(T_k - O_k)$$

$$w_{jk}^{(new)} = w_{jk}^{(old)} + \eta\delta_j O_j$$

where η is the learning rate of of the hidden layer neurons. Similarly, δ_j can be denoted by

$$\delta_j = O_j(1 - O_j)(T_j - O_j)$$

where T_j is the ideal response. Following the chain rule, δ_i for the i layer can be similarly obtained.

The elegant nature of the equations for both δ_i and δ_j emanate from the differentiable characteristic of the sigmoidal function. Thus, when the errors are propagated backwards, we get a formulation in terms of expected and calculated output patterns. Herein lies the power and simplification, which makes backpropagation so versatile. The equation for changing the weights between the input and hidden layers can similarly be represented as

$$w_{ji}^{(new)} = w_{ji}^{(old)} + \eta_2\delta_i x_i.$$

The process of weight correction can be speeded up by adding a momentum term, which takes into account the correction applied in the faster iterations. It may be pointed out that the backpropagation performs gradient descent in E. By adjusting the weights after each sample patterns is presented, the network converges to a fixed weight vector. McClelland and Rumelhart had shown how by selecting small learning rate, a good approximation of gradient descent can be achieved through the sequence of small movements [16].

8.11.3 Kohonen's Self-Organizing Feature Map

The essential constituents of Kohonen's neural network model are as follows [17]:

- An array of neurons receiving coherent inputs simultaneously, and computing a simple output function
- A mechanism for comparing the neuron outputs to select the neuron producing maximum output
- A local interaction between the selected neuron and its neighbors.
- An adaptive mechanism that updates the interconnection weights

The self-organizing feature map (SOFM) is an unsupervised learning network [17], which transforms p-dimensional input patterns to a q-dimensional (usually $q = 1$ or 2) discrete map in a topologically ordered fashion. Input points that are close in p-dimension are also mapped closely on the q-dimensional lattice. Each lattice cell is represented by a neuron that has a p-dimensional adaptable weight vector associated with it. With every input its match with each weight vector is computed. Then the best matching weight vector and some of its topological neighbors are adjusted so that the similar patterns are clustered together. Initially, the process starts with a large neighborhood; with passage of iteration, the neighborhood size is reduced gradually. At a given time instant, within the neighborhood, the weight vector associated with each neuron is updated differently. The strength of interaction between the winner and a neighboring node is inversely related to the distance (on the lattice) between them.

Consider the self-organizing network given in Figure 8.7. Let an M-dimensional input pattern be simultaneously incident on each of an $N \times N$ array of neurons. The output of the ith neuron is defined as

$$\eta_i(t) = \sigma \left[[\boldsymbol{m}_i(t)]^T \boldsymbol{x}(t) + \sum_{k \in S_i} w_{ki} \, \eta_k(t - \triangle t) \right], \tag{8.1}$$

where $\boldsymbol{x}$ is the M-dimensional input vector incident on it along the connection weight vector $\boldsymbol{m}_i$, k belongs to the subset S_i of neurons having interconnections with the ith neuron, w_{ki} denotes the fixed feedback coupling between the kth and ith neurons, $\sigma[.]$ is a suitable sigmoidal output function, t denotes a discrete time index, and T stands for the transpose.

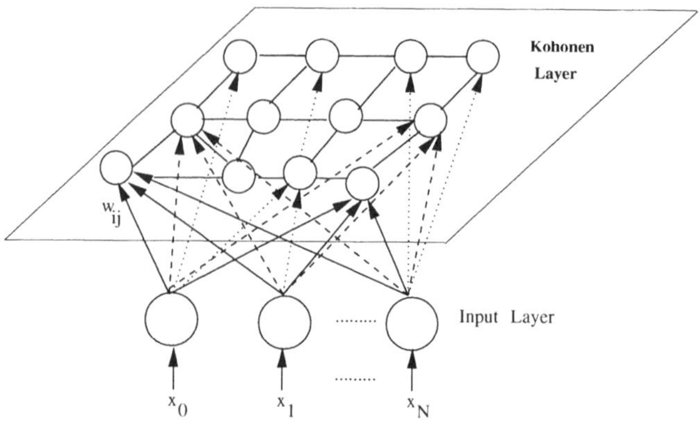

Fig. 8.7 Kohonen self-organizing feature map.

Initially the components of the $\boldsymbol{m}_i$ values are set to small random values lying in the range $[0, 0.5]$ or even sometimes $[-0.1, 0.1]$. If the best match between vectors $\boldsymbol{m}_i$ and $\boldsymbol{x}$ occurs at neuron c, then we have

$$\|\boldsymbol{x} - \boldsymbol{m}_c\| = \min_i \|\boldsymbol{x} - \boldsymbol{m}_i\|, \quad i = 1, 2, \ldots, N^2, \tag{8.2}$$

where $\|.\|$ indicates the Euclidean norm.

The weight updation is given as [17]

$$\boldsymbol{m}_i(t+1) = \begin{cases} \boldsymbol{m}_i(t) + \alpha(t)\,(\boldsymbol{x}(t) - \boldsymbol{m}_i(t)) & \text{for } i \in N_c \\ \boldsymbol{m}_i(t) & \text{otherwise,} \end{cases} \tag{8.3}$$

where $\alpha(t)$ is a positive constant that decays with time and N_c defines a topological neighborhood around the maximally responding neuron c, such that it also decreases with time. Different parts of the network become selectively sensitized to different inputs in an ordered fashion so as to form a continuous map of the signal space. After a number of sweeps through the training data, with weight updating at each iteration obeying Eq. (8.3), the asymptotic values of $\boldsymbol{m}_i$ cause the output space to attain proper topological ordering. This is basically a variation of *unsupervised* learning.

8.11.4 Counterpropagation Neural Network

A counterpropagation network, suggested by Hecht - Nielson is a combination of Kohonen's self organizing network and Grossberg's outstar network [22]–[24]. The counterpropagation network architecture is as shown in Figure 8.8.

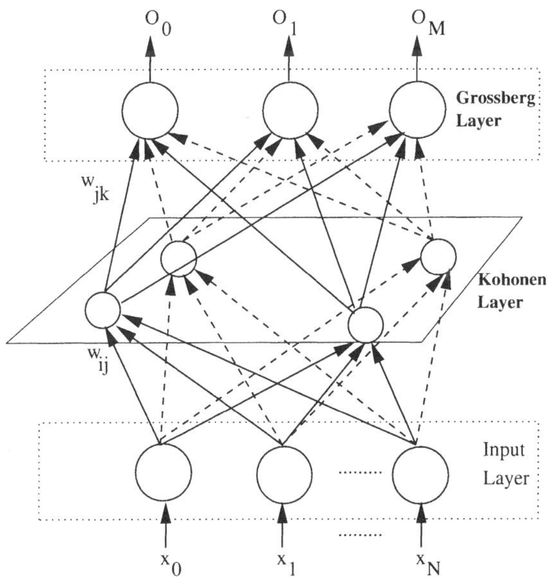

Fig. 8.8 Architecture of a Counterpropagation Neural Network.

The counterpropagation neural network employs a supervised learning mechanism where the input and output vectors propagate in a counter-flow manner and hence the name counterpropagation network. The network has an input layer in which there are fanout connections of the input vector, $X = \{x_1, x_2, \cdots, x_N\}$ to a middle (hidden) layer which is a two-dimensional Kohonen's layer. The final Grossberg outstar layer with M number of processing elements provides the approximation of the output vector $O = \{O_1, O_2, \cdots, O_M\}$.

The training and operation of the counterpropagation network are carried out in two phases. In the first phase, the Kohonen's layer is trained by the self-organization principle as explained in the previous section. The ith node in the Kohonen's layer is chosen such that $\|W_i - X\| \leq \|W_j - X\|$, for all $j \neq i$. The weight vectors of the ith node in the Kohonen's layer are then updated as

$$W_i^{new} = W_i^{old} + \alpha \left(X - W_i^{old}\right) Z_i.$$

Here the learning constant α is adaptively changed as the training progresses. The parameter Z_i is chosen such that $Z_i = 1$ when i is the winner node and $Z_i = 0$ otherwise. The output of the Kohonen's layer is next fed to the Grossberg layer, which is trained as

$$U_i^{new} = U_i^{old} + \beta \left(Y - U_i^{old}\right) Z_i.$$

As the learning progresses after presentation of a large number of training samples, W_i vector is updated in such a way that they are approximately equiprobable in the nearest neighbor sense. The network functions as a look-up table after the training is over. When a test pattern $\hat{X}$ is presented, the network finds the ith vector W_i closest to $\hat{X}$ and produces the output Y associated with this weight vector. The network is model independent and data driven. The crisp counterpropagation network may be generalized to a fuzzy counterpropagation network by extending the Kohonen's layer to a fuzzy self-organized feature map [22].

8.11.5 Global Features of Networks

Keeping in view the above discussion, there is an underlying commonality in the functioning of all the different neural networks. These are

- All the neural networks has inherent parallelism in their functioning.

- The functions of the basic functional units i.e., the neurons are more or less identical in a particular type of network.

- The parameters of a neural network undergo incremental changes which eventually settles down to a steady value after the network converges.

- The level of activation of a particular neuron depends exclusively on the current state of the neuron and those surrounding it and which are directly interconnected with it.

8.12 SUMMARY

Recognition and classification of image patterns are important tasks in low level computer vision and image processing due to its diverse applications. The image patterns are key to classification and interpretation of images. In this chapter, we have discussed various supervised and unsupervised pattern classification techniques. These techniques find wide applications in diverse areas of image analysis, viz., segmentation, pixel classification, and interpretation of images. Amongst the supervised classifiers, we have discussed Baysian classification which is a parametric model and K-nearest neighbor classifier which is a nonparametric model of classification. We have also described three different neural network architectures namely, multilayer perceptron with error backpropagation, Kohonen's self organizing feature map, and counterpropagation neural networks. All these three networks perform quite well in pixel classification, object detection, and image segmentation.

REFERENCES

1. R. O. Duda, P. E. Hart, and David G. Stork, *Pattern Classification*, 2nd ed., Wiley, New York, 2000.

2. K. S. Fu, *Syntactic pattern recognition and applications*, Prentice-Hall, Englewood Cliffs, NJ, 1982.

3. R. C. Gonzalez and M. G. Thomason, *Syntantic Pattern Recognition: An Introduction*, Addison-Wesely, Reading, MA, 1978.

4. T. Pavlidis, *Structural Pattern Recognition*, Springer Verlag, New York, 1977.

5. P. Perner, P. Wang, and A. Rosenfeld (Eds.), *Advances in Structural and Syntactical Pattern Recognition*, Springer Verlag 1996.

6. J. Schuermann, *Pattern Classification: A Unified View of Statistical and Neural Approaches*, Wiley, New York, 1996.

7. T. M. Cover and P. E. Hart, "Nearest neighbour Pattern Classification," *IEEE Trans on Information Theory*, Vol. IT-13(1), January 1967, 21–27.

8. A. K. Majumdar and A. K. Ray, "Syntactic Pattern Recognition", *Pattern Recognition from Classical to Modern Approaches*, Ed. by S. K. Pal & A. Pal, World Scientific, 2001, 185- 229.

9. P. Perner and M. Petrou (Eds.), *Machine Learning and Data Mining in Pattern Recognition*, Springer Verlag, New York. 1999.

10. H. Freeman, "On the encoding of arbitrary geometric configuration," *IEEE Trans. Elec. Computers*, Vol. EC-10, 1961, 260–268.

11. A. K. Ray, A. K. Majumdar and B. Chatterjee, "A Formal Power Series Approach to the Construction of Minimal Fuzzy Automata," *Information Sciences*, 55(1-3), 1991, 189-207.

12. E. Jungert, "Extended Symbolic Projections as a Knowledge Structure for Spatial Rasoning and Planning," *Pattern Recognition*, New York, Springer Verlag, 1988.

13. S. Mitra and T. Acharya, *Data Mining: Multimedia, Soft Computing and Bioinformatics*, Wiley, Hoboken, New Jersey, 2003.

14. S. Haykim, *Neural Networks: A comprehensive foundation*, 2nd ed., Prentice Hall, NJ, 1999.

15. Brian D. Ripley, *Pattern Recognition and Neural Networks*, Cambridge University Press, January 1996.

16. D. E. Rumelhart and J. L. McClelland, eds., *Parallel Distributed Processing: Explorations in the Microstructures of Cognition*, vol. 1, MIT Press, Cambridge, MA, 1986.

17. T. Kohonen, *Self-Organization and Associative Memory*, Springer–Verlag, Berlin, 1989.

18. W. S. McCulloch and W. Pitts, "A logical calculus of the idea immanent in nervous activity," *Bulletin of Mathematical Biophysics*, vol. 5, 1943, 115–133.

19. D. O. Hebb, *The Organization of Behaviour*, Wiley, New York, 1949.

20. F. Rosenblatt, "The perceptron: A probabilistic model for information storage and organization in the brain," *Psychological Review*, vol. 65, 1958, 386–408.

21. M. Minsky and S. Papert, *Perceptrons: An Introduction to Computational Geometry*, MIT Press, Cambridge, MA, 1969.

22. B. Mannan and A. K. Ray, "Crisp and fuzzy competitive learning networks for supervised classification of multispectral IRS scenes," *Int J. Remote Sensing*, 24(17), 2003, 3491–3502.

23. R. Hecht-Nielson, "Counterpropagation Networks," *Applied optics*, 26, 1987, 4979–4984.

24. R. Hecht-Nielson, "Applications of Counter-propagation Networks," *Neural Networks*, 1, 1988, 131–139.

9

Texture and Shape Analysis

9.1 INTRODUCTION

The human vision system perceives scenes having variations of intensity and color, which form certain repeated patterns, called *texture*. In a textured image, a set of local statistics or attributes vary slowly or remain approximately periodic. These attributes, which are repetitive or quasi-repetitive patterns, dominate any textured scene.

A texture is an ensemble of repetitive subpatterns, which follow a set of well defined placement rules. These sub patterns themselves are made up of more fundamental units, called primitives. Such characterization of textures are generally applicable mainly to deterministic type of textures, such as, line arrays, checker boards, hexagonal tiling, etc. There are images, such as satellite images of the earth surface, which apparently do not possess such basic pattern primitives which are repeated in the overall pattern.

There is a large diversity of textural patterns and their rendering mechanisms in nature. In his theory of textons, Julesz formalized the way in which human perception is able to discriminate textures in a preattentive way [1]–[4]. He concluded that blobs called *textons*, together with their geometric and intensity attributes, form the primitives on which texture perception is based.

The attributes and utilities of textures can be summarized as

- Textures are repetitive patterns, which characterize the surfaces of many classes of objects. Thus classification of object patterns becomes easy if the textures present in the image are identified and differentiated from each other.

- Textures provide vital information about the arrangement of the fundamental elements of an image.
- The attributes of a texture may be described in qualitative terms such as *coarseness*, *homogeneity*, *orientation* of image structure, and *spatial relationships* between image intensities or tones. Texture analysis is the quantification and use of such image properties which aid in texture discrimination.

In view of its multifaceted properties, Texture analysis finds a lot of applications in many areas, e.g., medical image analysis, automatic surface inspection, remote sensing etc.

9.1.1 Primitives in Textures

A *primitive* is a connected set of pixels, characterized by a set of attributes. The simplest primitive is a single pixel with its gray tone attribute. In general, a connected set of pixels all having the same gray tone or all having same edge direction forms a primitive. Some local attributes, other than gray tone may be considered while defining a primitive. Other attributes includes measures of shape of connected regions and homogeneity of its local property. For example, a connected set of resolution cells can be associated with its length or elongation of its shape or variance of its local property. Primitives can be generated from image data by various neighborhood operators.

9.1.2 Classification of textures

Based on the attributes, textures are of two types:

1. Microtextures
2. Macrotextures

This classification is based on the size of the primitives that constitute the textures. Both microtexture and macrotexture are composed of primitive elements with specific shapes, sizes and placements. Two important attributes of such textures are coarseness and directionality. Coarseness relates to the size of the texture elements. If the primitive element size is large, then the texture is termed a coarse texture or macro texture and if the size is small the resultant texture is a fine or micro texture. Directionality corresponds to the orientation of the texture elements and to their spatial arrangement.

Based on the feature selection and classification philosophy, texture understanding methods are divided into three major groups:

- spatial methods
- structural methods
- statistical methods

9.1.2.1 Spatial Methods In spatial methods, a texture is modeled as a *surface*, from which spatially compact and visually perceptual features like lines, edges, orientation, etc., are extracted over a larger area. These methods look for such repetitive perceptual features, and creates a vector of these perceptual features in the texture. Various orthonormal transforms may be employed to characterize these features. Based on the spatial relationships among the primitives, textures may be classified as *weak* and *strong* textures. Weak textures are those which have weak spatial relationship amongst the primitives, while strong textures have non-random spatial interactions between the primitives

9.1.2.2 Structural methods In structural characterization, a texture is viewed as made up of many primitive textural elements, called *texel*, arranged according to some specific placement rules. In structural texture analysis, the texture is considered as the repetition of some primitives with a certain rule of placement. Human beings have the ability to perceive the structural characteristics of some textures. The traditional Fourier spectrum analysis is often used to determine these primitives and the displacement rule. For example, the first few most prominent components in the spectrum can be used to characterize a given texture.

9.1.2.3 Statistical methods In statistical methods a texture is modeled as a random field and a statistical probability density function model is fitted to the spatial distribution of intensities in the texture. Typically, these methods measure the interactions of small number of pixels. High-order statistics like Markov Models, may be used as invariant statistical texture classifiers. In these methods, the aim is to characterize the stochastic properties of the spatial distribution of gray levels in an image. The most common features used in practice are the measures derived from the spatial gray tone cooccurrence matrix.

9.2 GRAY LEVEL COOCCURRENCE MATRIX

A well-known statistical tool for extracting second-order texture information from images is the *gray level cooccurrence matrix* (GLCM). It was possibly Julesz who had first proposed the conjecture that the second order statistics is sufficient for human discrimination of textures [2]–[4].

Originally introduced by Haralick et al. [5], GLCM measures second-order texture characteristics which play an important role in human vision, and has been shown to achieve a similar level of classification performance.

The GLCM of an $N \times N$ image $f(i, j)$, containing pixels (with dynamic range G) with gray levels $\{0, 1, \cdots, G-1\}$ is a two-dimensional matrix $C(i, j)$, where each element of the matrix represents the probability of joint occurrence of intensity levels i and j at a certain distance say d and an angle θ. Thus there may be multiple number of cooccurrence matrices depending on various values of d and θ. A number of texture characterization and segmentation methods use GLCM [6]–[9] to classify various texture classes.

Several texture measures may be directly computed from the gray level cooccurrence matrix. Haralick et al. suggested a set of 14 features, which can be used to classify texture images [5]. Some of the important features like angular second moment, contrast, entropy, inverse difference moment, and so on, are described below. The following features may be computed from the normalized cooccurrence matrix $C_{norm}(i,j)$, which may be obtained by dividing each element in $C(i,j)$ by total number of pixel pairs.

(A) *Angular second moment*: The *angular second moment* (ASM) feature is defined as

$$F_1 = \sum \sum C(i,j),$$

where $C(i,j)$ represents the joint probability of occurrence of pixels with intensities i and j, and L is the number of distinct gray levels. This is a measure of local homogeneity in the image. Its value is high when the image has very good homogeneity. In a non homogenous image, where there are many gray level transitions, the ASM assumes lower values.

(B) *Contrast*: The *contrast* feature F_2 is given by

$$F_2 = ij(i-j)^2 \sum \sum C(i,j).$$

This measure shows the amount of local variation and is the opposite of homogeneity. Thus when high values concentrate along the diagonal of the cooccurrence matrix, the contrast feature yields a measure of the difference moment of cooccurrence matrix. It is a measure of the amount of local variations present in the image. For uniform images, the value of contrast is zero, which is the minimum value for the contrast. As the variations in the image increase, the value of contrast also increases.

(C) *Correlation*: This measure analyzes the linear dependency of gray levels of neighboring pixels. Typically this measure is high, when the scale of local texture is larger than the distance. The correlation feature (F_3) is a measure of gray tone linear dependencies in the image.

(D) *Entropy*: The entropy of a texture (F_4) is measured as

$$F_4 = \sum \sum ijC(i,j) \log C(i,j).$$

This measure yields a measure of complexity and complex textures tend to have high entropy.

(E) Inverse Difference Moment (IDM): The IDM feature is given by

$$F_5 = \sum \sum \frac{C(i,j)}{[1+(i-j)*2]}$$

where i and j are two different pixel intensities. The IDM feature is the inverse of the contrast of the cooccurrence matrix. It is a measure of the amount of local uniformity present in the image. If the image is spatially uniform, the value of inverse difference moment will be maximum, approaching 1. In case of an image having large number of adjacent pixels, where $|i - j|$ is very small, value of inverse difference moment will be high. Thus IDM is high in those images having very less contrast and low for images having very high contrast.

Although there are a number of other potential features which can be extracted from the cooccurrence matrix, it has been observed that all the features proposed in [5] are not effective in all types of textures and there is a need to extract application-specific appropriate features. This approach increases the feature discriminatory power with a reduction in the classification error.

9.2.1 Spatial Relationship of Primitives

Once we have the locations and attributes of a set of primitives, we can construct a spatial relationship (e.g., adjacency relationship) among the primitives.

The following features extracted from spatial relationships of the texture primitives are effectively used in texture discrimination.

- **Edge per unit area:** To compute the *edge per unit area* around a pixel in an image, the magnitudes of the edge gradients of the pixel in a neighborhood is first computed by a standard edge operator. Mean of the edge gradient magnitudes in the neighborhood is called the *edge per unit area* associated with the pixel.

- **Run-length:** The gray level *run-length* primitive in its one-dimensional form is a maximum collinear connected set of pixels all having the same gray level. Properties of the primitives can be lengths of run and angular orientation of run.

- **Maximal component run-length:** The *maximal component run-length* is a maximally connected set of pixels all having the same gray level. This feature have properties such as size of run-length, average gray level, maximum or minimum diameter and its angular orientation.

- **Relative Extrema Density:** *Extrema* is defined along a horizontal scan as:

  1. **Relative minimum:** In any row of pixels, a pixel i having gray level $g(i)$ is a *relative minimum* if $g(i)$ satisfies both

$$g(i) \leq g(i+1) \text{ and } g(i) \leq g(i-1) \tag{9.1}$$

2. **Relative maximum:** A pixel is a *relative maximum* if $g(i)$ satisfies both

$$g(i) \geq g(i-1) \text{ and } g(i) \geq g(i-1) \tag{9.2}$$

9.2.2 Generalized Cooccurrence

Strong texture measures take into account the cooccurrence between the texture primitives. This is often referred in literature as the *generalized cooccurrence* relationship [6, 10].

To define the concept of *generalized cooccurrence*, it is necessary to first decompose an image into its primitives. Let Q be the set of all primitives on the image. Then we need to measure primitive properties such as mean gray tone, variance of gray tones, regions, size shape etc. Let T be the set of all the properties and f be a function of assigning to each primitive in Q a property of T Finally we need to specify a spatial relation between primitives such as distance or adjacency. Let $S \leq QXQ$ be the binary relation pairing all primitives which satisfy the spatial relation.

9.3 TEXTURE SPECTRUM

The *texture spectrum*, a statistical way of describing texture feature of an image, was first conceived by He and Wang [11]–[13]. In this method a texture unit represents the local texture information for a given pixel and its neighborhood, and the global texture of an image is characterized by its texture spectrum. A brief review of the texture spectrum as proposed originally is presented as follows.

The basic concept is that a texture image can be represented as a set of essential small units termed as texture units, which characterize the local texture information for a given pixel and its neighborhood. The statistics of all the texture units over the entire image reveal the global texture aspects.

In a square raster digital image, each pixel is surrounded by eight neighboring pixels. The local texture information for pixel is then extracted from the neighboring pixels, which form the elements of the 3×3 window with the pixel under consideration as the central one. It can be noted that the eight neighborhoods represents the smallest complete unit from which texture spectrum information can be obtained.

Given a neighborhood of 3×3 pixels, which are denoted by a set containing nine elements, $V = \{V_0, V_1, \cdots, V_8\}$, where V_i ($i = 0, 1, \cdots, 8$) represents the gray level of the ith element in the neighborhood with V_0 representing the gray level of the central pixel. It is important to note that the eight pixels in the neighborhood are always taken in some order and the subscripts might denote the direction in which a particular neighborhood pixel lies.

The corresponding texture unit TU of the pixel is then defined by a set containing eight elements. Thus, $TU = \{E_1, E_2, \cdots, E_8\}$, and E_i $(i = 1, 2, \cdots, 8)$ is determined by the following description:

$$E_i = \begin{cases} 0 & \text{when } V_i < V_0 \\ 1 & \text{when } V_i = V_0 \\ 2 & \text{when } V_i > V_0 \end{cases}$$

where the element E_i occupies the same position as the pixel i. As each element of the TU has one of the three possible values, the combination of all the eight elements results in $3^8 = 6561$ possible texture units in total. There is no unique way to label and order all these texture units. The proposed algorithm labels these texture units by the following formula:

$$N_{TU} = 3^{i-1}E_i, \quad i = 1, 2, \cdots, 8$$

where N_{TU} represents the texture unit number.

Each texture unit number describes the local texture aspect of a given pixel, which are the relative gray level relationships between the central pixel and its neighbors. Thus, the statistics on frequency of occurrence of all the texture unit numbers over a large region of an image should reveal texture information. With this background, the term *texture spectrum* is defined as the frequency distribution of all the texture unit (numbers) with the abscissa indicating the texture unit number N_{TU} and the ordinate the frequency of its occurrence.

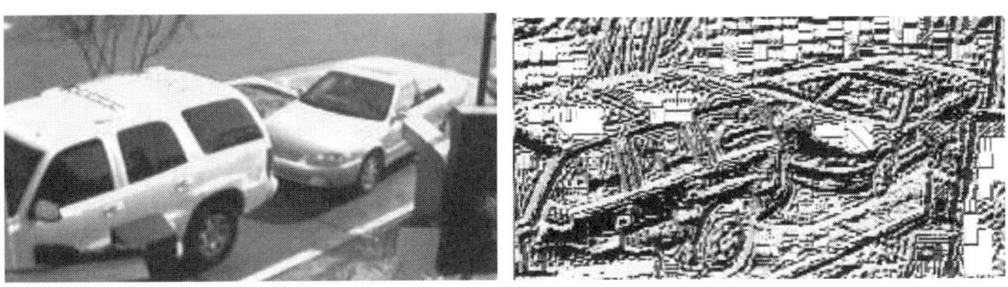

(a) (b)

Fig. 9.1 (a) An input image, (b) texture spectrum of the image.

The results of texture spectrum is shown in Figure 9.1.

9.4 TEXTURE CLASSIFICATION USING FRACTALS

Another interesting way of looking at a textured image is like this: Suppose we change the resolution of the textured image, then this will lead to the change in their properties. If now we attempt to measure the change of the properties at various resolutions, then we will be able to sufficiently characterize the nature of the textured surface. If now we measure the areas of gray level surfaces at various resolutions, then we will observe that the area will decrease at coarse resolution. This is because the finer details inside the area will disappear when we capture the image at a course resolution. Thus the rate of decrease of areas as the images are captured at lower resolutions or scales will characterize the nature of the texture. Textures may thus be classified based on the change in their properties with changing resolution. Fractal properties of the picture are computed from the rate of this decrease in area, and are used for texture comparison and classification. There have been attempts to understand this relationship by the researchers of the computer vision community. The advent of the concepts of fractals has enhanced this understanding [14, 15].

9.4.1 Fractal Lines and Shapes

Mandelbrot has first suggested the technique of length measurement of a coastline [14]. Using a yardstick of length δ, if n is number of steps the yardstick takes to cover the entire coastline, the approximate length of the coastline is $n\delta$. It may be noted here that as we choose smaller and smaller length of the scale, the observed length increases.

An alternative strategy to compute the length of the curve is to consider all the points which are located within a distance of say δ, on all sides from each point on the curve. The set of all such points thus forms a strip of width 2δ. The length of the coastline can thus be computed as the area covered by this strip divided by 2δ. Needless to say that as $\delta \rightarrow 0$, the length increases substantially. Here also the measured length is a function of the scale δ.

The third strategy may be to cover the coastline with a number of disks of radius δ. The sum of the areas covered by all these disks divided by 2δ yields the approximate length of the coastline.

The length of the coastline L estimated by any of the three methods mentioned above is a function of the scale δ. The higher is the value of δ the less is the estimation of the length. This estimated length $L(\delta)$ is related to the *fractal dimension* d of the line by the following equation

$$L(\delta) = Fe^{1-d},$$

where F is a constant for particular type of the line. It is the true distance between two points in case of a straight line. For a straight line the actual distance is exactly equal to the estimated distance and thus the *fractal dimension* of a straight line is 1. Although the estimated length of the line $L(\delta)$ is proportional to the scale δ, one chooses the fractal dimension d of a coastline is independent of the scale δ. If we plot $L(\delta)$ versus δ, we will get a straight line with slope $(1 - d)$. The fractal dimension is actually calculated from the slope of this line.

9.4.2 Fractals in Texture Classification

In an image each pixel may be conceived as a point in a three dimensional space, where the pixel location is given by its two-dimensional coordinates and the pixel intensity forms the third dimension. Thus each pixel in a texture image may be considered as a cuboid with its length and width equal to the pixel dimensions, and height equal to pixel intensity. The total exposed surface of forms a measure of the surface area of the texture. The area may be calculated at different integer ruler sizes L, by averaging adjacent pixels to generate a new image with pixel size L. Assuming the texture is a continuous self-similar fractal, then surface area and ruler size follow the relation

$$A(L) \propto L(2 - D),$$

where D is the fractal dimension of the image. The fractal dimension D can be inferred from the slope of $A(L)$ versus L in the logarithmic scale, i.e., $\{\log_2(A(L)),\ \log_2(L)\}$.

Extending the concepts of Mandelbrot as presented above, the fractal dimension of such an image may be computed by different techniques. One of these techniques, known as "box counting" method, is based on computing the number of boxes of dimension $L \times L \times L$ (where L is the scale size) to cover the entire space. This is true only if the entire 3D space is statistically self-similar. The number of boxes $N(L)$ of size $L \times L \times L$ needed to cover the set obeys the power law

$$N(L) \propto L^{-D},$$

where D is the fractal dimension, and $N(L)$ is the number of boxes used for covering the space. The fractal dimension D may be estimated from the slope of a linear fit to the data $\{\log_2 L,\ -\log_2(N(L))\}$ by computing a number of $N(L)$ for several L's. Sometimes it becomes easier to scale the pixel intensity before estimating the *fractal dimension*. The box counting method described above provides a good estimate and it works well for optimum box sizes for which the discrete boxes more closely approximates the continuous surface.

9.4.3 Computing Fractal Dimension Using Covering Blanket method

In this method we again consider each pixel in the texture to be a point in discrete 3-D space, using integer pixel intensity as the third dimension. Peleg et. al. have presented an algorithm to compute the fractal dimension of a texture surface [16]. The basic philosophy of *covering blanket method* of fractal dimension computation is to cover the surface of the two-dimensional plot of the intensity image with two blankets each of thickness, one above the image surface and the other below the surface. Effectively thus we may conceive to have a single blanket with thickness 2δ, wrapped around the image intensity surface $f(i,j)$. Let us consider that the upper covering surface is denoted as $U_\delta(i,j)$ and the lower covering surface is denoted as $L_\delta(i,j)$. Initially we may assume that both the upper and lower covering blankets coincide with the image surface $f(i,)$, i.e., $U_\delta^0(i,j) = L_\delta^0(i,j) = f(i,j)$, where $U_\delta^0(i,j)$ and $L_\delta^0(i,j)$ are the initial values of the upper and lower covering blankets $U_\delta(i,j)$ and $L_\delta(i,j)$ respectively. The two blanket surfaces are defined below.

1. Upper blanket surface

$$U_\delta(i,j) = \max\left\{U_{\delta-1}(i,j) + 1, \max\left\{U_{\delta-1}(m,n)\right\}\right\},$$

   where $|(m,n) - (i,j)| \leq 1$.

2. Lower blanket surface

$$L_\delta(i,j) = \min\left\{L_{\delta-1}(i,j) + 1, \min\left\{L_{\delta-1}(m,n)\right\}\right\},$$

   where $|(m,n) - (i,j)| \leq 1$.

In the above formulation, only those image points (m, n) have been chosen in the 3×3 neighborhood around the central (i,j)th pixel, whose distances are less than or equal to one from the (i,j)th pixel location. This implies that while computing the upper and lower blanket surfaces we have chosen only the immediate four connected neighbors from the (i,j)th pixel location. In reality, we could as well choose the eight connected neighbors in the neighborhood of the (i,j)th pixel. Thus all those points (x, y) are included in the covering blanket which satisfy the condition

$$L_\delta(x,y) \leq f(x,y) \leq U_\delta(x,y).$$

It may be noted here that any blanket with thickness δ will automatically include the points covered by the blanket with thickness $\delta - 1$. The volume of the blanket is computed from the upper blanket and lower blanket as follows:

$$V_\delta = \sum \left(U_\delta(i,j) - L_\delta(i,j)\right).$$

The surface area can be computed from the difference of the two volumes of the blankets measured with radius δ and $\delta - 1$ and may be given as

$$A_\delta = \frac{V_\delta - V_{\delta-1}}{2}.$$

The division by 2 is necessitated to take into account both the upper as well as lower layers. By taking the difference between V_δ and $V_{\delta-1}$, we have isolated those features which have changed from scale $\delta - 1$ to scale δ. The above formulation of the surface area yields good results for both fractal as well as nonfractal surfaces.

The above area A_δ is related to the fractal dimension D as

$$A_\delta = Fe^{2-D}.$$

If we plot A_δ versus δ on a log-log scale, the slope of the resultant straight line is $2-D$, where D is the fractal dimension of the image surface. For nonfractal surfaces this slope is not a straight line. Peleg et. al [16] computed a fractal signature for each gray level image surface by finding the slope of the best fitting straight line through three points $[\log_2(\delta-1), \log_2 A_{\delta-1}]$, $[\log_2 \delta, \log_2 A_\delta]$, and $[\log_2(\delta+1), \log_2 A_{\delta+1}]$. If the object is a fractal object, the fractal signature so computed should be equal to $2-D$ for all values of δ. In Figure 9.2b), we shown the segmented description of five distinctly different textures representing different fractal surfaces.

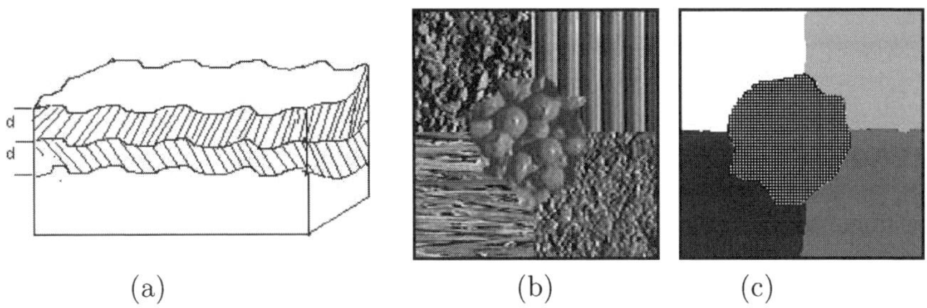

Fig. 9.2 (a) Covering blanket to compute fractal dimension, (b) A multi-textured image, (c) segmented image using fractal dimension

9.5 SHAPE ANALYSIS

Understanding complex objects using texture, color, motion, etc. from the pixel statistics have been widely investigated. Like textures, the form or the shape is another fundamental unit of perception and recognizing objects using their global shape has been gaining immense importance in diverse application areas like biomedical image analysis, video surveillance, biometrics, and so on.

Shape is the ensemble of all the geometrical information of an object which do not change even when the location, scale and orientation of the object are changed. Thus shape is invariant to Euclidean similarity transformations. There are various ways to describe shapes. The techniques of shape analysis may be categorized as (1) contour-based shape analysis and (2) region-based shape analysis.

In contour based shape analysis methods, a shape is represented by a coarse discrete sampling of the object contour. A set of landmark points are extracted from this contour. Since there exists variability amongst the shape patterns, shape preserving transformations, i.e., rigid translation, rotation and scale invariant transforms are necessary before establishing the equivalence amongst the shape classes.

In region based approach, on the other hand, invariant shape features are extracted from the interior as well as boundary pixels inside a region, and classification is carried out using these features.

To accomplish the shape analysis task, statistical shape analysis is being widely investigated [17]–[19].

9.5.1 Landmark Points

An interesting way to describe a shape pattern is by defining a finite set of *landmark points* inside the object. Dryden & Mardia define landmarks as points of correspondence on each object that matches between and within the population [17] and classified these points into three subgroups:

1. *Anatomical landmark points*: These points are assigned by an expert that corresponds between organisms in some biologically meaningful way. For example the joins of tissues or bones may be considered as *anatomical landmarks.* Landmark points have been successfully used for medical image analysis [17, 20]
2. *Mathematical landmark points*: These are the points located on an object, which obey some interesting mathematical or geometrical property, such as, high curvature or an extremum point.
3. *Pseudo-landmark points*: These are a set of constructed points on an object either on the outline or between other landmark points and may be appropriately defined by the user. Continuous curvilinear shapes may be approximated by a large number of pseudo landmark points. Usually pseudo landmarks may be chosen as a set of equispaced points along the outline between the pair of other landmarks.

Landmark points may be described as nodes or vertices of a polygon enclosing the shape pattern. They may be a set of fiducial points on say biometric patterns like fingerprints. They also may represent some important key points or marker locations in a remotely sensed image, e.g., the intersection between two roads.

9.5.2 Polygon as Shape Descriptor

An efficient representation of a planar shape may be the concatenation of the ordered pair of an n-point polygon in k dimensions. This can be achieved by concatenating each dimension into a $k \times n$-vector. The vector representation of a two-dimensional planar shapes may thus be represented as

$$x = [\{x_1, x_2, \cdots, x_n\}, \ \{y_1, y_2, \cdots, y_n\}].$$

The location, scale and rotational invariant shape representation may be achieved by establishing a coordinate reference with respect to position, scale, and rotation, to which all the shape patterns should be aligned.

Quite often we need to align shapes with one-to-one point correspondence. Thus given two shapes, the shape alignment procedure involves four steps:

Step 1. Compute the centroid of each shape.

Step 2. Rescale each shape to have equal size.

Step 3. Align with respect to position the two shapes at their centroids.

Step 4. Align with respect to orientation by rotation.

9.5.3 Dominant points in Shape Description

One scheme to detect a set of dominant points on the curve is to determe the curvature at each point, and computing the resultant cumulative curvature for all the points starting with the last detected dominant point [21].

Dominant points are those points along an image outline that store a lot of important information about the shape of the image. These points are usually points of high curvature. It does not mean that the points of small curvature are devoid of such information. In many cases, the cumulative curvature due to a large number of such points occurring consecutively is significant. A number of algorithms have been suggested for finding curvature extrema on a digital curve. In general, there are two approaches to the problem.

- Method I: to detect the dominant points directly through angle or corner detection schemes.
- Method II: to obtain a piecewise linear polygonal approximation of the digital curve depending on certain restrictions on the extent to which the shape has been preserved. Dominant points then correspond approximately to the intersections of adjacent line segments of the polygon. These points are also known as the vertices or break points of the closed curve (polygon).

9.5.4 Curvature and Its Role in Shape Determination

For a smooth curve on a real Euclidean plane, curvature is defined as the change in slope as a function of arc length and can be expressed in terms of first- and second-order derivatives. It is known that points having high curvature are rich in information content regarding the shape of the curve. As a result of this fact, several dominant point detection algorithms use techniques for direct measurement of discrete curvature or its functions (also called measures of significance). Since shape and curvature are intimately related, it is important to find the curvature accurately. Keeping in mind that the curve being used is a digitized version of a smooth curve, precise determination of curvature is a challenge. After the determination of digital curvature at each point, the next part is to detect the dominant points for which several schemes have been suggested. A lot of schemes use the curvature or its functions and then determine the dominant points using a threshold.

9.5.5 Polygonal Approximation for Shape Analysis

Given a two-dimensional image the problem of approximating its shape from its polygonal representation has received paramount importance during the last decades. Such a polygonal representation finds a number of applications in diverse areas such as chromosome analysis, industrial machine part classification, character recognition, biometric data analysis, etc. The outline of a two-dimensional object usually characterizes the fundamental features of the object patterns. Any closed planer curve may be approximated by a polygon in any desired accuracy so that its representation can have a smooth appearance, which is a sequence of straight line segments. Identifying such a curve which passes through or near a set of given points is the problem of polygonal approximation.

It has been observed from the human visual information system, that some dominant points along an object contour are rich in information content and they are sufficient to characterize the shape of the object. An approach to apply the basic philosophy of the human understanding and recognition procedure of complex two-dimensional curves leads to the detection of dominant points on the curves which when joined by straight line segments can approximate the shape to any desired degree of accuracy.

There are many algorithms for the detection of dominant points on digital curves. Piecewise linear approximation of the planar curve allows for a variable number of segments. After an arbitrary initial choice, the segments are split or merged in order to derive the error norms under a prespecified bound.

A parallel algorithm based on the determination of average curvature of the points on the curve by determining the region of support of each point, may be found in [21]. Since the level of detail represented at each point on the digital curve varies, a smoothing factor based on the local properties of the curve has to be used to find the curvature at each point. This smoothing factor is determined by the region of support. The advantage of the algorithm is that it requires no smoothing parameter.

9.6 ACTIVE CONTOUR MODEL

Segmentation of monochrome images uses basic properties of gray-level values to detect the isolated points, lines and edges. Alternately segmentation can also be performed by thresholding, region growing, region splitting and merging. Quite often they produce spurious edges and gaps which do not necessarily correspond to boundary objects. The limitation of these methods is due to their complete reliance on the information contained in the local neighborhood of the image. They ignore both model-based information and higher order organization of the image. Another problem associated with these methods is edge grouping. After extracting edges from the image, they have to be grouped or linked in order to determine boundaries, which often does not yield good results.

The application of prior knowledge, say geometrical knowledge, strengthens the visual interpretation of shape via the stabilizing influence of prior expectations of the shapes that are likely to be seen. The seminal work of Kass et al. on *snakes* provides a novel approach to visual analysis of shape [22]. Snakes are active contour models which are widely used in detecting object boundary shape as well as for tracking a moving object in an image sequence. It is an elastic contour which is fitted to features detected in an image. The nature of its elastic energy draws it more or less strongly to certain preferred configurations, representing prior information about shape which is to be balanced with evidence from an image. Thus Snake is an energy-minimizing spline guided by external constraint forces and influenced by image forces that pull it towards features such as lines and edges. Snakes lock onto nearby edges, localizing them accurately [22, 23]. A number of approaches have been proposed for shape analysis using active contours [22]–[24].

These models utilize deformable contours, which conform to various object shapes and motion. Snakes have been used for edge and curve detection, segmentation, shape modeling and visual tracking. It is a parametric curve, and its properties are specified through a function called energy functional. A partial differential equation controlling the snake causes it to evolve so as to reduce its energy. The motion of the snake is caused by simulated forces acting upon it.

The edge map of the image can be viewed as a landscape on which the parametric curve can slither. A force acts upon the curve and moves across the landscape trying to reach the energy equilibrium. The model driving it across the landscape has following two components

1. The first component enables the snake to preserve the original shape, to develop a corner, etc.

2. The other component instructs it where to go.

It is important that the curve clings to the boundary of a specific object in the scene. The boundary can be recognized as low values of the negative edge map, so that the equilibrium equation should be set up in such a way that the curve tends to minimize the term involving the negative edge map. Since we know the general shape of the object in question, we may design the evolution equation in such a way that the snake easily can embrace the object. It is elastic, stiff, and is able to develop a corner. In addition, if the inertia is attributed to a snake it acquires dynamic behavior which can be used to apply prior knowledge of motion, not just of shape.

9.6.1 Deformable Template

Deformable templates constitute another important approach to object estimation. This approach employs prior knowledge about the shape of the object in a direct manner. This prior shape information is specified as a sketch, binary template or a parametric prototype. The a priori information is then encoded either in the form of edge information from a binary template or the parameter vector. That information does not need to be exact in the sense that it matches the boundaries of the image exactly.

We may say that the difference between snakes and deformable templates is that snakes are form-free energy minimizing functions. In snakes model, there is no global structure of the curve except for some general regularization constraints such as continuity and smoothness of the boundary. On the other hand, parametric deformable templates control deformation using a set of parameters, which are capable of encoding a specific shape. This type of model is used when more specific shape information is available, which can be described either by a binary temple or a set of parameters.

The prototype template describes one and the most likely instance of the object shape. We apply a parametric transformation to the prototype and deform its boundaries varying the deformation parameters in order to capture a large variety of possible instances. If the object in interest is of biological nature, the form will resemble, but still vary from individual to individual. Those small variations can be captured by the random deformations of the prototype, so that a deform template may match the object of interest better then the original template. Objects may also be noise corrupted or degraded in some way so that the original shape is lost. In that case too a deformed template may match the object better that the original one.

Initially boundaries of the object in an image is extracted using an appropriate edge detector. Matching is next performed on all the objects by aligning the templates in the database with the image in question using some potential energy function.

9.6.1.1 Basic Snake Model The basic snake model, as proposed in [22] is given by

$$E_{snake} = \int_0^l \{E_{int}(v(s)) + E_{image}(v(s)) + E_{cont}(v(s))\}\, ds$$

In the basic snake model, the position of the snake on the image is represented parametrically by a planar curve $v(s) = (x(s), y(s))$, E_{int} represent the internal energy of the spline due to bending, E_{image} represents image forces pushing the snake towards the desired object. E_{cont} gives rise to external constraint forces.

The internal spline energy can be written as:

$$E_{int} = \frac{\alpha(s)\,|v_s(s)|^2 + \beta(s)\,|v_{ss}(s)|^2}{2}$$

where $v_s(s)$ is the first derivative and $v_{ss}(s)$ is the second derivative of $v(s)$ with respect to s. The spline energy is composed of a first-order term controlled by $\alpha(s)$ and a second-order term controlled by $\beta(s)$. The first-order term makes the snake act like a membrane and the second-order term makes it act like a thin plate. Adjusting the weights $\alpha(s)$ and $\beta(s)$, controls the relative importance of the membrane and thin-plate term.

In a digital image, the image energy function E_{image} is derived from the image so that it takes on its smaller values at the feature of interest, such as boundaries. Given a gray-level image $f(x, y)$, viewed as a function of continuous position variables (x, y), typically external energies designed to lead an active contour towards edges are:

$$E^1_{image}(x, y) = -|\nabla f(x, y)|^2$$

and

$$E^2_{image}(x, y) = -|\nabla \left[G_\sigma(x, y) * f(x, y)\right]|^2$$

where $G_\sigma(x, y)$ is a two-dimensional Gaussian function with standard deviation σ and ∇ is the gradient operator.

It is easy to see from these definitions that larger σ's will cause the boundaries to become blurry. Such large σ 's are often necessary, however in order to increase the capture range of the active contour. A snake has it weakness associated with short capture range, which would not enable user to randomly place the starting points of snake. Also, a concavity kind of edge becomes highly challenging to snake deformation. However, many other further improved models have also been developed.

A balloon model is added on another term to snake's external fore. It is a unit vector perpendicular to snake's curve at a point. The specialty of this force is to inflate snake from within like a balloon. So, if a snake is placed inside of the image, it can expand itself until it hits the boundary of that image. Hence, the edge is traced. This ability helped us greatly when the object is too close to the boundary of a picture and cannot be enclosed from outside. Also, balloon has a better adoption to concaved edges than traditional snake. From an initial oriented curve we add to the previous forces a pressure force pushing outside as if we introduced air inside. The force F now becomes:

$$F = k_1 n(s) - k \frac{\nabla P}{\|\nabla P\|}$$

where P is the potential associated with the external force, $n(s)$ is the unit vector to the curve at point $v(s)$ and k_1 is the amplitude of this force. If the sign of k_1 is changed it will have effect of deflation instead of inflation.

The results of boundary tracking by Snake algorithm is shown in Figure 9.3. Although balloon is a better model, however, it still has short capture range which leads to limitations of snake. A better snake model is the *gradient vector flow* (GVF) snake. GVF snake keeps the strong and useful part of the traditional snake: the internal force. But, instead of image gradient, it

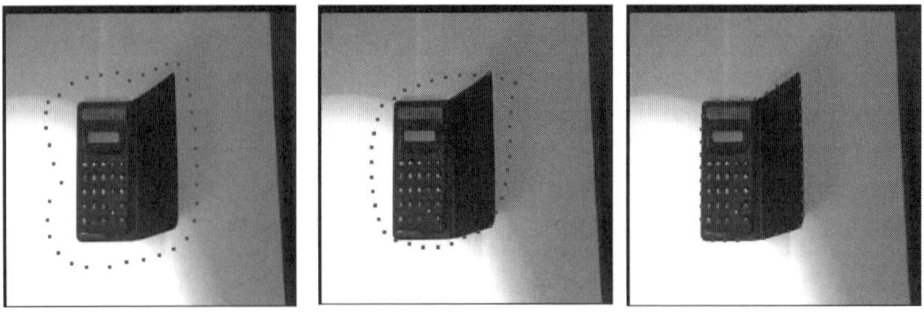

(a) (b) (c)

Fig. 9.3 Results of active contour modelling by SNAKE - (a) original Image, (b) contour after 10 iterations, (c) final contour.

created its own force field: GVF force field. Snake or active contour models are good tools for depicting the exact edges and provide higher level of usage. No matter how capable a snake's abilities are, its performance is very much rely on the intuitive user's interpretation of images in order to provide it with a good guidance towards its destiny.

9.7 SHAPE DISTORTION AND NORMALIZATION

While capturing two-dimensional images, based on the camera placement there are four possible basic forms of planar object shape distortions–rotation, scaling, translation, and skewing [25, 26]. A good shape descriptor should be invariant to these distortions. It is thus important to normalize the shape patterns in its original and various distorted forms, such as scaled, rotated, translated or skewed forms, so that they all,more or less, resemble similar to each other. When an appropriate set of features are extracted only after such a normalization, shape classification yields much better accuracy. We will consider shape normalization of a binary image $f(x, y)$ in which $f(x, y) = 1$ indicates that (x, y) is an object pixel, otherwise it is a background pixel. Such a normalization algorithm to normalize the shapes, called shape compacting [25], involves the following steps:

1. computing the shape dispersion matrix M,

2. aligning the coordinate axes with the eigen vectors of M, and

3. rescaling the axes using the eigenvalues of M.

9.7.1 Shape Dispersion Matrix

For a given shape, its dispersion matrix M reveals the variances and covariances amongst the pixels in the image. The dispersion matrix is a key element in the normalization process. The alignment of the coordinate axis uses the dispersion matrix M and takes care of the rotation of the object. Rescaling the coordinate axes is integral component of shape compacting and it uses the dispersion matrix. The basic philosophy of shape normalization process is that after the normalization operation, the shape will have a dispersion matrix equal to an identity matrix multiplied by a constant. This is an indication that the shape is in its most compact form.

To compute the dispersion matrix first we calculate the shape centroid

$$\bar{x} = \frac{\sum_x \sum_y x \cdot f(x,y)}{\sum_x \sum_y f(x,y)}, \quad \bar{y} = \frac{\sum_x \sum_y y \cdot f(x,y)}{\sum_x \sum_y f(x,y)}.$$

The shape dispersion matrix M is a 2 by 2 matrix

$$M = \begin{bmatrix} m_{1,1} & m_{1,2} \\ m_{2,1} & m_{2,2} \end{bmatrix}$$

where

$$m_{1,1} = \left(\frac{\sum_x \sum_y x^2 \cdot f(x,y)}{\sum_x \sum_y f(x,y)} \right) - \bar{x}^2, \quad m_{2,2} = \left(\frac{\sum_x \sum_y y^2 \cdot f(x,y)}{\sum_x \sum_y f(x,y)} \right) - \bar{y}^2,$$

$$m_{1,2} = m_{2,1} = \left(\frac{\sum_x \sum_y x \cdot y \cdot f(x,y)}{\sum_x \sum_y f(x,y)} \right) - \bar{x} \cdot \bar{y},$$

If we consider each object pixel as a data point, the shape can be viewed as a cluster of pixels. The shape dispersion matrix M computed above is exactly the covariance matrix of the cluster. It has already been discussed that a set of principal components may be selected from the covariance matrix, which is used to decouple the set of correlated features. It is also necessary to scale the features so that the clusters become compact. The shape dispersion matrix essentially performs the same function; it normalizes a shape by making it compact.

9.7.2 Shifting and Rotating the Coordinate Axes

The origin of the coordinate system is shifted to the center of the shape and then the coordinate system is rotated according to the eigenvectors of the dispersion matrix M.

The matrix M has two eigenvectors E_1 and E_2 corresponding to the eigenvalues λ_1 and λ_2. The two normalized eigenvectors E_1 and E_2 of M are computed as follows:

$$\lambda_1, \lambda_2 = \frac{m_{1,1} + m_{1,2} \pm \sqrt{(m_{1,1} + m_{1,2})^2 + 4m_{1,2}^2}}{2}$$

and

$$E_1 = \begin{bmatrix} e_{1x} \\ e_{1y} \end{bmatrix} = \begin{bmatrix} \frac{m_{1,2}}{\sqrt{(\lambda_1 - m_{1,1})^2 + m_{1,2}^2}} \\ \frac{\lambda_1 - m_{1,1}}{\sqrt{(\lambda_1 - m_{1,1})^2 + m_{1,2}^2}} \end{bmatrix}$$

$$E_2 = \begin{bmatrix} e_{2x} \\ e_{2y} \end{bmatrix} = \begin{bmatrix} \frac{m_{1,2}}{\sqrt{(\lambda_2 - m_{1,1})^2 + m_{1,2}^2}} \\ \frac{\lambda_2 - m_{1,1}}{\sqrt{(\lambda_2 - m_{1,1})^2 + m_{1,2}^2}} \end{bmatrix}$$

Now we can construct a matrix R from E_1 and E_2 by:

$$R = \begin{bmatrix} E_1^T \\ E_2^T \end{bmatrix} = \begin{bmatrix} e_{1,x} & e_{1,y} \\ e_{2,x} & e_{2,y} \end{bmatrix}$$

Since M is real and symmetric, E_1 and E_2 are orthogonal to each other. Furthermore, they are normalized to unit length. Thus, R is an orthonormal matrix.

We now transform the coordinate system by first translating the origin to the shape center and then multiplying the coordinates with matrix R. Now, each object pixel location (x, y) will have a new location (x', y') given by:

$$\begin{bmatrix} x' \\ y' \end{bmatrix} = R \cdot \begin{bmatrix} x - \bar{x} \\ y - \bar{y} \end{bmatrix}$$

Since R is an orthonormal matrix, the geometric interpretation of the transformation by R is pure coordinate rotation. The new coordinate axes are in the same directions as E_1 and E_2.

The dispersion matrix D of the translated and rotated shape is given by:

$$D = \begin{bmatrix} \lambda_1 & 0 \\ 0 & \lambda_2 \end{bmatrix}$$

9.7.3 Changing the scales of the bases

In the previous step, we have rotated the coordinate system so that the new X-axis points in the direction in which the shape is most dispersed. The effect of the rotation on the dispersion matrix is that now it is a diagonal matrix. Since our objective is to have a shape whose dispersion matrix is a scaled identity matrix, in this last step we will change the scales of the two axes according to the eigenvalues λ_1 and λ_2. That is, for an object pixel location (x', y'), the new location (x'', y'') is obtained through a transformation defined by W:

$$\begin{bmatrix} x'' \\ y'' \end{bmatrix} = W \cdot \begin{bmatrix} x' \\ y' \end{bmatrix} = \begin{bmatrix} k/\sqrt{\lambda_1} & 0 \\ 0 & k/\sqrt{\lambda_2} \end{bmatrix} \cdot \begin{bmatrix} x' \\ y' \end{bmatrix}$$

where k is a system-wide constant.

Since W is a diagonal matrix, the effect of the above step on the shape is to change the scales of the two coordinate basis vectors so that the shape is in

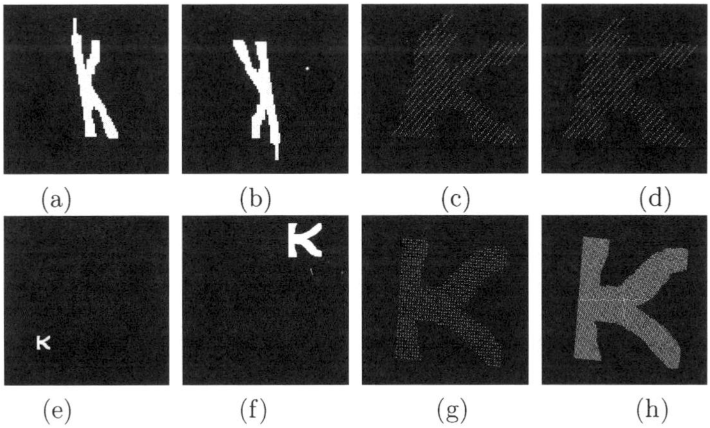

Fig. 9.4 Normalized images by shape dispersion matrix: (a)-(b) rotated image of letter K, (c)-(d) normalized results, (e)-(f) scaled and translated version of letter K image, (g)-(h) normalized results

its most compact form and with a normalized size. The results of translation, rotation, and scale invariance compact form generation has been shown in Figure 9.4.

9.8 CONTOUR-BASED SHAPE DESCRIPTOR

The contour-based shape descriptors may be classified as *Fourier shape descriptor* [27, 28], *wavelet-based shape descriptors* [29], *triangulation-based descriptors*, *minimum bounding circles* or *ellipse-based shape descriptors*, etc. Here we describe the Fourer based shape descriptor only.

9.8.1 Fourier based shape descriptor

In this method, we first compute the object's shape signature. Then we compute the discrete fourier transform on that shape signature. The shape signature of an object may be derived from either the boundary points of the object or they may be computed from the curvature or the radii information of the boundary points. For example, the shape signature of an arbitrarily shaped object may be given as $\{f_1, f_2, \cdots, f_n\}$, where each f_i represents the distance from the centroid to the ith point of the boundary. These points may be captured at an interval of equal angles from the centroid, such that these points are uniformly sampled along the boundary. The shape signature as suggested above is, however immune to noise. Even a small change in the location of the object boundary may lead to a large change in the overall shape representation. On the other hand, if we compute the Fourier transform on the shape signature, then the fourier coefficients captures the general shape characteristics in a better way. The fourier descriptors so computed are, however, not invariant to rotation and scaling. By appropriate normalization of the boundary to make it of a standard size and orientation,the fourier descriptor preserves all the shape attributes of an object. The magnitude information of $F(u)$ is invariant to rotation.

9.8.1.1 Normalized Fourier Descriptor The Fourier descriptor is computed by taking the DFT of the tangents of the sequence of points along the boundary of the shape. These descriptors are invariant to translation and rotation, but they are not scale invariant. By appropriately normalizing the Fourier descriptor, we may make it scale invariant which will be called as normalized Fourier descriptors. The normalized Fourier descriptors are computed by normalizing the length of the shape contour to 1.

Also the scale normalization can be achieved by creating a new feature vector $\frac{|F(1)|}{|F(0)|}, \frac{|F(2)|}{|F(0)|}, \cdots, \frac{|F(n)|}{|F(0)|}$, which are scale invariant features. The major limitation of these descriptors is that they do not take into consideration the interior of the shape, such as the holes within them. Also occluded objects, or complex objects consisting of multiple number of disconnected regions cannot be described by these descriptors.

9.8.1.2 Shape Features The polygonal approximation of an object as obtained using the technique described in the previous section should be free from noise and redundant points. The shape features of the object can be extracted using the vertices of the polygon. Let us assume that we have set of vertices $V_0, V_1, \cdots, V_{N-1}$ of the polygon represented by (x_0, y_0), (x_1, y_1), $\cdots$, (x_{N-1}, y_{N-1}). The following shape features are useful to discriminate between two shape classes.

1. Shape *Compactness* is defined as

$$C = \frac{4\pi A}{P^2},$$

where A is the area of the polygon and P is its perimeter. If the shape is a circular one, its compactness will be equal to 1. If, however, the space is a very thin and long bar, its compactness will be close to 0.

2. *Eccentricity* is defined as

$$E = \frac{\mu_{2,0} + \mu_{0,2} - \sqrt{(\mu_{2,0} - \mu_{0,2})^2 + 4\mu_{1,1}^2}}{\mu_{2,0} + \mu_{0,2} + \sqrt{(\mu_{2,0} - \mu_{0,2})^2 + 4\mu_{1,1}^2}},$$

where $\mu_{p,q} = \sum\sum(x - \bar{x})^p(y - \bar{y})^q$ is the (p, q) order central moment of the shape and $(\bar{x}, \bar{y})$ is the centroid of the object. The eccentricity actually represents the ratio of the minor axis to major axis of the best fitting ellipse of the shape.

3. Solidity $= \frac{A}{H}$, where A is the area of the polygon and H is the convex hull area of the polygon approximating the shape. If the shape is convex, its solidity will be closer to 1.

9.9 REGION BASED SHAPE DESCRIPTORS

The region based shape descriptors take into consideration all the interior and bounadary pixels of the shape. The most prominent region shape descriptor is based on the moments. Various types of moments have been used for moment based shape classification [30]–[33]. These are

- geometric moment
- Hu's invariant moments
- Zernike moments
- Radial Chebyshev Moments

The Geometric and invariant moments have been discussed in Chapter 11.

The seven moments of Hu are invariant to translation, rotation and scaling. However, their basis is not orthogonal and hence these moments have high degree of information redundancy. The higher order Hu's moments are also very sensitive to noise. Also since the basis involves powers of p and q, the dynamic range of the moment values of the objects belonging to the same shape class is very high. They are normalized with respect to the spread or the dynamic range.

9.9.1 Zernike moments

Keeping in view the above problems associated with the invariant Moments, it was found necessary to have moments in which the basis functions are orthogonal [30]. Teague introduced two different continuous-orthogonal moments–Zernike and Legendre moments, based on the orthogonal Zernike and Legendre polynomials,respectively. Several studies have shown the superiority of Zernike moments over Legendre moments due to their better feature representation and Rotation invariance: The magnitudes of Zernike moments are invariant to rotation. They are robust to noise and can take care of minor variations in shape. Since the basis is orthogonal,they have minimum information redundancy.

9.9.2 Radial Chebyshev Moments (RCM)

Mukundan et al. has suggested the use of discrete orthogonal moments to eliminate the problems associated with the continuous orthogonal moments [33]. They introduced Chebyshev moments based on the discrete orthogonal Chebyshev polynomial and showed that Chebyshev moments are superior to geometric, Zernike, and Legendre moments in terms of image reconstruction capability. They have introduced radial Chebyshev moments, which possess rotational invariance property

9.10 GESTALT THEORY OF PERCEPTION

The Gestalt theory suggests how we perceive the objects around us and how we discriminate the objects in the foreground from the entire background. A set of laws, popularly known as Gestalt Laws guides us in our understanding of human perception in discriminating between the objects and the background. Some of the Gestalt Laws, which define the human perception process are shown below.

1. A set of objects which are close together, i.e., in proximity with each other in space are perceived as grouped together.

2. Objects which are similar to one another in terms of shape, size, or color are grouped together.

3. Objects that form closed units tend to be perceived as together.

4. Elements forming continuous lines or curves appear to be grouped together.

5. There is a minimum contrast which needs to be detected by an observer for discriminating the spatial frequencies in a picture.

9.11 SUMMARY

Textures and shapes are extremely important features in human as well as machine vision and understanding systems. These areas have been gaining paramount importance in recent times. In this chapter, we have provided glimpses of some of the techniques for texture feature extraction using gray level co-occurrence matrices, fractal dimension, etc. Active contour modelling techniques have been discussed for shape analysis and image processing.

REFERENCES

1. B. Julesz, "Visual texture discrimination," *IRE Trans. On Inf. Theory*, Vol-8, 1962, 84-92.

2. B. Julesz, E. N. Gilbert, L. A. Shepp, and H. L. Frisch, "Inability of humans to discriminate between visual textures that agree in second-order statistics -revisited," *Perception*, 2, 1973, 391–405.

3. B. Julesz, "Experiments in the visual perception of textures," *Scientific America*, 232, 1975, 34–43.

4. B.Julesz, "A theory of pre-attentive texture discrimination based on first order statistics of textons," *Biol. Cybernet.*, 41, 1981, 131–138.

5. R. M. Haralick, K. Shanmugam, and I. Dinstein, "Textural features for image classification," *IEEE Trans. Syst. Man Cybern.*, SMC-3, 1973, 610–621.

6. R. W. Conners and C. A. Harlow, "A theoretical comparison of texture algorithms," *IEEE Transactions on systems man and cybernetics*, 21(1), January 1991, 252–261.

7. S. W. Zucker and D. Terzopoulous "Finding structure in cooccurrence matrices for texture analysis," *Computer graphics and image processing*, Vol-12, 1980, 286–308.

8. C. R. Dyer, T. H. Hong and A. Rosenfield, "Texture classification using gray level cooccurrence based on edge maxima," *IEEE Transactions on Systems, Man and Cybernetics*, Vol-10, 1980, 158–163.

9. C. V. Jawahar and A. K. Ray, "Incorporation of gray level imprecision in representation and processing of digital images," *Pattern Recognition Letters*, 17, 1996, 541–546.

10. L. S. Davis, S. A. Johns, and J. K. Aggarwal, "Texture analysis using generalized cooccurrence matrices," *IEEE Trans. Patt. Anal. Mach. Intell.*, PAMI-1(3), 1979, 251–259.

11. D. C. He and L. Wang, "Texture unit, texture spectrum and texture analysis," *IEEE Trans. Geoscience and Remote Sensing*, 28(4), 1990, 509–512.

12. D. C. He and L. Wang, "Texture features based on texture spectrum", *Pattern recognition*, 24, 1991, 391–399.

13. D. C. He and L.Wang, "Unsupervised textural classification of images using the texture spectrum," *Pattern recognition*, 25(6), 1992, 247-255.

14. B. Mandelbrot, *The Fractal Geometry of Nature*, Freeman Co., San Francisco, 1982.

15. A. P. Pentland, "Fractal based description of natural scenes," *IEEE Trans. on Pattern Analysis and Machine Intelligence*, 6, 661–674, 1984.

16. S. Peleg, J. Naor, R. Hartley, and D. Avnir, "Multiple resolution texture analysis and classification," *IEEE Trans. on Pattern Analysis and Machine Intelligence*, PAMI-6(4), July 1994, 518–523.

17. I. L. Dryden and K. V. Mardia, *Statistical Shape Analysis*, Wiley, New York, 1998.

18. R. Bhattacharya and V. Patrangenaru, "Large sample theory of Intrinsic and extrinsic sample means on manifolds", *Ann. Statistics*, 31, 1–37, 2003.

19. A. Srivastava, S. H. Joshi, W. Mio, and X. Liu, "Statistical Shape Analysis: Clustering, Learning, and Testing," *IEEE Trans. on Pattern Analysis and Machine Intelligence*, PAMI-27(4), April 2005, 590–602.

20. J. T. Kent and K. V. Mardia, "Procrustes Tangent Projections and bilateral symmetry," *Biometrica*, 88, 2001, 469-485.

21. C. H. Teh and R. T. Chin, "On the detection of dominant points on digital curves," *IEEE Trans of Pattern Anal. and Machine Intel.*, 11(8), August 1989.

22. M. Kass, A. Witkin, and D. Terzopoulos, "Snakes: Active Contour Models," *Proceedings of First International Conference on Computer Vision*, London, 1987, 259–269.

23. D. J. Williams and M. Shah, "A Fast Algorithm for Active Contours and Curvature Estimation," *CVGIP: Image Understanding*, 55(1), January 1992, 14–26.

24. C. Xu and J. L. Prince, "Snakes, Shapes and Gradient Vector Flow," *IEEE Trans. on Image Processing*, 7(3), March 1998.

25. J.-G. Leu, "Shape Normalization Through Compacting," *Pattern Recognition Letters*, 10, 1989, 243–25.

26. M. E. Celebi and Y. A. Aslandogan, "A Comparative Study of Three Moment-Based Shape Descriptors" *Proc. of the IEEE International Conference on Information Technology: Coding and Computing*, Las Vegas, NV, April 2005.

27. E. Persoon and K.-S. Fu, "Shape Discrimination Using Fourier Descriptors" *IEEE Trans. On Systems, Man and Cybernetics*, SMC-7(3), 1977, 170-179.

28. H. Kauppinen, T. Seppanen, and M. Pietikainen, "An Experimental Comparison of Auto-regressive and Fourier-Based Descriptors in 2D Shape Classification" *IEEE Trans. on Pattern Analysis and Machine Intelligence*, 17(2), 1995, 201–207.

29. C.-H. Chuang and C.-C.J. Kuo, "Wavelet Descriptor of Planar Curves: Theory and Applications," *IEEE Trans. on Image Processing*, 5(1), 56–70, 1996.

30. M. R. Teague, "Image Analysis Via the General theory of Moments," *Journal of Optical Society of America*, 70(8), 1980, 920–930.

31. A. Khotanzad, "Rotation Invariant Pattern Recognition Using Zernike Moments," *Proceedings of the International Conference on Pattern Recognition*, 326–328, 1988.

32. H. Kim and J. Kim, "Region-based Shape Descriptor Invariant to Rotation, Scale and Translation," Signal Processing: Image Communication, 16, 2000, 87–93.

33. R. Mukundan and K. R. Ramakrishnan , *Moment Functions in Image Analysis: Theory and Applications*, World Scientific Publication Co., Singapore, 1998.

10

Fuzzy Set Theory in Image Processing

10.1 INTRODUCTION TO FUZZY SET THEORY

With the introduction of fuzzy set theory by L. A. Zadeh in 1965 [1], there has been a significant progress in the area of visual and cognitive sciences. Fuzzy sets have been extensively used in computer vision and machine intelligence since then. Fuzzy set deals with the imprecision and vagueness embedded in human understanding systems and provides an elegant framework for describing, analyzing, and interpreting the vague and uncertain events. The human vision system is essentially a fuzzy system, since we can understand and interprete the imprecise visual world around us. In this chapter we will provide glimpses of fuzzy sets and some of its applications in image processing.

10.2 WHY FUZZY IMAGE?

A gray image possesses ambiguity within pixels due to inherent vagueness or imprecision embedded in an image, rather than to the randomness in probabilistic sense. The imprecision in an image pattern is due to several factors.

- Ambiguity in gray values of an image
- Spatial (i.e., geometrical) ambiguity
- Imprecision in Knowledge base
- Combination of all the above factors

An image is a two-dimensional mapping of the three-dimensional world around us. During the process of capturing the two-dimensional image data, a lot of ambiguity arises in the gray values of the image. For example, in a digital image pixels of gray levels I and $I + \Delta I$ are two different intensity levels, yet they visually appear to be the same bright pixel levels. There exists an ambiguity or imprecision in describing whether a pixel is bright or dark.

Spatial ambiguity is caused by the imprecision in object boundaries, or the edges within the image. In a digital image, a pixel is either an edge or a no-edge. But the edges are not always precisely defined and thus ambiguity arises in practical images.

Fuzzy images are characterized by the degree to which each pixel belongs to a particular region. It is a mathematical tool to describe and interpret ill-defined attributes, such as, pixel gray level, edges, regions, etc. Also the primitives, such as corners, curves, etc. in an image may be described more logically using fuzzy sets.

In many image-processing applications, we employ expert knowledge to interpret an input scene. Examples are object recognition, region segmenetation, scene description, and so on. Fuzzy logic offer us powerful tools to represent and process human knowledge in the form of fuzzy if-then-else rules. Even the structural information in a pattern may be better described using fuzzy structural rules.

In view of the above discussion it may be observed that there is a necessity to use alternative tool, like fuzzy sets to describe, analyze and interpret images. This will provide a framework for representing and manipulating the uncertainties in the image, such as image regions, boundaries, etc.

10.3 INTRODUCTION TO FUZZY SET THEORY

The conventional set (or crisp set) theory is based on a binary valued membership, which implies that a particular element either belongs to a particular set or it does not belong to it. A crisp set is defined as one whose elements fully belong to the set and they possess well-defined common attributes, which can be measured quantitatively. In a crisp set the common attributes are equally shared by all the elements of the set.

In fuzzy sets, on the other hand, the degree of membership of an element to the set is indicated by a membership value, which signifies the extent to which the element belongs to the set. The membership value lies between 0 and 1, with membership 0 indicating no membership and 1 indicating full membership of the element to the set. In a crisp set, the membership values of its elements are either 0 or 1.

The membership of an element x in a fuzzy set is obtained by a membership function $\mu(x)$ that maps every element belonging to the fuzzy set X_F to the interval $[0,1]$. Formally this mapping is written as:

$$\mu(x) : X_F \longrightarrow [0:1].$$

The membership assignment is primarily subjective in the sense that the users specify the membership values. In many applications, the membership function is chosen based on an objective criterion. In some situations probability measures may be used for assigning the membership values. These membership functions are useful in modelling and quantifying various imprecise *linguistic*, *nonmathematical*, or *ambiguous* terms.

There are two commonly used ways to denote a fuzzy set. If μ is the membership function of a particular element x, then the fuzzy set X_F may be written as a set of fuzzy singletons:

$$X_F = \{(x, \mu(x))\}.$$

Alternatively the fuzzy set may be explicitly indicated as the union of all $x \mid \mu(x)$ singletons and thus can be written as:

$$X_F = \bigcup x \mid \mu(x) \text{ for all } x \in X$$

10.4 PRELIMINARIES AND BACKGROUND

In this section we will present some of the preliminaries and basic operations involved in fuzzy sets, which will be useful in understanding diverse applications of fuzzy sets [2, 3].

10.4.1 Fuzzification

Fuzzification is used to transform a crisp data set into a fuzzy data set or simply to increase the fuzziness of an existing fuzzy set. Thus for fuzzification we use a fuzzifier functions which may be dependent on one or more parameters.

Let us consider the set of all bright pixels in an image. Here the intensity of the pixel is qualified by the linguistic term *bright*. The pixel intensity I is a variable, which is called a linguistic variable, since it can assume linguistic values *bright*, *dark*, etc. Such linguistic variables do not have any precise value but they convey imprecise concepts which we human beings can understand.

If x is a member of the fuzzy set of all *bright* pixels, then the extent to which it is bright is given by its membership function $\mu(x)$. One possible membership function can be written as:

$$\mu(x) = \frac{1}{1 + \left(\frac{x}{F_2}\right)^{-F_1}}, \tag{10.1}$$

where F_1 and F_2 are known as exponential and denominational fuzzifiers respectively [4]. As may be observed, the selection of the parameters, i.e., exponential and denominational fuzzifiers, is provided by the users and they may use subjective judgements for parameter selection.

The above fuzzification function was originally proposed by Zadeh [1]–[3] and has since been extensively used to model the membership function of a fuzzy element x by choosing suitable fuzzifier values [4, 5].

In Eqn. 10.1, F_2 denotes the crossover point across which the membership values will change quite appreciably and F_1 denotes the rate at which this change will occur. It may be noted that the exponential fuzzifier mentioned above is positive.

10.4.2 Basic Terms and Operations

Some of the definitions and important operations in fuzzy sets are given below. **Intersection and Union**: The *intersection* of two fuzzy sets $X = \{x \mid \mu_1(x)\}$ and $Y = \{y \mid \mu_2(y)\}$ is defined as the fuzzy set $Z = \{z \mid \mu(z)\}$ such that $z \in X$ and $z \in Y$ and $\mu(z) = \min[\mu_1(z), \mu_2(z)]$. Similarly the *union* of two fuzzy set is defined on the basis of the max function. The union of X and Y is defined as $Z = \{z \mid \mu(z)\}$, $z \in X$ and $z \in Y$, such that $\mu(z) = \max[\mu_1(z), \mu_2(z)]$.

Complement: The *complement* of a fuzzy set $X = \{x \mid \mu_1(x)\}$ is defined as the set $Z = \{z \mid \mu(z)\}$, $z \in X$ such that $\mu(z) = 1 - \mu_1(z)$.

Product: The product of two fuzzy sets $X = \{x \mid \mu_1(x)\}$ and $Y = \{y \mid \mu_2(y)\}$ is defined as the fuzzy set $Z = \{z \mid \mu(z)\}$ such that $z \in X$ and $z \in Y$, and $\mu(z) = \mu_1(z)\mu_2(z)$.

Power of a fuzzy set: The power of a fuzzy set $X = \{x \mid \mu_1(x)\}$ raised to the power α is defined as the fuzzy set $Z = \{z \mid \mu(z)\}$ such that $z \in X$ and $z \in Y$, such that $\mu(z) = \mu_1(z)^\alpha$.

Equality: Two fuzzy sets $X = \{x \mid \mu_1(x)\}$ and $Y = \{y \mid \mu_2(y)\}$ are said to be *equal* if for all $x \in X \Rightarrow x \in Y$ and for all $y \in Y \Rightarrow y \in X$, such that $\mu_1(x) = \mu_2(y)$ and $\mu_1(y) = \mu_2(x)$.

Empty set: A fuzzy set is said to be an *empty set* if the membership values of all its elements is zero.

Normal set: A fuzzy set is called a *normal set* if the membership values of all its elements are one.

Alpha cut or Alpha level set: If $X = \{x \mid \mu_1(x)\}$ is a fuzzy set then the *alpha cut* or *alpha level set* of X is defined as the fuzzy set $Z = \{z \mid \mu(z)\}$ such that $z \in X$ and $\mu(z) = \mu_1(z)$ with the condition that $\mu_1(z) > \alpha$.

Linguistic Variables and Hedge: Fuzzy set theoretic techniques are mainly employed in ambiguous situations and statements. For example, it deals with adjectives like *bright*, *dark*, *large*, *small*, *medium*, or adverbs like *very*, *more* or *less*, etc. These words have no precise semantics and they cannot be represented using crisp mathematics. In fuzzy mathematics, such adjectives or adverbs may be adequately described in terms of linguistic variables and the ambiguity may be modeled to a large extent.

Example: Let us consider the set of bright pixels in an image, in which a particular pixel has membership value 0.9. If we want to compute the membership value of that pixel to the set of *very bright* pixels, its new membership value will be somewhat lower. We make a suitable transformation on its existing membership value (in the set of bright pixel) to obtain its new membership value in the set of very bright pixels. A popularly used transformation to qualify a fuzzy element by the adverb *very* is to square its existing membership value, which yields its new membership value 0.81, which is less than 0.9 as expected.

In Table 10.1, some of the linguistic modifiers and their corresponding operations on the membership values are listed.

Table 10.1 Membership function of linguistic hedges

Modifier	**Operation on the membership value, μ is transformed to μ'**
VERY	$\mu' = \mu^2$ (Concentration)
MORE OR LESS	$\mu' = \sqrt{\mu}$ (Dilation)
INTENSIFY	$\mu' = 2\mu^2$ if $\mu \leq 0.5$ and $\mu' = 1 - 2(1-\mu)^2$ if $\mu > 0.5$
PLUS	$\mu' = \mu^{1.25}$
MINUS	$\mu' = \mu^{0.75}$
NOT	$\mu' = 1 - \mu$

The above table summarizes the various ways in which ambiguous linguistic terms can be interpreted mathematically in fuzzy domain to obtain more meaningful quantitative expressions.

10.5 IMAGE AS A FUZZY SET

To solve the ambiguity in image description, we can define membership function $\mu(g)$, which gives satisfactorily the extent to which a particular pixel with gray value g is dark or bright. For example, to represent dark, the pixels toward the lower end of the gray level scale will have high membership values (near 1), which gradually transit to low values (near 0) as the pixel gray level changes from dark to light. Fuzzy sets may be satisfactorily applied to model ambiguous and imprecise image information. Here we have assumed that the imprecise image information concerns the darkness or lightness of a pixel.

Similarly the ambiguity resulting from the imprecision in an image, such as, the continuity of a line segment, shape of objects, similarity of regions, which cannot be represented suitably and uniquely in conventional mathematics may be taken care of using fuzzy sets.

With this background we can proceed to the formal way of representing an image as a fuzzy set.

Conventionally an image X of size $M \times N$ is represented as a two-dimensional matrix where the pixel at location $(i,\ j)$, $i = 0, \cdots, M-1$, $j = 0, \cdots, N-1$, is represented as x_{ij} . Thus for an n-bit gray scale image x_{ij} can take values from 0 to $L-1$, where $L = 2^n$ and mathematically

$$X = \bigcup_i \bigcup_j x_{ij}.$$

Definition of a fuzzy image includes the gray value x_{ij} along with a membership value $\mu(x_{ij})$ associated with it. This membership value $\mu(x_{ij})$ represents the extent to which a pixel at location $(i,\ j)$ belongs to a class, having a specific attribute. Thus in fuzzy set theory, an image X of size $M \times N$ having L levels of gray can be considered as an array of fuzzy singletons, each associated with a membership value or function. Thus a fuzzy image can be represented as:

$$X = \bigcup_i \bigcup_j x_{ij} \mid \mu(x_{ij}).$$

The membership function of the gray value essentially reflects the membership or belongingness of the pixel to a certain class.

10.5.1 Selection of the Membership Function

The assignment of the membership function may be performed by several ways.

- *Membership based on visual model*: The membership function may be assigned in accordance with the human visual perceptual model. We may model the variation of the membership values of the pixels in a linear fashion as the pixel gray value changes from 0 to $L-1$ (for an L level image). The human visual response to illumination, however, is not linear but closely exponential. When a shade is too dark or too light (or bright), the human eye fails to notice any slight change around it. Thus the membership values of the gray levels toward the extreme low or extreme high end of the entire scale should not change appreciably. However, the response of the human eye to different degrees of illumination is approximately linear in the mid range. The membership function may be kept more or less linear in that range of gray values.

- *Statistical Distribution*: The membership values of the pixels may be assigned on the basis of image statistics as a whole or on the basis of local information at a pixel calculated from the surrounding pixels. The probability density function of the Gaussian or gamma distribution may be used for assignment of membership values [6].

- *Gamma membership function*: The pdf of gamma distribution is given as:

$$f(x) = \frac{\left(\frac{x-m}{\beta}\right)^{\gamma-1} \exp\left(-\frac{x-m}{\beta}\right)}{\Gamma(\gamma)}, \quad x \geq \mu, \text{ and } \gamma, \beta > 0 \tag{10.2}$$

where γ is the shape parameter, m is the location parameter, β is the scale parameter and Γ is the gamma function, which is defined as

$$\Gamma(a) = \int_0^\infty t^{a-1} e^{-t} dt.$$

The case where $m = 0$ and $\beta = 1$ is called the standard gamma distribution, and the equation for the gamma distribution reduces to

$$f(x) = \frac{x^{\gamma-1} \exp(-x)}{\Gamma(\gamma)} \quad x > 0, \gamma > 0.$$

Now if $\beta = 1$, $\gamma = 1$, and $m \neq 0$, then the gamma distribution in the above equation becomes

$$f(x) = \exp(-(x - m)). \tag{10.3}$$

Gamma membership functions have been observed to yield very good results in image processing [6].

10.6 FUZZY METHODS OF CONTRAST ENHANCEMENT

The human eye does not respond to subtle differences in illumination. The purpose of contrast enhancement is to improve the overall visibility of the image gray levels. This may be achieved by transforming the image gray levels in such a way that the dark pixels appear darker and light pixels appear lighter. Such a transform increases the differences in gray level intensity and thus enables our vision system to discern these differences.

The contrast stretching algorithms may employ single-level or multilevel thresholding. For example, a threshold can be selected at a gray level x_I and any gray level falling below that is still reduced and any gray level falling above it is increased. Care is taken that the stretching *saturates* at the extreme ends. For example, if we employ linear stretching across a single threshold x_t, mathematically it can be represented as

$$\left.\begin{aligned} CON(x) &= (1+v_1)x, \quad x > x_t \\ &= x, \qquad\qquad\ \ x = x_t \\ &= (1-v_2)x, \quad x < x_t \end{aligned}\right\} \tag{10.4}$$

where $CON(x)$ is the contrast intensifier function which transforms a gray level x into its intensified values, and v_1 and v_2 are fractions between 0 and 1 which decide the percentage stretching of the gray value x across the selected threshold.

In similar ways linear stretching algorithms can also be devised with multiple thresholds, such that the above logic may be employed between a pair of consecutive thresholds.

Fuzzy methods for contrast enhancement employ a membership function $\mu(x)$ to determine the extent of *darkness* (or *brightness*) of a pixel gray value and then apply a suitable transformation on the membership value to generate the membership value of that pixel in the contrast intensified image.

10.6.1 Contrast Enhancement Using Fuzzifier

Several interesting image enhancement techniques have been suggested based on fuzzy sets [5, 7]. For an image whose gray level at the location (m, n) is given by x_{mn} the membership function is defined as

$$\mu(x_{mn}) = \left[1 + \frac{x_{max} - x_{mn}}{F_d}\right]^{-F_e} \tag{10.5}$$

where x_{max} is the maximum gray level available in the image and F_d and F_e are denominational fuzzifier and exponential fuzzifier respectively [5].

Eq. 10.5 shows that as $\mu(x_{mn}) \to 1$ when $x_{mn} \to x_{max}$, i.e., the highest gray level represents maximum brightness. The membership function, as in Eq. 10.5, represents the degree of brightness of a particular pixel.

It may be noted that the membership value always occurs within a range defined by $[\alpha,1]$. The minimum gray level that a pixel can assume is 0, which yields the lower limit of α. Thus α is obtained as

$$\alpha = \left[1 + \frac{x_{max}}{F_d}\right]^{-F_e}. \tag{10.6}$$

An appropriately chosen intensifier operator stretches the contrast in an image. It transforms membership values above 0.5 to higher or lower values. Intensifier operation becomes increasingly nonlinear as the membership values moves away from 0.5 to still higher values and those below 0.5 to still lower values. The intensifier operator may be chosen as :

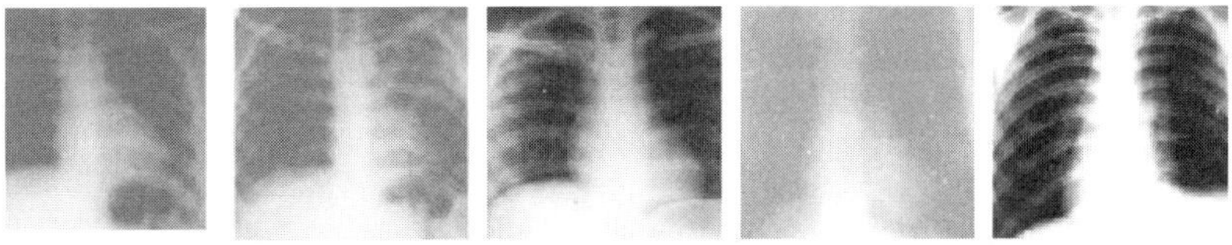

(a) (b) (c) (d) (e)

Fig. 10.1 (a) A chest X-ray image1, (b) contrast stretched image using crisp logic, (c) fuzzy contrast enhancement, (d) chest X-ray image2, (e) fuzzy constrast enhanced image2

$$\left.\begin{aligned} INT(\mu) &= 2\mu^2, & \mu \leq 0.5 \\ &= 1 - 2(1-\mu)^2, & \mu > 0.5 \end{aligned}\right\} \tag{10.7}$$

where $INT(\mu)$ is the intensification function on the membership values. The modified gray values are obtained by taking an inverse mapping function computed by the modified membership values, obtained after the original membership values are operated upon by the INT operator.

The aforementioned methods are applied to the image and results are shown in Figure 10.1. For easy comparison the result obtained by stretching is also included. The fuzzy based approach has been found to give visibly better results compared to the conventional linear contrast stretching method.

It can be seen that the performance of the conventional linear stretching is quite poor as it darkens the image uniformity and thus makes the vision stressed. The fuzzy based method shows better results and does not darken the image. Selecting two sets of F_d and F_e, shows the effect of parameter variation on this method. The selection has been entirely supervised.

10.6.2 Fuzzy Spatial Filter for Noise Removal

Spatial filtering utilizes an averaging procedure to generate the smoothened image. The weights usually chosen for the averaging are crisp and they do not depend on the pixel intensity in the image. Thus all regions of the image under an arbitrary neighborhood W are equally affected by averaging. The limitations of such a method are that they do not take into account

- the effect of the difference of gray levels between the central pixel and a neighboring pixel
- the diminishing influence of the pixels, which are situated at increasing distance from the central pixel.

These two principal limitations of the conventional spatial filtering are taken into consideration while designing *fuzzy spatial filter*. Fuzzy spatial filters which smoothen a given image by employing averaging techniques, decide the weights of their masks on the basis of the above considerations.

While averaging, the contribution of a particular pixel in a certain neighborhood would decrease as it becomes increasingly different from that of the central pixel. This principle leads us to define the fuzzy membership values of a pixel. These membership values, in turn, serve as the weights in the averaging process of the pixels in the neighborhood. Thus the weights used in this filter are not constant and depend on the neighborhood information of the central pixel.

10.6.3 Smoothing Algorithm

The membership function determines the extent to which a particular pixel in the neighborhood represents the central pixel. Thus as the pixel moves away from the central pixel its membership value would decrease. At the same time if it has a large difference of gray value from that of the central pixel its membership value would become lower than what it would have been if the difference were smaller.

Taking this into account the membership function can be modeled as a double bell-shaped curve. The membership function as it can be expected depends on two parameters, difference of gray values and distance [4, 7].

Let the gray level of the pixel at the location (m, n) be given by x_{mn}. If the central pixel gray level be given by x_c, then the membership value $\mu(x_{ij})$ of any pixel in the neighborhood of x_c is given by:

$$\mu(x_{ij}) = \exp\left[\frac{-(x_c - x_{ij})^2}{\alpha}\right] \exp\left[\frac{-d^2}{\beta}\right] \tag{10.8}$$

where d is the distance between the central pixel x_c and the neighboring pixel x_{ij}; α and β are two scaling factors which determine the extent of flatness of the membership function. It can be noticed that the membership function returns 1 only when the neighboring pixel and the central pixel coincide. That is to say $x_c = x_{ij}$. Thus the central pixel has the maximum contribution in determining its modified value.

In the averaging procedure the membership values themselves serve as the weights of the gray levels of the pixels in the neighborhood. In light of above discussions the algorithm can be presented concisely as follows:

- For a central pixel x_c determine a neighborhood W of some suitable size.
- For each pixel x_ij in the neighborhood determine the membership value μ_{ij}.
- The modified gray value of the central pixel can thus be given by

$$x'_c = \frac{\lfloor \sum\sum \mu_{ij}\, x_{jj} \rfloor}{\lfloor \sum\sum \mu_{ij} \rfloor}, \; (i,j) \in W, \; x_{ij} \neq x_c \tag{10.9}$$

10.7 IMAGE SEGMENTATION USING FUZZY METHODS

Fuzzy set theory has been extensively applied in the area of image thresholding and segmentation. A number of thresholding approaches have been suggested in the literature [8]–[13]. Several interesting fuzzy distance functions and similarity measures have been suggested in [8, 9, 14, 15]. These measures have been demonstrated to yield excellent results in image thresholding. The concepts of fuzzy first order and second order statistics, viz., fuzzy histogram and cooccurrence matrices have been presented in [16]–[18]. These measures yield excellent segmented description of images having bimodal or multimeodal histograms. The fuzzy second order statistics characterize textured images in a better way than their corresponding hard counterparts [16, 17]. A number of research reports indicate the superiority of fuzzy sets in segmenting an image over its crisp counterpart [12]–[18]. In the foregoing discussion in Chapter 7, we have noted that the selection of a set of appropriate thresholds is important in image segmentation. Often the image histogram does not show any well-defined peaks, and consequently the selection of proper thresholds becomes difficult. Similarly there is a possibility that the pixels belonging to the same object are partitioned in diverse groups because of poor threshold selection which does not take into account the effect of possible different illumination at different parts of the same object.

The problem of thresholding involves identifying an optimal threshold T and segmenting the scene into two meaningful regions–object (O) and background (B), i.e,

$$\begin{aligned} O &= \{(m,n)|I_{mn} \geq T\} \\ B &= \{(m,n)|I_{mn} < T\} \end{aligned} \tag{10.10}$$

in such a way that $O \bigcup B = G$ and $O \bigcap B = \phi$.

Algorithms employing such a formulation dichotomize the pixels deterministically into object and background classes and therefore fail to reflect the structural details embedded in the original gray distribution. Many of these schemes often yield poor results in some classes of images. An appropriate thresholding scheme that

- provides a soft thresholded description of the image based on fuzzy set theoretic concepts,
- can incorporate a wide range of object background size and scatter imbalances, and
- does not depend excessively on assumptions of gray distributions in the image

have been investigated in [16]–[18].

Fuzzy thresholding involves partitioning the image into two fuzzy sets $\tilde{O}$ and $\tilde{B}$ corresponding to object and background by identifying the membership distributions $\mu_{\tilde{O}}$ and $\mu_{\tilde{B}}$ associated with these regions. A natural extension into fuzzy setting has been proposed in [5], where an image is characterized by a monotonic membership function $\mu_{\tilde{O}}$ such that

$$\begin{aligned} \mu_{\tilde{O}}(I_{mn}) < 0.5 \text{ if } I_{mn} < T \\ \mu_{\tilde{O}}(I_{mn}) > 0.5 \text{ if } I_{mn} > T, \end{aligned} \tag{10.11}$$

and the cross over point of the membership function matches with the hard threshold. Background region, $\tilde{B}$, was considered as the fuzzy complement of object region, $\tilde{O}$, i.e.,

$$\mu_{\tilde{B}}(j) + \mu_{\tilde{O}}(j) = 1.0 \,\forall\, j \tag{10.12}$$

Huang and Wang proposed a fuzzy thresholding scheme which minimizes the fuzziness in the thresholded description and at the same time accommodates the variations in the gray-values in each of the regions [10]. They assigned memberships as

$$\mu_{\tilde{O}}(f_{mn}) = \begin{cases} \left[1 + \frac{\mu f_{mn}}{K}\right]^{-1} & \text{if } f_{mn} \geq T \\ 0 & \text{if } I_{mn} < T \end{cases} \tag{10.13}$$

where μ is the mean gray-value of $\tilde{O}$ and the parameter K controls the amount of fuzziness in the segmented description. A similar membership assignment was employed for $\tilde{B}$ also. They classified the pixels unequivocally into object or background regions with the help of the hard threshold T and thereby led to an abrupt discontinuity of membership distribution in the object and background regions, such that,

$$\begin{aligned} \tilde{O} \bigcup \tilde{B} \subseteq G \\ \tilde{O} \bigcap \tilde{B} = \phi \end{aligned} \tag{10.14}$$

The histogram representing the frequency of occurrence is modeled as

$$h_j = \rho_1 f\left(d\left(j, \mu_1\right), \sigma_1\right) + \rho_2 f\left(d\left(j, \mu_2\right), \sigma_2\right), j \epsilon L \tag{10.15}$$

Here, ρ_1 corresponds to the ratio of sizes of object and background regions in the image while another parameter $\rho_2 = \frac{\sigma_2}{\sigma_1}$ denotes the measure for ratio of scatter. When $\rho_1 = \rho_2 = 1.0$, the object and background regions have equal size and equal dispersion. A typical example of $f(.)$ is a Gaussian function with a total number of N_i pixels given by

$$f\left(d(\,j, v_i\,), \sigma_i\right) = \frac{N_i}{\sigma_i \sqrt{2\Pi}} \exp^{-d(j,v_i)^2/(2\sigma_i^2)} \tag{10.16}$$

Bayes decision theory provides a minimum error optimal threshold T, where both object and background distributions provide equal gray-density. If the histogram is bimodal, and the minimum error threshold corresponds to the valley of histograms. This does not imply either that the histogram valley, always corresponds to the optimal threshold T or that the histograms generated by the model are always bimodal.

Various fuzzy set theoretic distance and divergence measures have been suggested in a number of applications, such as, threshold selection, image segmentation by fuzzy divergence, region extraction, and region extraction from color images [6, 8, 9].

10.8 FUZZY APPROACHES TO PIXEL CLASSIFICATION

Fuzzy approaches to pixel classification have found applications in problems where (1) precise knowledge about the pattern classes is not available, (2) large number of pattern samples are not available for statistical estimation of parameters, (3) patterns have partial membership to different classes.

In the hard clustering approach, each sample is considered to belong to only one cluster and the clusters are considered as disjoint sets of data. The k-means algorithm partitions a given set of samples into k classes by iteratively re-computing the partition. In each iteration, the means of all the classes are computed and a sample is assigned to the class, whose mean is nearest from the pattern sample. Mathematically, k-means algorithm minimizes the sum of within cluster scatters and provides accurate results when classes are compact and well separated.

In practice, however, there are many situations where the clusters are not disjoint and a sample pattern or a pixel in an image may belong to different clusters. Such a situation cannot be handled by crisp clustering techniques. For example, a particular pixel on the river bed in an aerial image may belong partly to *water* class and partly to the *wet-soil* class. In fuzzy segmentation such problems can be handled quite well and several algorithms are available for segmentation of images, where the class separation is not well defined [8, 12, 18, 19].

Fuzzy approaches to supervised pattern classification and clustering may be found in [19]. The basic postulates of fuzzy clustering is that a member may have partial memberships grades in several fuzzy clusters. A membership value in the interval $[0, 1]$ is assigned to each sample in every cluster, based on certain measurements.

In fuzzy clustering a pattern is assigned with a degree of belongingness to each cluster in a partition. Here we will present the most popular fuzzy clustering algorithm, known as fuzzy c-means algorithm.

10.9 FUZZY C-MEANS ALGORITHM

Fuzzy c-means clustering algorithm [19, 20], a generalization of the hard c-means algorithm yields extremely good results in image region clustering and object classification. As in hard k-means algorithm, fuzzy c-means algorithm is based on the minimization of a criterion function. The following criterion function may be chosen.

$$J(U,V) = \sum_{i=1}^{c} \sum_{k=1}^{n} u_{ik}^{m} |x_k - v_i|^2 \tag{10.17}$$

where,

- $x_1, x_2, \cdots, x_n$ are the n sample points
- $V = \{v_1, v_2, \cdots, v_c\}$ are the cluster centers
- $U = [u_{ik}]$ is $c \times n$ matrix, where u_{ik} is the membership value of the kth input sample x_k in the ith cluster. The membership values satisfy the following conditions:

$$\left.\begin{array}{l} 0 \leq u_{ik} \leq 1 \\ \sum_{i=1}^{c} u_{ik} = 1 \\ 0 < \sum_{k=1}^{n} x_k < n \end{array}\right\}$$

- $m \in [1, \infty]$ is an exponent weight factor. There is no fixed rule for choosing the exponent weight factor. However, in many applications $m = 2$ is a common choice. In case of crisp clustering m may be chosen as 1.

The above three conditions imply the followings:

- The membership values of each sample x_k to a particular cluster should lie between 0 and 1
- each sample x_k must belong to at least one cluster and the sum of the membership values to each cluster should be one
- Each class must have at least one sample and all the samples cannot belong to a particular class.

It may be pointed here that if the samples have crisp membership to each cluster, then u_{ik} assumes the values 0 or 1 for each sample, depending upon its belongingness to each cluster. The objective function in this case is the sum of the squared Euclidean distances between each input sample and its corresponding cluster center, weighted by the fuzzy membership values.

The algorithm iteratively updates the cluster centers using the expression:

$$V_i = \frac{\sum_{k=1}^{n} u_{ik}^{m} x_k}{\sum_{k=1}^{n} u_{ik}^{m}}, \quad i = 1, 2, 3, \cdots, c \tag{10.18}$$

The fuzzy membership function of the k th sample x_k to the ith cluster is given by the following;

$$u_{ik} = \frac{\left[\frac{1}{(x_k - v_i)^2}\right]^{\frac{1}{m-1}}}{\sum_{j=1}^{c}\left[\frac{1}{(x_k - v_i)^2}\right]^{\frac{1}{m-1}}}, \; i = 1, 2, \cdots, c; \; k = 1, 2, \cdots, n \tag{10.19}$$

It can be noted that the weight factor m, reduces the influence of small membership values.

The FCM algorithm is thus summarized as follows:

1. Initialize $U^{(0)}$ randomly or based on some approximation. Initialize $V^{(0)}$ and calculate $U^{(0)}$. Set the iteration counter $t = 1$. Select the number of cluster centers c and choose the value of m.

2. Compute the cluster centers. Given $U^{(t)}$, calculate $V^{(t)}$ according to the Eq. 10.18.

3. Given $V^{(t)}$ update the membership values to $U^{(t+1)}$ according to Eq. 10.19.

4. Stop iteration if $|u_{ik}^{(t+1)} - u_{ik}^{(t)}| \leq \varepsilon$, where ε is small positive number.

5. Increment iteration counter to $t = t + 1$. Go to step 2.

It may be noted that, while applying the FCM to image clustering or segmentation the data points or the sample points $x_1, x_2, \cdots, x_n$ are the pixel gray values. Thus n represents the total number of pixels in the image.

10.10 FUSION OF FUZZY LOGIC WITH NEURAL NETWORKS

Fusion of fuzzy logic with neural networks has been suggested by a number of researchers [21]–[23]. The advantages of fuzzy logic combined with the attractive properties of neural networks yield better performance of a pattern classifier. In a fuzzy neural network, neuron functions, their input-outputs, and the network parameters are fuzzified. These networks can handle fuzzy input output relationships which are often encountered in many image processing applications. Several fuzzy neural networks have been used in diverse applications of image processing [21]–[23].

Fuzzy MLP with error backpropagation system uses fuzzy inputs represented in the form of quantitative, linguistic, or a combination of these. Further, the system can learn the input output relationships and generate rules which can subsequently be used by a conventional expert system. A logical extension of multilayer perceptron, can be used to learn the input output fuzzy relationship in the form of if-then-else rules. The output is presented with a certain amount of confidence level. The novel feature of the network is to query for missing information. The neurons are programmed to do min-max and product probabilistic sum operators to represent the nodes at the hidden and output layers respectively. Backpropagation algorithm is suitably modified for logical operations. The rules generated from the fuzzy neural networks can be applied to rule based systems to overcome the problem of rule extraction from experts. Thus neuro-fuzzy networks have been found to be useful for generating rule base for expert systems.

Adaptive Resonance Theory based neural networks have evolved from the biological theory of cognitive information processing. Grossberg had first proposed this network in 1976.

10.10.1 Fuzzy Self Organising Feature Map

The Self Organising Feature Map, proposed by Kohonen, can be generalized To design a Fuzzy Kohonens network, which yields better performance compared to its crisp counterpart. The basic objective of this network is to cluster the input patterns, in such a way that the total Eucledian distance between each pattern and its nearest cluster centroid is minimum. Here the concept of winner is a fuzzy one.Each input pattern evokes partial response in more than one node and the weights of more than one node can be updated proportional to the response.The learning takes place according to the membership of the winner. The algorithm for Fuzzy Kohonen [22] is shown below.

Step 1: Initialize the neurons with random weights $w_I(0)$, $i = 1, \cdots, N$.

Step 2: Compute the distances $d(x_k, w_i)$ from the input pattern x_k to each of the competing neurons w_i.

Step 3: Compute the membership of the winner neuron based on the distance measure $d(x_k, w_i)$.

Step 4: Update the weight associated with each neuron.

The weight updataion is performed in accordance to the following updation rule.

$$w_i(t+1) = w_i(t) + \alpha(t) z_i(t)[x_k - w_i(t)] \tag{10.20}$$

where z_i is the fuzzy scaling function given by

$$z_i = (\mu_{ik})^{f_m} \tag{10.21}$$

where

$$\mu_{ik} = \frac{\left(\frac{1}{D_{ik}}\right)^{\frac{1}{f_m - 1}}}{\sum_{p=1}^{c} \left(\frac{1}{D_{pk}}\right)^{\frac{1}{f_m - 1}}} \tag{10.22}$$

and $D_{ik} = d(x_k, w_i)$. The scaling function z_i depends on the fuzzy generator f_m which is a real number greater than 1. Interestingly, the closer is the neuron to the input patter the larger is its win membership. During the learning process, all the neurons orient themselves towards the input pattern to gradually decreasing levels proportional to their distances.

10.11 SUMMARY

In this chapter we have introduced the fundamentals of Fuzzy set theory and some of their applications in Image Processing. The concepts of imprecision in images have been addressed here and the strategies adopted in image enhancement. Segmentation and pixel classisifaction using fuzzy set theory have been presented. Several membership assignment schemes have been discussed in detail. Various contrast enhancement strategies along with histogram based fuzzy enhancement and thresholding have been suggested. Fuzzy thresholding yields better results than their corresponding crisp counterparts. Also the pixel classifiction using fuzzy neural networks perform better than the crip networks. To fuzzy neural networks, viz., fuzzy self organizing feature map and fuzzy counterpropagation networks have been presented in this chapter.

REFERENCES

1. L. A. Zadeh, "Fuzzy Sets," *Information and Control*, 8, 1965, 338–353.

2. A. Kaufmann, *Introduction to the theory of fuzzy subsets: Fundamental theoretical elements*, Academic Press, Vol-1, New York, 1980.

3. G. J. Klir and B. J. Yaun, *Fuzzy sets and fuzzy logic: Theory and Application*, Prentice Hall of India, 1997.

4. J. C. Bezdek and S. K. Pal, *Fuzzy models for pattern recognition*, IEEE Press, Piscataway, NY, 1992.

5. S. K. Pal and R. A. King, "Image enhancement using smoothing with fuzzy sets", IEEE Transactions on systems man and cybernetics, Vol. SMC-11, July 1981, 494-=505.

6. T. Chaira and A. K. Ray, "Fuzzy Approach to color region extraction," *Pattern Recognition Letters*, 12(24), 2003, 1943-1950.

7. J. D. Fahnestock and R. A. Schowengerdt, "Spatially varying contrast enhancement using local range modification," *Optical Engineering*, 22(3), 1983, 378–381.

8. T. Chaira and A. K. Ray, "Threshold selection using fuzzy set theory," *Pattern Recognition Letters*, 25, 2004, 865–874.

9. T. Chaira, T and A. K. Ray, "Fuzzy measures for color image retrieval," *Fuzzy sets and systems*, 150(3), 2005, 545–560.

10. L. K. Huang and M. J. Wang, "Image thresholding by minimizing the measure of fuzziness," Pattern Recognition, 28(1), 1995, 41–51.

11. H. D. Cheng and H. H.Chen, "Image segmentation using fuzzy homogeneity criterion," *Information Science*, 98, 1997, 237–262.

12. A. D. Brink, "Thresholding of digital images using two-dimensional histogram and fuzzy entropy principle," *IEEE Trans. on Image Processing*, 9(4), 2000.

13. C. Chang, K. Chen, J. Wang, and M.L.G. Althouse, "A relative entropy based approach in image thresholding," *Pattern Recognition*, 27, 1994, 1275–1289.

14. J. Fan and W. Xie, "Distance measure and induced fuzzy entropy," *Fuzzy Sets and System*, 104, 1999, 305–314.

15. W. J. Wang, "New similarity measures on fuzzy sets and fuzzy elements," *Fuzzy sets and systems*, 85, 1997, 305–309.

16. C. V. Jawahar and A. K.Ray, "Fuzzy Statistics of Digital Images," *IEEE Signal Processing Letter*, 3, 1996, 225–227.

17. C. V. Jawahar and A. K. Ray, "Techniques and Applications of Fuzzy Statistics in Digital Image Analysis," *Fuzzy Theory Systems: Techniques and Applications*, Ed. C.T. Leondes, Academic Press, 12, 1999, 759–778.

18. C. V. Jawahar, P. K. Biswas, and A. K. Ray, "Analysis of Fuzzy Thresholding Schemes", *Pattern Recognition*, 33, 2000, 1339–1349.

19. J. Bezdek, *Pattern Recognition with Fuzzy Objective Function Algorithms*, Plenum press, New York, 1981.

20. N. R. Pal and J. C. Bezdek, "On cluster validity for the fuzzy C-means model," *IEEE Trans. Fuzzy Systems*, 3(3), 1995, 370–379.

21. J. J. Buckley and Y. Hayashi, "Fuzzy Neural Networks - A survey," *Fuzzy Sets and Systems*, 66, 1994, 1–13.

22. B. Mannan and A. K. Ray, "Crisp and fuzzy competitive Learning Networks for Classification of Multispectral IRS scenes", *Int J.Remote Sensing*, 24(17), 2003, 3491-3502.

23. S. Mitra and T. Acharya, *Data Mining: Multimedia, Soft Computing and Bioinformatics*, Wiley, Hoboken, NJ, 2003.

11

Image Mining and Content-Based Image Retrieval

11.1 INTRODUCTION

Tremendous amount of visual information in the form of image and video data are distributed all over the world like any other nonvisual data such as numeric and nonnumeric data, speech, voice, text, etc. *Data mining* is the field of study to extract valuable information from very large data set by discovering patterns and knowledge embedded into the data. *Image mining* broadly deals with extraction of the valuable information embedded in large image and video databases [1]. This promising field is still in the early stage of its development. Most of the work in this area has been restricted mainly in the development of *content-based image retrieval* (CBIR) systems.

Retrieval of a query image from a large database of images is an important task in the area of computer vision and image processing. The advent of large multimedia collection and digital libraries has led to an important requirement for development of search tools for indexing and retrieving information from them. A number of good search engines are available today for retrieving the text in machine readable form, but there are not many fast tools to retrieve intensity and color images. The traditional approaches to searching and indexing images are slow and expensive. Thus there is continued need to develop efficient algorithms in image mining and CBIR.

The indexing and retrieval of images usually seeks to find semantic information. Retrieving this semantic information and automatic segmentation of the image into objects using machine vision techniques is a nontrivial problem today. Many image attributes such as color, shape, and texture are having direct correlation with semantics embedded in the image. For example, the skin of tigers has a unique texture wherever they are found and using this image attribute, it may be possible to index and retrieve the images of tigers.

Image retrieval using similarity measures is an elegant technique used in *content-based image retrieval* (CBIR). Ideally a CBIR system should automatically extract the semantic information about the images for a specific application area.

To a very large extent, the low-level image features such as color, texture, and shape are widely used for CBIR. While attempting the task of image retrieval, we identify the mutual correspondence between two images in a set of database images using similarity relations. The content-based query system processes a query image and assigns this unknown image to the closest possible image available in the database. In view of the above discussion, it may be concluded that selection and extraction of low-level image features constituting the image and subsequent similarity-based matching and classification are the two issues in content-based image retrieval [1]–[8].

11.2 IMAGE MINING

Traditional data mining techniques have been developed mainly for structured datatypes. The image datatype does not belong to this structured category, suitable for interpretation by a machine, and hence the mining of image data is a challenging problem. Content of an image is visual in nature and the interpretation of the information conveyed by an image is mainly subjective, based on the human visual system.

Image data have been used for machine vision, based on extraction of desired features from an image and interpretation of these features for particular applications. This is a challenging area of study, and it has been extensively explored in pattern recognition and machine vision for quite some time. Although interpretation of the image content by the human visual system is a natural and apparently effortless procedure, it remains a mystery how the human brain processes this information. Hence modeling the process of human interpretation of the semantic content of images is still a research challenge. As a result, it is difficult to define a single set of algorithms or functionalities that can claim to comprise a complete set of image mining tools.

The research and development of mining image data is relatively new, and has become an emerging field of study today. Most of the activities in mining image data have been in the search and retrieval of images based on the analysis of similarity of a query image or its feature(s) with the entries in the image database. The image retrieval systems can be broadly categorized into two categories based on the type of searches.

In the first category, the images are described based on user-defined texts [3, 4]. The images are indexed and retrieved based on these rudimentary *descriptions*, such as their size, type, date and time of capture, identity of owner, keywords, or some text description of the image. As a result, this is often called *description based* or *text-based image retrieval* process. The image indices are predefined based on these descriptions, and they are searched on these indices when a query is posed as

Find the images from an image database which matches with the given set of descriptions, e.g. images captured on January 8 to June 30, 2005 and size bigger than 100 Kbyte.

The text-based descriptions of the images are usually typed manually for each image by human operators, because the automatic generation of keywords for the images is difficult without incorporation of visual information and feature extraction. As a result, this is a very labor-intensive process and is impractical in today's multimedia information age. Moreover, since the description of images is very much subjective, the automated process to generate a text-based description for indexing of the images could be very inaccurate and incomplete.

In the second category, the query can be posed as

Find the images similar to a given query image.

This second category of similarity-based image retrieval process is the *content-based image retrieval* (CBIR) [5]–[9]. In CBIR systems, the images are searched and retrieved based on the visual content of the images. Based on these visual contents, desirable images features can be extracted and used as index or basis of search.

There are, in general, three fundamental modules in a content-based image retrieval system:

1. Visual content or feature extraction

2. Multidimensional indexing and

3. Retrieval

The images in an image database are indexed-based on extracted inherent visual contents (or features) such as color, texture, pattern, image topology, shape of objects, and their layouts and locations within the image, etc. An image can be represented by a multidimensional vector of the extracted features from the image. The feature vector actually acts as the *signature* of the image. This feature vector can be assumed to be associated to a point in the multidimensional space. As an example, an image can be represented by an N-dimensional feature vector whose first n_1 components may represent *color*, the next n_2 components may represent *shape*, the following n_3 components may represent some image *topology*, and finally n_4 components may represent *texture* of the image, so that there are $N = n_1 + n_2 + n_3 + n_4$ components.

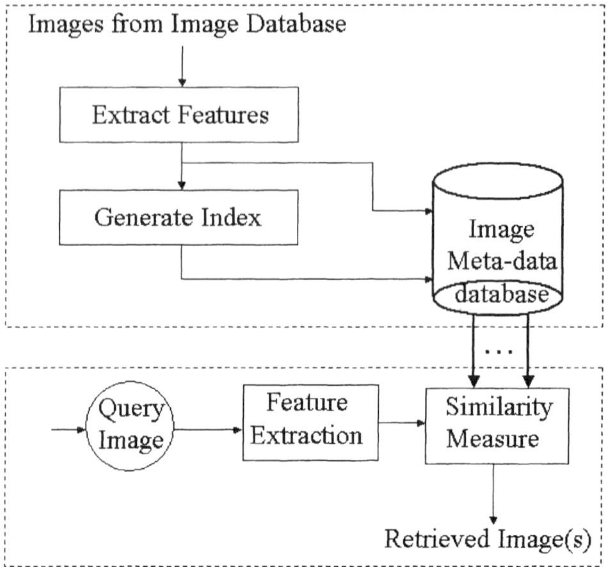

Fig. 11.1 Architecture of a content-based image retrieval system.

As a result, an *example image* can be simply used as a query using visual content-based indexing.

The query image can be analyzed to extract the visual features and can be compared to find matches with the indices of the images stored in the database. The extracted image features are stored as metadata, and images are indexed based on this metadata information. This metadata information comprises some measures of the extracted image features. The feature vectors of similar images will then be clustered in the N-dimensional space. Retrieving *similar* images to a *query* image then boils down to finding the indices of those images in the N-dimensional search space whose feature vectors in the N-dimensional space are within some threshold of proximity to the point of the query image. This indexing structure is popularly known as *multidimensional access structure* (MAS) [10].

The architecture for a possible content-based image retrieval system is shown in Figure 11.1. The CBIR systems architecture is essentially divided into two parts. In the first part, the images from the image database are processed offline. The features from each image in the image database are extracted to form the metadata information of the image, in order to describe the image using its visual content features. Next these features are used to index the image, and they are stored into the metadata database along with the images. In the second part, the retrieval process is depicted. The query image is analyzed to extract the visual features, and these features are used to retrieve the similar images from the image database. Rather than directly comparing two images, similarity of the visual features of the query image is measured with the features of each image stored in the metadata database as their signatures. Often the similarity of two images is measured by computing the distance between the feature vectors of the two images. The retrieval systems return the first k images, whose distance from the query image is below some defined threshold.

Several image features have been used to index images for content-based image retrieval systems. Most popular among them are color, texture, shape, image topology, color layout, region of interest, etc. We discuss some of these features in greater detail in the following sections.

11.3 IMAGE FEATURES FOR RETRIEVAL AND MINING

11.3.1 Color Features

Color is one of the most widely used visual features in content-based image retrieval [11]–[14]. While we can perceive only a limited number of gray levels, our eyes are able to distinguish thousands of colors and a computer can represent even millions of distinguishable colors in practice. Color has been successfully applied to retrieve images, because it has very strong correlations with the underlying objects in an image. Moreover, color feature is robust to background complications, scaling, orientation, perspective, and size of an image.

Although we can use any color space for computation of a color histogram HSV (hue, saturation, value), HLS (hue, lightness, saturation), and CIE color spaces (such as CIELAB, CIELUV) have been found to produce better results as compared to the RGB space. Since these color spaces are visually (or perceptually) uniform compared to the RGB, they are found to be more effective to measure color similarities between images.

11.3.1.1 Color Histogram This is the most commonly used color feature in CBIR [11, 12]. Color histogram has been found to be very effective in characterizing the global distribution of colors in an image, and it can be used as an important feature for image characterization. To define color histograms, the color space is quantized into a finite number of discrete levels. Each of these levels becomes a bin in the histogram. The color histogram is then computed by counting the number of pixels in each of these discrete levels. There are many different approaches to quantize a color space to determine the number of such discrete levels [11]–[13].

Using the color histogram, we can find the images that have similar color distribution. One can think of the simplest measure of similarity by computing the distance between two histograms. Let us consider that $H^{(1)} = \{h_1^{(1)}, h_2^{(1)}, \cdots, h_K^{(1)}\}$ and $H^{(2)} = \{h_1^{(2)}, h_2^{(2)}, \cdots, h_K^{(2)}\}$ are two feature vectors

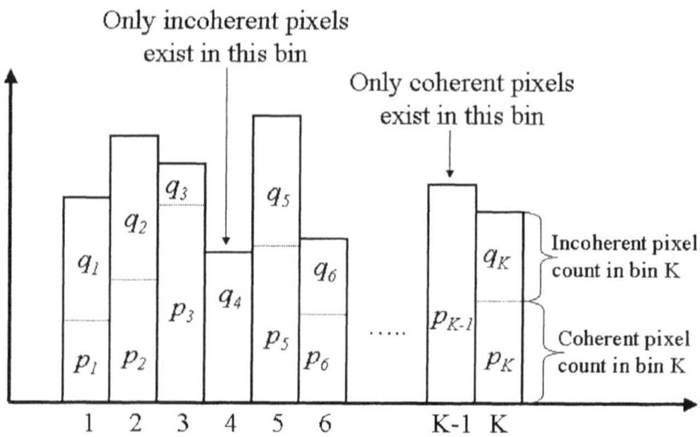

Fig. 11.2 Partitioning the color histogram with coherence and noncoherence pixel counts.

generated from the color histograms of two images, where $h_j^{(1)}$ and $h_j^{(2)}$ are the count of pixels in the jth bin of the two histograms respectively, and K is the number of bins in each histogram. We can define a simple distance between two histograms as

$$d_1 = \sum_{j=1}^{K} |h_j^{(1)} - h_j^{(2)}|. \tag{11.1}$$

There is another distance measure between two histograms, popularly known as *histogram intersection*. The histogram intersection is the total number of pixels common to both the histograms. This can be computed as

$$I(H^{(1)}, H^{(2)}) = \sum_{j=1}^{K} \min(h_j^{(1)} - h_j^{(2)}). \tag{11.2}$$

The above equations can be normalized to maintain the value of the distance measure in the range $[0, 1]$. To normalize $I(H^{(1)}, H^{(2)})$, it is divided either by the total number of pixels in one of the histograms or by the size of the image.

11.3.1.2 Color Coherence Vector One problem with the color histogram-based similarity measure approach is that the global color distribution doesn't reflect the spatial distribution of the color pixels locally in the image. This cannot distinguish whether a particular color is sparsely scattered all over the image or it appears in a single large region in the image.

The color coherence vector-based [14] approach was designed to accommodate the information of spatial color into the color histogram. Here we can classify each pixel in an image, based on whether it belongs to a large uniform region. For example, we can consider a region to be uniformly colored if it consists of the same color and the area of the region is above a certain threshold (say, 2%) of the whole image area. We refer to the pixels in these regions as *coherent* pixels.

In this approach, each histogram bin is divided into two parts. One contains the count of pixels belonging to a large uniformly colored region and the other contains the same colored pixels belonging to a sparse region. Let us consider that the ith bin of the histogram contains p_j coherent pixels and q_j incoherent pixels. Using this partition, the *color coherence vector* of an image can be expressed as $\{(p_1, q_1), (p_2, q_2), \cdots, (p_K, q_K)\}$. We show this in Figure 11.2. It should be noted that $\{p_1 + q_1, p_2 + q_2, \cdots, p_K + q_K\}$ is the original color histogram without this distinguishing power. The color coherence vectors provide superior retrieval results with this additional distinguishing capability, as compared to the global color histogram method [14].

11.3.1.3 Color Moment This is a compact representation of the color feature to characterize a color image [12]. It has been shown that most of the color distribution information is captured by the three low-order moments. The first-order moment (μ) captures the mean color, the second-order moment (σ) captures the standard deviation, and the third-order moment captures the skewness (θ) of color. These three low-order moments (μ_c, σ_c, θ_c) are extracted for each of the three color planes, using the following mathematical formulation.

$$\mu_c = \frac{1}{MN} \sum_{i=1}^{M} \sum_{j=1}^{N} p_{ij}^c, \tag{11.3}$$

$$\sigma_c = \left[\frac{1}{MN} \sum_{i=1}^{M} \sum_{j=1}^{N} (p_{ij}^c - \mu_c)^2 \right]^{\frac{1}{2}}, \tag{11.4}$$

$$\theta_c = \left[\frac{1}{MN} \sum_{i=1}^{M} \sum_{j=1}^{N} (p_{ij}^c - \mu_c)^3 \right]^{\frac{1}{3}}, \tag{11.5}$$

where p_{ij}^c is value of the cth color component of the color pixel in the ith row and jth column of the image. As a result, we need to extract only nine parameters (three moments for each of the three color planes) to characterize the color image. Weighted Euclidean distance between the color moments of two images has been found to be effective to calculate color similarity [12].

11.3.1.4 Linguistic Color Tag The global color distribution using color histogram does not take advantage of the fact that the adjacent histogram bins might actually represent roughly the same color, because of the limited ability of the human perceptual system. Moreover, only a limited number of color shades are sufficient for visual discrimination between two images. To take advantage of this, a color matching technique based on *linguistic tags*, to identify a color with a name, has been proposed recently [15]. The concept behind this technique is to construct *equivalence classes* of colors, which are identified by *linguistic tags* (or color name such as pink, maroon, etc.), that perceptually appear the same to the human eye but are distinctly different from that of neighboring subspaces. Using this approach, the dimensionality of color features is significantly reduced. This also helps to reduce computations for color similarity measures. A color palette of 15 colors corresponding to 15 equivalent classes was developed by Iqbal and Aggarwal using this technique, and was applied effectively in color similarity measures to distinguish color images [15].

11.3.2 Texture Features

Texture is a very interesting image feature that has been used for characterization of images, with application in content-based image retrieval. There is no single formal definition of *texture* in the literature. However, a major characteristic of texture is the *repetition of a pattern or patterns over a region* in an image. The elements of patterns are sometimes called *textons.* The size, shape, color, and orientation of the textons can vary over the region. The difference between two textures can be in the degree of variation of the textons. It can also be due to spatial statistical distribution of the textons in the image. Texture is an innate property of virtually all surfaces, such as bricks, fabrics, woods, papers, carpets, clouds, trees, lands, skin, etc. It contains important information regarding underlying structural arrangement of the surfaces in an image. When a small area in an image has wide variation of discrete tonal features, the dominant property of that area is *texture.* On the other hand, the *gray tone* is a dominant property when a small area in the image has very small variation of discrete tonal features. Texture analysis has been an active area of research in pattern recognition since the 1970s [16, 17].

A variety of techniques have been used for measuring textural similarity. In 1973, Haralick et al. proposed co-occurrence matrix representation of texture features to mathematically represent gray level spatial dependence of texture in an image [16]. In this method the co-occurrence matrix is constructed based on the orientation and distance between image pixels. Meaningful statistics are extracted from this co-occurrence matrix, as the representation of texture. Since basic texture patterns are governed by periodic occurrence of certain gray levels, co-occurrence of gray levels at predefined relative positions can be a reasonable measure of the presence of texture and periodicity of the patterns.

Several texture features such as *entropy*, *energy*, *contrast*, and *homogeneity*, can be extracted from the co-occurrence matrix of gray levels of an image. The gray level co-occurrence matrix $C(i\,j)$ is defined by first specifying a displacement vector $d_{x,y} = (\delta x,\ \delta y)$ and then counting all pairs of pixels separated by displacement $d_{x,y}$ and having gray levels i and j. The matrix $C(i\,j)$ is normalized by dividing each element in the matrix by the total number of pixel pairs. Using this co-occurrence matrix, the texture features metrics are computed as follows.

$$Entropy = -\sum_i \sum_j C(i,j) \log C(i,j), \tag{11.6}$$

$$Energy = \sum_i \sum_j C^2(i,j), \tag{11.7}$$

$$Contrast = \sum_i \sum_j (i-j)^2 C(i,j), \tag{11.8}$$

$$Homogeneity = \sum_i \sum_j \frac{C(i,j)}{1+|i-j|}. \tag{11.9}$$

Practically, the co-occurrence matrix $C(i,\ j)$ is computed for several values of displacement $d_{x,y}$, and the one which maximizes a statistical measure is used.

Tamura et al. proposed computational approximations to the texture features, based on the psychological studies in visual perception of textures [17]. The texture properties they found visually meaningful for texture analysis are *coarseness*, *contrast*, *directionality*, *linelikeness*, *regularity*, and *roughness*. These texture features have been used in many content-based image retrieval systems [5, 6]. Popular signal processing techniques have also been used in texture analysis and extraction of visual texture features. Wavelet transforms have been applied in texture analysis and classification of images, based on multiresolution decomposition of the images and representing textures in different scales [18]–[21]. Among the different wavelet filters, Gabor filters were found to be very effective in texture analysis.

11.3.3 Shape features

Shape is another image feature applied in CBIR. Shape can roughly be defined as the description of an object minus its position, orientation and size. Therefore, shape features should be invariant to *translation*, *rotation*, and *scale*, for an effective CBIR, when the arrangement of the objects in the image are not known in advance. To use *shape* as an image feature, it is essential to segment the image to detect object or region boundaries; and this is a challenge. Techniques for shape characterization can be divided into two categories.

The first category is *boundary-based*, using the outer contour of the shape of an object. The second category is *region-based*, using the whole shape region of the object. The most prominent representatives of these two categories are *Fourier descriptors* [22] and *moment invariants* [23]. The main idea behind the *Fourier descriptors* is to use the Fourier-transformed boundaries of the objects as the shape features, whereas the idea behind *moment invariants* is to use region-based geometric moments that are invariant to translation and rotation. Hu identified seven normalized central moments as shape features, which are also scale invariant. We provide expressions for these seven invariants below.

11.3.3.1 Moment invariants Let $F(x, y)$ denote an image in the two-dimensional spatial domain. *Geometric moment* [1] of order $p + q$ is denoted as

$$m_{p,q} = \sum_x \sum_y x^p y^q F(x, y), \tag{11.10}$$

for $p,\ q = 0, 1, 2, \ldots$. The central moments are expressed as

$$\mu_{pq} = \sum_x \sum_y (x - x_c)^p (y - y_c)^q F(x, y), \tag{11.11}$$

where $x_c = \frac{m_{1,0}}{m_{0,0}}$, $y_c = \frac{m_{0,1}}{m_{0,0}}$, and $(x_c,\ y_c)$ is called the center of the region or object. Hence the *central moments*, of order up to 3, can be computed as

$$\left.\begin{array}{lcl}
\mu_{0,0} & = & m_{0,0} \\
\mu_{1,0} & = & 0 \\
\mu_{0,1} & = & 0 \\
\mu_{2,0} & = & m_{2,0} - x_c m_{1,0} \\
\mu_{0,2} & = & m_{0,2} - y_c m_{0,1} \\
\mu_{1,1} & = & m_{1,1} - y_c m_{1,0} \\
\mu_{3,0} & = & m_{3,0} - 3x_c m_{2,0} + 2m_{1,0} x_c^2 \\
\mu_{1,2} & = & m_{1,2} - 2y_c m_{1,1} - x_c m_{0,2} + 2y_c^2 m_{1,0} \\
\mu_{2,1} & = & m_{2,1} - 2x_c m_{1,1} - y_c m_{2,0} + 2x_c^2 m_{0,1} \\
\mu_{0,3} & = & m_{0,3} - 3y_c m_{0,2} + 2y_c^2 m_{0,1}.
\end{array}\right\}. \tag{11.12}$$

The *normalized central moments*, denoted $\eta_{p,q}$, are defined as

$$\eta_{p,q} = \frac{\mu_{p,q}}{\mu_{0,0}^{\gamma}}, \tag{11.13}$$

where

$$\gamma = \frac{p + q + 2}{2} \tag{11.14}$$

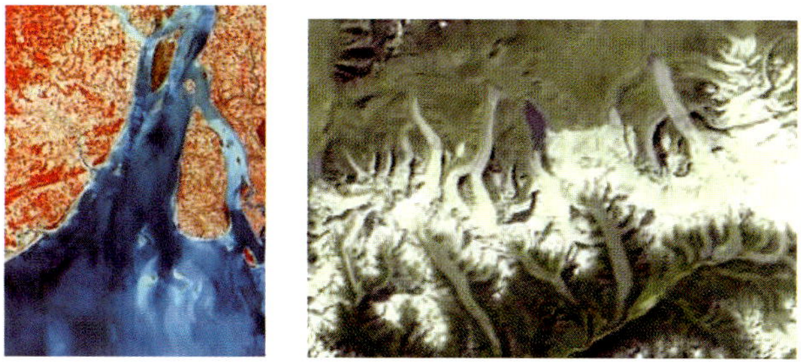

(a) (b)

Fig. 1.1 Example of a remotely sensed image of (a) delta of river Ganges, (b) Glacier flow in Bhutan Himalayas. *Courtesy: NASA/GSFC/METI/ERSDAC/JAROS, and U.S./Japan ASTER Science Team.*

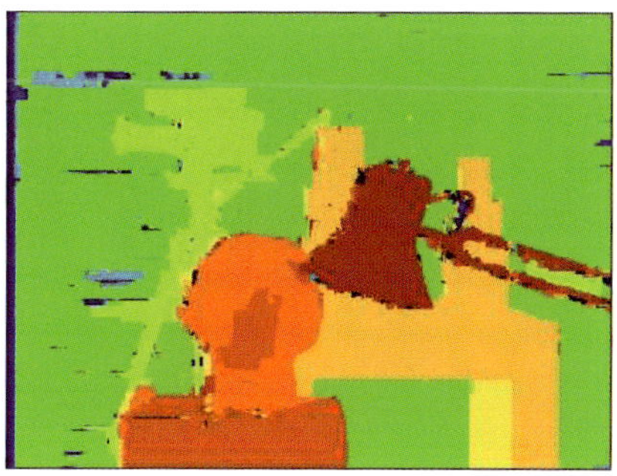

Fig. 2.13 The depth map image from a range camera.

Fig. 11.5 Result of color image retrieval by fuzzy Similarity measure.

(a) (b)

Fig. 13.1 Remotely sensed images from MODIS

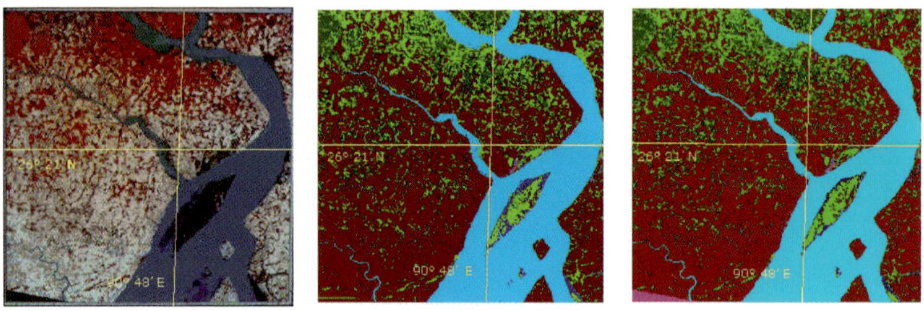

(a) (b) (c)

Fig. 13.2 Satellite Image (a) original, (b) segmented by multilayered perceptron (error backpropagation), (c) segmented by Fuzzy self-organizing feature map.

Fig. 14.2 Results of particle filter based tracking - moving human figure encapsulated in rectangular bounding boxes at an interval of 10 frames.

(a)

(b)

Fig. 3.4 (a) Original Image and (b) result of fuzzy assignment-based adaptive interpolation.

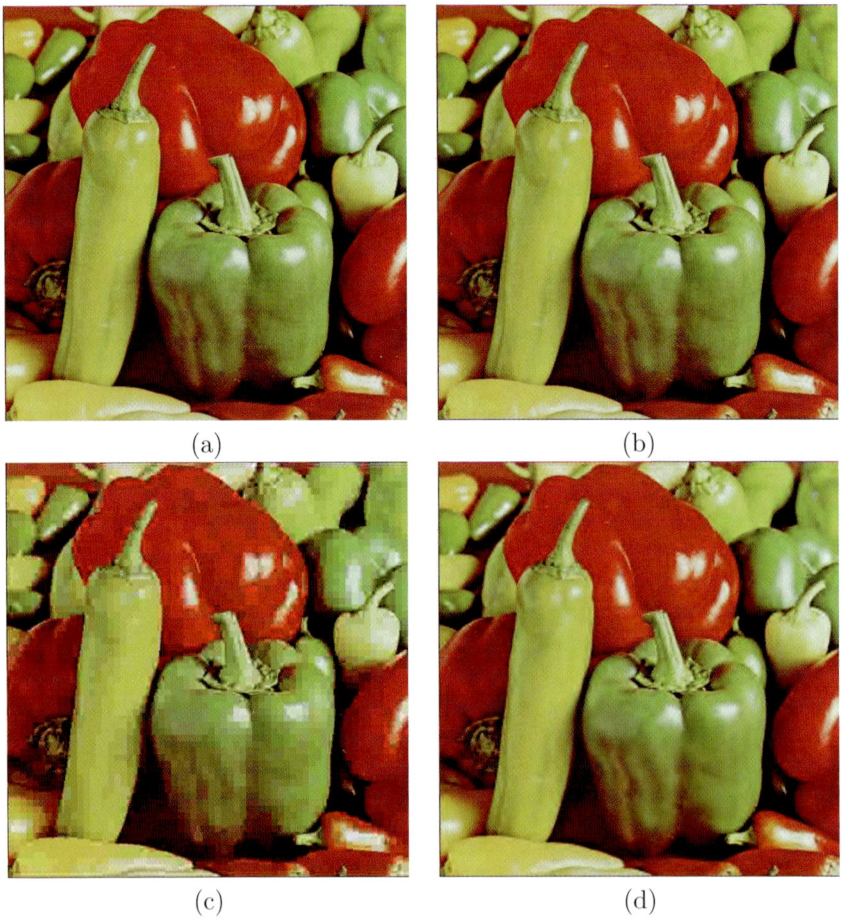

Fig. 15.6 (a) Original Peppers image, (b) compressed with baseline JPEG using quality factor 75 (1.57 bit/pixel), (c) compressed with baseline JPEG using quality factor 10 (0.24 bit/pixel), and (d) compressed with the new JPEG2000 standard using the same bit rate (0.24 bit/pixel).

for $p + q = 2, 3, \ldots$. A set of seven *transformation invariant moments* can be derived from the second- and third-order moments as follows:

$$\left.\begin{array}{rcl}\phi_1 &=& (\eta_{2,0} + \eta_{0,2}) \\ \phi_2 &=& (\eta_{2,0} - \eta_{0,2})^2 + 4\eta_{1,1}^2 \\ \phi_3 &=& (\eta_{3,0} - 3\eta_{1,2})^2 + (3\eta_{2,1} - \eta_{0,3})^2 \\ \phi_4 &=& (\eta_{3,0} + \eta_{1,2})^2 + (\eta_{2,1} + \eta_{0,3})^2 \\ \phi_5 &=& (\eta_{3,0} - 3\eta_{1,2})(\eta_{3,0} + \eta_{1,2})[(\eta_{3,0} + \eta_{1,2})^2 - 3(\eta_{2,1} + \eta_{0,3})^2] \\ && +(3\eta_{2,1} - \eta_{0,3})(\eta_{2,1} + \eta_{0,3})[3(\eta_{3,0} + \eta_{1,2})^2 - (\eta_{2,1} + \eta_{0,3})^2] \\ \phi_6 &=& (\eta_{2,0} - \eta_{0,2})[(\eta_{3,0} + \eta_{1,2})^2 - (\eta_{2,1} + \eta_{0,3})^2] \\ && +4\eta_{1,1}(\eta_{3,0} + \eta_{1,2})(\eta_{21} + \eta_{03}) \\ \phi_7 &=& (3\eta_{2,1} - \eta_{0,3})(\eta_{3,0} + \eta_{1,2})[(\eta_{3,0} + \eta_{1,2})^2 - 3(\eta_{2,1} + \eta_{0,3})^2] \\ && +(3\eta_{1,2} - \eta_{3,0})(\eta_{2,1} + \eta_{0,3})[3(\eta_{3,0} + \eta_{1,2})^2 - (\eta_{2,1} + \eta_{0,3})^2]. \end{array}\right\} . \tag{11.15}$$

This set of normalized central moments is invariant to translation, rotation, and scale changes in an image.

In addition to the geometric moments, the circularity, aspect ratio, symmetricity, and concavity are also used for segmentation and shape detection in images. We exclude detailed discussions on these features here.

The problem with shape-based CBIR system development is that the shape features need very accurate segmentation of images to detect the object or region boundaries. Image segmentation is an active area of research and most of the segmentation algorithms are still computationally very expensive for online image segmentation. Robust and accurate segmentation of images still remains a challenge in computer vision. As a result, shape feature-based image retrieval has been mainly limited to image databases where objects or regions are readily available.

11.3.4 Topology

A digital image can be represented by one or more topological properties [1], which typically represent the geometric shape of an image. The interesting characteristic of topological properties is that when changes are made to the image itself, such as stretching, deformation, rotation, scaling, translation, or other rubber-sheet transformations, these properties of the image do not change. As a result, topological properties can be quite useful in characterization of images and can be used as a signature of an image content to use in content-based image retrieval.

One topological property of a digital image is known as *Euler number* [1]. The *Euler number* is usually computed in a binary image. However, it can be extended to characterize gray-tone images as well by defining a vector of Euler numbers of the binary planes of the gray-tone image. This has been called the *Euler vector* [24]. The *Euler number* is defined as the difference between number of *connected components* and number of *holes* in a binary image. Hence if an image has C connected components and H number of

Fig. 11.3 Sample binary images with (a) Euler number 0 and (b) Euler number 2.

holes, the *Euler number* E of the image can be defined as

$$E = C - H. \tag{11.16}$$

Euler number of the binary image in Figure 11.3(a) is 0 because the image has one connected component and one hole, whereas the Euler number of Figure 11.3(b) is 2 because it has seven connected components and five holes. The binary image of letter B will have Euler number -1 because it has one connected component and two holes.

Euler number remains invariant despite the transformation of the image due to translation, rotation, stretching, scaling, etc. For some classes of digital images, Euler numbers have strong discriminatory power. In other words, once the Euler number for a particular digital image is known, the digital image may be readily distinguished from other digital images in its class. This implies that the Euler number may be used for more efficient searching or matching of digital images. For example, the Euler number may be used in medical diagnosis such as the detection of malaria infected cells. As the Euler number of an infected cell is often different from that of a normal cell, the malaria infected cells may be identified by calculating the Euler number of each cell image. Euler number may also be used for image searching, such as in a database of logo images.

11.3.4.1 Euler Vector Above usage of Euler number has been extended to gray-tone images, by defining a vector of Euler numbers [24]. This vector is called the *Euler vector*. Intensity value of each pixel in an 8-bit gray-tone image can be represented by an 8-bit binary vector b_i, $i = 0, 1, \cdots, 7$, that is, $(b_7, b_6, b_5, b_4, b_3, b_2, b_1, b_0)$, where $b_i \in \{0, 1\}$. The ith bit-plane is formed with b_i's from all the pixels in the gray-tone image. To define Euler vector, we retain the first four most significant bit-planes corresponding to (b_7, b_6, b_5, b_4), because they contain most of the information of the image. This 4-bit binary vector is converted to its corresponding reflected gray code (g_7, g_6, g_5, g_4), where $g_7 = b_7$, $g_6 = b_7 \bigotimes b_6$, $g_5 = b_6 \bigotimes b_5$, $g_4 = b_5 \bigotimes b_4$, and $\bigotimes$ denotes the binary XOR (modulo-2) operation.

11.3.4.2 Definition Euler vector of a gray-tone image is a 4-tuple (E_7, E_6, E_5, E_4), where E_i is the Euler number of the bit-plane formed with reflected gray codes g_i of all the pixels in the image.

Gray code representation of intensity values offers a distinct advantage over standard binary representation in this particular context. Euler vector is found to be more insensitive to noise and other changes, when the gray code is used. This happens because two consecutive numbers have unit Hamming distance in gray-code representation and, for most of the cases, a small change in intensity values cannot affect all the 4 bit planes simultaneously in gray representation. Euler vector can be used as a quick combinatorial signature of an image and has been used for image matching in content-based image retrieval [24].

11.3.5 Multidimensional Indexing

Multidimensional indexing is an important component of content-based image retrieval. Development of indexing techniques has been an active area in database management, computational geometry, and pattern recognition. However, the notion of indexing has subtle differences in different communities. The notion of indexing in multimedia data mining and content-based image retrieval is different from its notion in the traditional database management systems. In traditional database management systems (particularly for relational databases), the indexing refers to the access structure of the database files in terms organization of the records. Indexes are specified based on one or more attributes of the records in order to process queries based on those attributes. These record and file structures are well organized and supported by an access structure such as hashing, B-tree, etc. In the information retrieval community, the indexing mechanism is concerned with the process to assign terms (or phrases or keywords or descriptors) to a document so that the document can be retrieved based on these terms. The indexing in content-based image retrieval or mining multimedia data is similar to the notion adopted in the information retrieval. The primary concern of indexing is to assign a suitable description to the data in order to detect the information content of the data. As we explained in the previous sections, the descriptors of the multimedia data are extracted based on certain features or feature vectors of the data. These content descriptors are then organized into a suitable access structure for retrieval.

The key issues in indexing for content-based image retrieval are (1) reduction of high dimensionality of the feature vectors, (2) finding an efficient data structure for indexing, and (3) finding suitable similarity measures.

In CBIR, the dimensionality of feature vectors is normally very high. Today the dimensionality is typically of the order of 10^2. With the exploration of multimedia content, this order may grow in future. Before indexing, it is very important to reduce the dimensionality of the feature vectors. The most popular approach to reduce high dimensionality is application of the principal component analysis, based on singular decomposition of the feature matrices. The theory behind singular value decomposition of a matrix and generation of principal components for reduction of high dimensionality has been discussed in detail in Section 4.5. The technique has also been elaborated in Section ?? with regard to text mining. This can be applied to both text and image datatypes in order to reduce the high dimensionality of the feature vectors and hence simplify the access structure for indexing the multimedia data.

After dimensionality reduction, it is very essential to select an appropriate multidimensional indexing data structure and algorithm to index the feature vectors. There are a number of approaches proposed in the literature. The most popular among them are multidimensional binary search trees [25], R-tree [26], variants of R-tree such as R^*-Tree [27], SR-tree [28], SS-tree [29], Kd-tree [30], etc. All these indexing methods provide reasonable performance for dimensions up to around 20, and the performance deteriorates after that. Moreover, most of these tree-based indexing techniques have been designed for traditional database queries such as point queries and range queries, but not for similarity queries for multimedia data retrieval. There have been some limited efforts in this direction. Multimedia database indexing particularly suitable for data mining applications remains a challenge. So exploration of new efficient indexing schemes and their data structures will continue to be a challenge for the future.

After indexing of images in the image database, it is important to use a proper similarity measure for their retrieval from the database. Similarity measures based on statistical analysis have been dominant in CBIR. Distance measures such as Euclidean distance, Mahalanobis distance, Manhattan distance, and similar techniques have been used for similarity measures. Distance of histograms and histogram intersection methods have also been used for this purpose, particularly with color features.

Another aspect of indexing and searching is to have minimum disk latency while retrieving similar objects. Chang et al. proposed a clustering technique to cluster similar data on disk to achieve this goal, and they applied a hashing technique to index the clusters [31]. In spite of lots of development in this area, finding new and improved similarity measures still remains a topic of interest in computer science, statistics, and applied mathematics.

11.3.6 Results of a Simple CBIR System

Table 11.1 CBIR match performance (%)

Image Group	A	B	$A+B$	$A+B+C$
AIRPLANE	71.3	69.68	73.39	73.23
CAR	78.90	74.11	78.30	82.56
FLOWER	51.22	38.64	43.64	43.48
ANIMAL	25.44	48.45	48.89	49.56
Overall	62.70	38.90	64.80	66.55

A: Moment invariants only.
B: Circularity, symmetricity, aspect ratio and concavity.
C: Energy, entropy, contrast and homogeneity.

A simple image matching scheme for an experimental content-based image retrieval system that uses moment, shape, and texture features extracted from the images, has been discussed in [32]. The shape features have been extracted using moment computation (feature set A) of the regions, combined with circularity, aspect ratio, concavity, and symmetricity metrics (feature set B). The texture of the images has been generated using the energy, entropy, contrast, and homogeneity measures as image features (set C). We present here results of this system using around 290 samples taken from an image database containing several classes of images, including animals, cars, flowers, etc. Distances of the query image from the database images have been computed simply as $\sum_i \mid f_i - f'_i \mid$, where f_i and f'_i are the values of the ith feature of the database image and query image respectively. The top 10 closest images have been taken as the query result, excluding the query image itself if it is present in the database.

Table 11.2 CBIR weighted match performance (%)

Image Group	All Features (No weight)	All features (with weight)
AIRPLANE	73.23	69.84
CAR	82.56	86.75
FLOWER	43.48	77.03
ANIMAL	49.56	72.44
Overall	66.55	77.79

Tables 11.1 and 11.2 provide the results, where each database image is used as a query image to find the top 10 from the database. From Table 11.1, we

observe that the overall performance can be improved by mixing the different feature sets for image query.

Further improvement can be achieved by assigning weights to each feature as follows:

$$\sum_i W_{gi} \mid f_i - f_i' \mid,$$

where W_{gi} is the weight of the ith feature for image group g and the database image belongs to image group g. Here W_{gi} is computed as $\frac{1}{1 + \sigma_{gi}}$, where σ_{gi} is the standard deviation of the ith feature value of the images in group g. Table 11.2 indicates that this improves the matching performance.

The summary of the results is presented in Figure 11.4 for eight query images. In each row the first column indicates the query image. For all rows the five best matches are shown, with the query image (as all of them are present in the database) naturally coming first as the best match.

11.4 FUZZY SIMILARITY MEASURE IN AN IMAGE RETRIEVAL SYSTEM

In an image retrieval system, a dababase image containing a set of shape, texture and color features is created. The features of the query image are next matched with the features of each image in the database. Different similarity functions which are essentially mappings between a pair feature vectors and a positive real-valued number yield a measure of the visual similarity between the two images. As an example, let us consider the color histogram as the feature. In cases where the overall color distribution of an image is more important regardless of the spatial arrangement in the image, then indexing using global color distribution will be more useful; otherwise, the local distribution of colors may be used as features. A global color histogram represents M number of intensity histograms, where M is the number of colors (usually $M = 3$ for an RGB image). Some of the fuzzy logic-based similarity measures between two images are explained below [33].

- **Fuzzy similarity measure based on Min-Max ratio:** The similarity between two images A and B is given by

  $$S_1(A, B) = \frac{\sum_{N=1}^{L} \min\{\mu_A(N), \mu_B(N)\}}{\sum_{N=1}^{L} \max\{\mu_A(N), \mu_B(N)\}},$$

  where $\mu_A(N)$ and $\mu_B(N)$ are fuzzy membership values of the Nth gray level of the histogram A_N and B_N of the two images A and B respectively. For an identical pair of images, the similarity value will be 1.

- **Similarity measure based on normalized absolute difference:** The similarity measure based on normalized absolute difference between

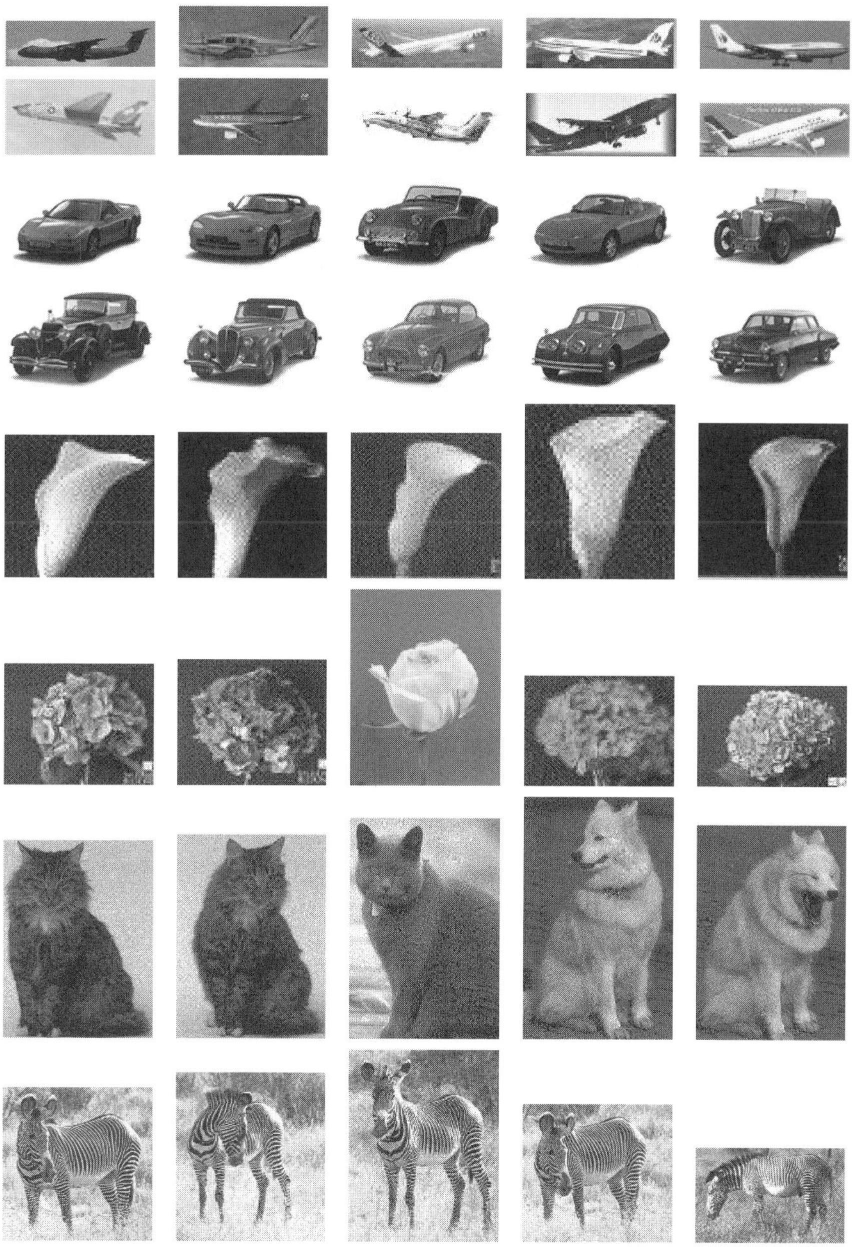

Fig. 11.4 Results of image matching in the CBIR system.

the fuzzy membership values of gray levels, computed from the histograms, is as below:

$$S_2(A,B) = 1 - \frac{\sum_{N=1}^{L} |\mu_A(N) - \mu_B(N)|}{\sum_{N=1}^{L} |\mu_A(N) + \mu_B(N)|}.$$

The variant of the above measure can be expressed as

$$S_3(A,B) = 1 - \frac{\sum_{N=1}^{L} |\mu_A(N) \bigcup \mu_B(N) - \mu_A(N) \bigcap \mu_B(N)|}{\sum_{N=1}^{L} |\mu_A(N) + \mu_B(N)|},$$

where intersection represents minimum and union represents maximum of two membership values. It can be shown that if $A_N \subseteq B_N \subseteq C_N$, corresponding to the histogram A_N, B_N, and C_N respectively for $N = \{0, 1, \cdots, L-1\}$ of three images A, B, and C then $S_1(A,B) \geq S_1(A,C)$ and $S_1(B,C) \geq S_1(A,C)$. This is an interesting property of the fuzzy similarity measure which has been stated earlier. Also it is easy to show that the similarity measures $S(A,B)$ between two images A and B are symmetric.

- **Similarity measure based on Tversky's model:** A feature contrast model has been characterized as a set of binary point features by Tversky [33]. Here a feature set has been considered as a set of logic predicates which are true for a set of stimuli (such as the binary features of an image). Let a and b be two stimuli and A and B are the sets of the features. A similarity measure $s(a,b)$ between a and b has been suggested as having the following properties.

  – Matching: $s(a,b) = F(A \bigcap B, A - B, B - A)$

  – Monotonicity: $s(a,b) > s(a,c)$, whenever $A \bigcap C \subseteq A \bigcap B$, $A - B \subseteq A - C$, $B - A \subseteq C - A$.

  A function that satisfied matching and monotonicity is called a matching function.

  Similarity based on Tversky's model has been extended to the similarity between gray and color images in [33] based on the generalized Tversky's index (GTI). Assuming, A_N and B_N are the histograms of two images A and B, the GTI is given as

$$GTI(A,B;\alpha,\beta) = \frac{f(A_N \bigcap B_N)}{f(A_N \bigcap B_N) + \alpha f(A_N - B_N) + \beta f(B_N - A_N)}, \tag{11.17}$$

  where $f(.)$ is a positive increasing function and α and β are two nonnegative parameters. The values of α and β provide the relative importance of the distinctive features in the similarity measures. Interestingly, the GTI measure explained above provides a set theoretic index for similarity assessment based on human perception. The same GTI measure as shown in Eq. 11.17 can be expressed in terms of the fuzzy membership values of the gray levels of images A and B as follows:

$$GTI(A, B; \alpha, \beta) = \frac{\sum_{N=1}^{L} \min_N\{\mu_A, \mu_B\}}{\sum_{N=1}^{L} \left(\min_N\{\mu_A, \mu_B\} + \alpha \min\{\mu_A, 1 - \mu_B\} + \beta \min_N\{1 - \mu_A, \mu_B\}\right)}, \tag{11.18}$$

where μ_A and μ_B are the membership values of the gray levels of the histograms of two images A and B. When the two histograms are similar, the GTI index will be high with nonnegative α and β values.

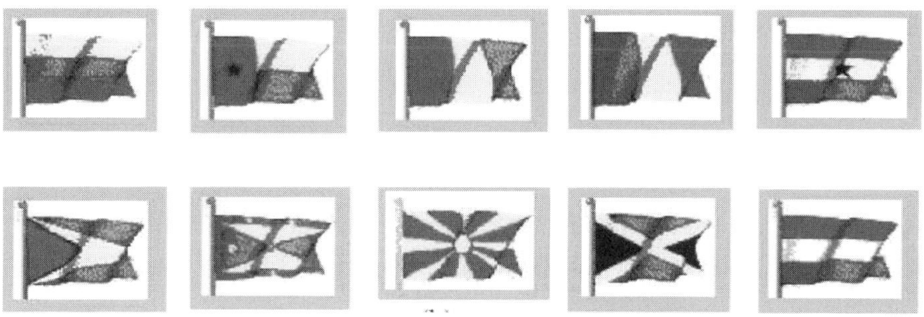

Fig. 11.5 Result of color image retrieval by fuzzy Similarity measure.

The results of color image retrieval using a fuzzy similarity measure is shown in Figure 11.5. The color version of Figure 11.5 is shown in the color pages section. The top leftmost image has been used as the query image to search in a database of color images of flags of various nations. The resulting matched images are shown in Figure 11.5.

11.5 VIDEO MINING

Currently text-based search engines are commercially available, and they are predominant in the *World Wide Web* for search and retrieval of information. However, demand for search and mining multimedia data based on its content description is growing. Search and retrieval of contents is no longer restricted to traditional database retrieval applications. As an example, it is often required to find a video clip of a certain event in a television studio. In the future the content customers will demand to search and retrieve video clips based on content description in different forms. It is not difficult to imagine that one may want to mine and download the images or video clips containing the presence of Mother Teresa from the Internet or search and retrieve them from a video archival system. It is even possible to demand for retrieval of a video which contains a tune of a particular song.

In order to meet the demands for retrieval of audio-visual contents, there is a need of efficient solution to search, identify, and filter various types of audio-visual content of interest to the user using non-text-based technologies. Recognizing this demand, the MPEG (Moving Picture Expert Group) standard committee, under the auspices of the International Standard Organization, is engaged in a work item to define a standard for multimedia audio-visual content description interface [34]. JPEG2000 is the new standard for still-picture compression and has been developed in such a way that metadata information can be stored in the file header for access and retrieval by users as well [32]. There is a mode in the JPEG2000 standard which particularly focuses on compressing moving pictures or video and its content description.

All these developments will influence effective mining of video data in the near future. Video mining is yet to take off as a mainstream active area of research and development by the data mining community. The development has so far been restricted to retrieval of video content only. However, there are ample opportunities that data mining principles can offer in conjunction with the video retrieval techniques toward the successful development of video data mining. In order to influence our readers in this direction, we present here a brief description of the MPEG7 standard for multimedia content description interface and a general discussion on a possible video retrieval system. Application of data mining techniques on top of this development is left to our readers for their imagination.

11.5.1 MPEG7: Multimedia Content Description Interface

There is a wrong notion that MPEG7 is another video compression standard. However, adoption of data compression principles is essential to define this standard for compact description of the multimedia contents. The goal of the MPEG7 is not to define another video compression standard. The purpose of this standardization activity is to specify a standard set of descriptors that can be used to identify content and permit search for particular multimedia content in a multimedia information system.

The goal of MPEG7 is to define a standard set of descriptors that can be used to describe various types of multimedia information, as well as the relationship between the various descriptors and their structures. In principle these descriptors will not depend on the way the content is available, either on the form of storage or on their format. For example, video information can be encoded with any compression scheme (MPEG1, MPEG2, MPEG4, JPEG, JPEG2000, or any other proprietary algorithm) or it can be uncompressed in its raw format without any encoding. It is even possible to generate a description of an analog video, or a picture drawn on paper. The audio-visual data description in MPEG7 may include still pictures, video, graphics, audio, speech, three-dimensional models, and information about how these data elements are combined in the multimedia presentation.

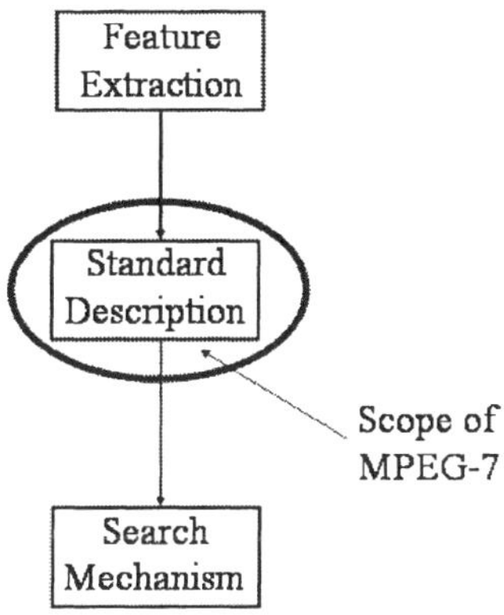

Fig. 11.6 Scope of the MPEG7 standard.

The top-level scope of the MPEG7 standard is shown in Figure 11.6. The block diagram emphasizes that only the audio-visual description of multimedia data is meant to be standardized. The standard neither defines nor deals with the mechanism for extraction of features from the multimedia data, nor is it connected with its encoding or search and retrieval mechanism. Accordingly, the MPEG committee was chartered to standardize the following elements as described in the requirement document for MPEG7 work item [35]:

1. A set of *descriptors*: A *descriptor* is a representation of a feature, such as color, shape, texture, image topology, motion, or title, to name a few. The descriptor defines the syntax and semantics of representation of the feature.

2. A set of *description schemes*: A *description scheme* specifies the structure and semantics of the relationships between its components, which may be both *descriptors* and *description schemes* as well.

3. A set of *coding schemes* for the descriptors.

4. A *description definition language* (DDL) to specify the description schemes and descriptors.

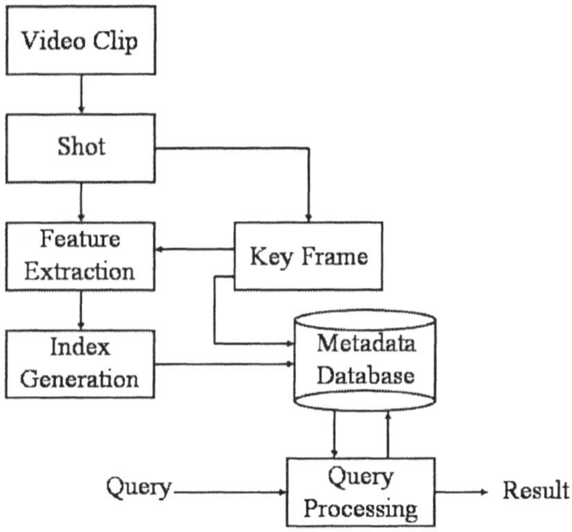

Fig. 11.7 Video retrieval systems architecture.

The goal of MPEG7 work item is a tall order. The final output of the standard is yet to be seen.

11.5.2 Content-Based Video Retrieval System

Content-based image retrieval techniques can be extended, in principle, to video retrieval systems. However, this is not very straightforward because of the temporal relationship of video frames and their inherent structure. A video is not only a sequence of pictures, it represents the actions and events in a chronological order to convey a story and represent a moving visual information. In other words, one may argue that each video clip can be considered as a sequence of individual still pictures and each individual frame can then be indexed and stored using the traditional content-based image retrieval techniques. Again, this is not very practical given the number of frames in a good-quality video clip of even a few minutes. This also does not capture the story structure, which is a collection of actions and events in time.

A generic video retrieval system, which can fit in MPEG7 model, is shown in Figure 11.7. As shown in this figure, a video clip is first temporarily segmented into video *shots*. A *shot* is a piece of a video clip (i.e., a group of frames or pictures), where the video content from one frame to the adjacent frames does not change abruptly. One of these frames in a shot is considered to be a *key frame*. This key frame is considered to be a representative for the picture content in that shot. Sequence of key frames can define the sequence of events happening in the video clip. This is very useful to identify the type and content of the video. Detection of shots and extraction of the key frames from video clips is a research challenge [36, 37].

As an individual still picture, each key frame can be segmented into a number of objects with desired meaningful image features such as shape, texture, color, topology, and many others as defined in Section 11.2 for content-based image retrieval. The semantic relationship between these individual features or feature vectors defines an object of interest to the user. We can apply similar feature extraction techniques and generate index structures for the key frames using the feature vectors, as described for content-based image retrieval. These indexed feature data and the corresponding key frames are stored in the *metadata* database. Collection of this metadata information describes the content of the video clip.

In a video retrieval system, the *query processing* depends on the applications. It can be similar to content-based image retrieval, as described in Section 11.2, or it can be more complex depending upon the type of query processing. In its simplest form, a picture can be supplied to the video retrieval system as a *query* image. This query image is then matched with each and every key frame stored in the metadata database. The matching technique is repeated, based on the extraction of different features from the query image, followed by the matching of these features with the stored features of the key frames in the database. Once a match of the query picture with a key frame is found, the piece of video can be identified by the index of the key frame along with the actual shot and the video clip.

Success of the MPEG7 will influence the future development of image and video mining.

11.6 SUMMARY

Discovery knowledge and mining new information from large image databases is an interesting area of study today in this era of internet and multimedia computing and communication. Development of image mining has been influenced by the research on content-based image retrieval. In this chapter, we have presented two approaches in content-based image retrieval based on both crisp classical feature extraction techniques as well as fuzzy similarity based techniques for image matching. Here we have discussed various features based on color, texture, shape, etc. for feature-based matching and retrieval. We have shown some results and explained how the content based image retrieval can be used as image and video mining tools.

REFERENCES

1. S. Mitra and T. Acharya. Data Mining: Multimedia, Soft Computing and Bioinformatics. Wiley, Hoboken, NJ, 2003.

2. Y. P. Tan, "Content-based Multimedia Analysis and Retrieval," in *Information Technology: Principles and Applications*, Ed. A. K. Ray and T. Acharya, 233–259, Prentice Hall India, New Delhi, 2004.

3. H. Tamura and N. Yokoya, "Image database systems: A survey," *Pattern Recognition*, 17, 1984, 29–43.

4. S.-K. Chang and A. Hsu, "Image information systems: Where do we go from here?," *IEEE Transactions on Knowledge and Data Engineering*, 4, 1992, 431–442.

5. M. Flickner, H. Sawhney, W. Niblack, J. Ashley, Q. Huang, B. Dom, M. Gorkani, J. Hafine, D. Lee, D. Petkovic, D. Steele, and P. Yanker, "Query by image and video content: The QBIC system," *IEEE Computer Magazine*, 28, 1995, 23–32.

6. A. Pentland, R. Picard, and S. Sclaroff, "Photobook: Content-based manipulation of image databases," *International Journal of Computer Vision*, 18, 1996, 233–254.

7. A. Gupta and R. Jain, "Visual information retrieval," *Communications of the ACM*, 40, 1997, 71–79.

8. J. R. Smith and S.-F. Chang, "Visually searching the web for content," *IEEE Multimedia Magazine*, 4, 1997, 12–20.

9. W.-Y. Ma and B. S. Manjunath, "Netra: A toolbox for navigating large image databases," *Multimedia Systems*, 7, 1999, 184–198.

10. H. Samet, *Application of Spatial Data Structures: Computer Graphics, Image Processing and GIS*, Addison-Wesley, Reading, MA, 1990.

11. M. Swain and D. Ballard, "Color indexing," *International Journal of Computer Vision*, 7, 1991, 11–32.

12. M. Stricker and M. Orengo, "Similarity of color images," in *Proceedings of SPIE Storage and Retrieval for Image and Video Databases III*, vol. 2185 (San Jose, CA), February 1995, 381–392.

13. H. J. Zhang and D. Shong, "A scheme for visual feature-based image indexing," in *Proceedings of SPIE Conference on Storage and Retrieval for Image and Video Databases III*, vol. 2185 (San Jose, CA), February 1995, 36–46.

14. G. Pass and R. Zabith, "Histogram refinement for content-based image retrieval," in *Proceedings of of IEEE Workshop and Applications of Computer Vision*, 1996, 96–102.

15. Q. Iqbal and J. K. Aggarwal, "Combining structure, color and texture for image retrieval: A performance evaluation," in *Proceedings of International Conference on Pattern Recognition* (Quebec City, Canada), 2002.

16. R. M. Haralick, J. Shanmugam, and I. Dinstein, "Texture feature for image classification," *IEEE Transactions on Systems, Man, and Cybernetics*, 3, 1973, 610–621.

17. H. Tamura, S. Mori, and T. Yamawaki, "Texture features corresponding to visual perception," *IEEE Transactions on Systems, Man, and Cybernetics*, 8, 1978, 460–473.

18. T. Chang and C. J. Kuo, "Texture analysis and classification with tree-structured wavelet transform," *IEEE Transactions on Image Processing*, 2, 1993, 429–441.

19. A. Laine and J. Fan, "Texture classification by wavelet packet signatures," *IEEE Transactions on Pattern Recognitions and Machine Intelligence*, 15, 1993, 1186–1191.

20. W.-Y. Ma and B. S. Manjunath, "A comparison of wavelet features for texture annotation," in *Proceedings of IEEE Int. Conf. on Image Processing II* (Washington, D.C.), 1995, 256–259.

21. B. S. Manjunath and W.-Y. Ma, "Texture features for browsing and retrieval of image data," *IEEE Transactions on Pattern Analysis and Machine Intelligence*, 18, 1996, 837–842.

22. E. Persoon and K. Fu, "Shape discrimination using Fourier descriptors," *IEEE Transactions on Systems, Man and Cybernetics*, 7, 1977, 170–179.

23. M. K. Hu, "Visual pattern recognition by moment invariants," in *Computer Methods in Image Analysis* (J. K. Aggarwal, R. O. Duda, and A. Rosenfeld, eds.), IEEE Computer Society, Los Angeles, 1977.

24. A. Bishnu, P. Bhunre, B. B. Bhattacharya, M. K. Kundu, C. A. Murthy, and T. Acharya, "Content-based image retrieval: Related issues with Euler Vector," in *Proceedings of the IEEE Int. Conf. on Image Processing* (Rochester, NY), pp. II:585–II:588, September, 2002.

25. J. L. Bentley, "Multidimensional binary search trees used for associative searching," *Communications of the ACM*, 18, 1975, 509–517.

26. A. Guttman, "R-trees: A dynamic index structure for spatial searching," in *Proceedings of ACM SIGMOD* (Boston), June 1984, 47–57.

27. N. Beckmann, H.-P. Kriegel, R. Schneider, and B. Seeger, "The R^*-tree: An efficient and robust access method for points and rectangles," in *Proc. of ACM SIGMOD*, May 1990.

28. N. Katamaya and S. Satoh, "The SR-tree: An index structure for higher-dimensional nearest neighbor queries," in *Proc. of the Int. Conf. on Management of Data, ACM SIGMOD*, May 1997, 13-15.

29. A. W. David and J. Ramesh, "Similarity indexing with SS-tree," in *Proc. of the 12th Int. Conf. on Data Engineering* (New Orleans), February 2002, 516–523.

30. M. Overmars, M. D. Berg, M. V. Kreveld, and O. Schwarzkopf, *Computational Geometry: Algorithms and Applications.* Springer-Verlag, New York, 1997.

31. E. Chang, C. Li, J. Z. Wang, P. Mork, and G. Wiederhold, "Searching near-replicas of images via clustering," in *Proc. of the SPIE Symposium of Voice, Video, and Data Communications* (Boston), September 1999, 281–292.

32. T. Acharya and P. S. Tsai. *JPEG2000 Standard for Image Compression: Concepts, Algorithms, and VLSI Architectures*, Wiley, Hoboken, New Jersey, 2004.

33. T. Chaira and A. K. Ray, "Fuzzy Measures for Color Image Retrieval," *Fuzzy Sets and Systems*, 150(3), 2005, 545–560.

34. MPEG7: ISO/IEC JTC1/SC29/WG211, N2207, Context and objectives, March, 1998.

35. MPEG7: ISO/IEC JTC1/SC29/WG211, N2727, MPEG7 requirements, MPEG Seoul Meeting, 1999.

36. B. C. O'Connor, "Selecting key frames of moving image documents: A digital environment for analysis and navigation," *Microcomputers for Information Management*, 8, 1991, 119–133.

37. A. Nagasaka and Y. Tanaka, "Automatic video indexing and full-search for video appearances," in *Visual Database Systems* (E. Knuth and I. M. Wegener, eds.), Elsevier Science Publishers, Vol. II, Amsterdam, 1992, 113-127.

12

Biometric And Biomedical Image Processing

12.1 INTRODUCTION

Some of the major applications of image processing that we have witnessed in the last two decades are in the areas of biometric and biomedical image processing. The human vision system comes across a large set of biometric features and biomedical images and recognizes them without any conscious effort. To impart this capability to a machine is, however, difficult. The biometric identification systems are useful in several applications such as commercial and law enforcement applications, especially in criminal identification, security system, videophone, credit card verification, photo-IDs for personal identification, etc. Recognition of human faces, fingerprints, signatures, and many other such biometric images constitute an important area of research in the field of computer vision.

Similarly there are different types of biomedical non-evasive imaging modalities such as X-ray, computed tomography (CT), magnetic resonance imaging (MRI), ultrasound images, and many others, which are used in the medical field for disease diagnosis and treatment planning. These imaging modalities reflect the state of the internal anatomy and dynamic body functions. It is important to understand the principal imaging modalities and the processing techniques to enhance, filter, segment, and interpret such images. The radiation in diverse forms utilized in these imaging techniques interact with various tissues to produce images from which the anatomical and structural information of various organs are extracted. The study of science and technology of such information transformation is essentially the studies of biomedical imaging. In this chapter we have presented some of these imaging modalities and their analysis and processing techniques.

12.2 BIOMETRIC PATTERN RECOGNITION

Human face and human signature represent some of the most common biometric patterns that our visual system encounters daily. We present here the classification techniques of these two biometric features. A lot of interest has been generated in automated face recognition and a number of implementation approaches have been proposed [1]–[3].

The major strategies used in face identification are either based on features or they are based on face space, such as *Eigenface* or *Fisherface.*

Most of the the feature based methods extract features from front view of the face and sometimes also from side face profiles. An automatic face recognition system employing both front and side views of the face is more accurate, since it takes advantage of the explicit information inherently available in both the views of the human face. Face recognition approaches employ diverse techniques like neural nets, elastic template matching, Karhunen-Loeve expansion, algebraic moments, iso-density lines, etc. [2]–[5]. Each of these methods has its advantages and limitations. The feature extraction and matching techniques for face recognition are presented in the next section.

12.2.1 Feature Selection

A set of landmark points are first identified from the front and side views of the human face, which are then used for feature measurement based on area, angle and distances between them. The combined set of features extracted from both the views is usually very effective to distinguish faces and provides more reliability over systems using features only from a single view because the side profile features provide additional structural profile information of the face, not visible from the frontal images.

Extraction of features from the front view may be performed from the edge images [5]. The template matching may be used for extraction of the eyes from the face image, while features such as nose, lips, chin, etc., may be extracted from the horizontal and vertical edge maps of a human face.

12.2.2 Extraction of Front Facial Features

Correlation-based technique extracts several front facial components such as eyes, eyebrows, eye points, etc. A set of eye templates are initially chosen. The facial image $f(i,j)$ is convolved with a set of appropriately chosen templates $T(m,n)$, represented by the following filter operation:

$$F(i,j) = \sum\sum T(m,n) * f(i+m, j+n).$$

This convolution process generates a set of energy measures at the output of the filter. The position of the eye is determined from the output of the convolution filter. Using translation, scale, and rotation invariant affine transform, we can detect the eyes from the convolution filter.

Once the two eyes have been detected, the eyebrow positions can be located within a small search region above the eye-center. Subsequently, a set of eyebrow templates can be matched along the column of the iris in each half of the eyebrow template window to detect the eyebrows above the left and right eyes. Some of the front facial features are invariant features, which do not change with facial expression, whereas others are variant features. Some of the front facial points are shown in Figure 12.1. Among the front facial features, eyes have a significant role in the recognition process.

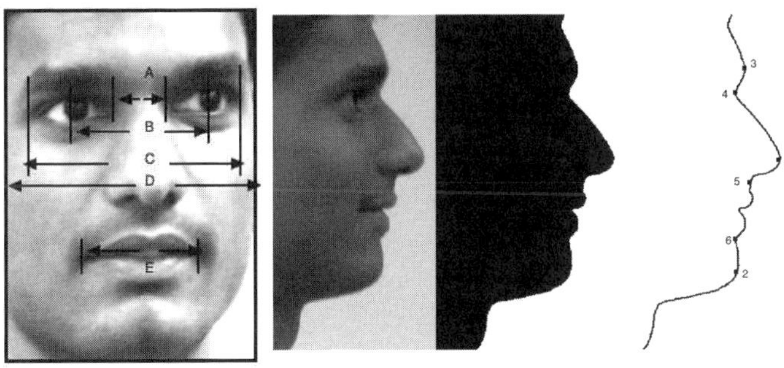

(a) (b) (c) (d)

Fig. 12.1 (a) Some of the Fiducial feature points of front face, (b) side view, (c) binarized side view, (c) contour of the side profile with fiducial points marked on it.

Some of the invariant and variant features from the front face view. are illustrated as follows:

1. Invariant features
   - Distance between left and right iris centers
   - Distance between two inner eye points
   - Distance between two outer eye points
   - Distance form eye-center (mid point of distance between two iris centers) to nose tip
2. Variant features:
   - Distance between left iris center and left eyebrow (same column)
   - Distance between right iris center and right eyebrow (same column)
   - Face width at the nose tip

12.2.3 Extraction of side facial features

12.2.3.1 Selection of fiducial marks The outline profile of the side view of a human face yields a set of good features for identification. The selected landmark points are (1) *nose* point, (2) *chin* point, (3) *forehead* point, (4) *bridge* point, (5) *brow* point, (6) *lip-bottom-curve* point.

The digitized photograph of the side face is transformed to side outline profile and the landmark points are then extracted. The side face image, its binarized version, side profile, and the selected landmark points are shown in Figure 12.1. Seven distance measures and one area measure are now extracted from the landmark points obtained from each side profile.

12.2.3.2 Distance and Area measures A set of distance measures computed from the extracted landmark points are (1) *Nose* to *Forehead* distance, (2) *Nose* to *Bridge* distance, (3) *Nose* to *Nose Bottom* distance, (4) *Brow* to *Bridge* distance, (5) *Brow* to *Chin* distance, (6) *Nose* to *Chin* distance, (7) *Nose Bottom* to *Lip Bottom curve point* distance. Area of the profile on the left portion of the line joining forehead and chin point is taken as the area measure feature.

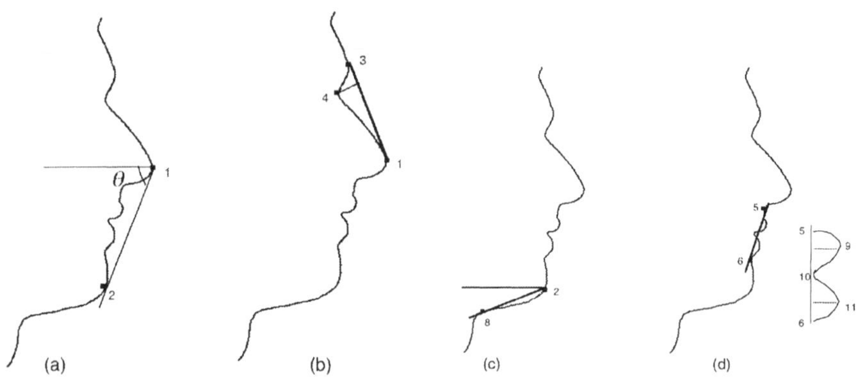

Fig. 12.2 (a) Nose to Chin distance, (b) Nose to forehead distance (c) Chin angle (d) lip distance

Some of the side facial features are shown in Figure 12.2. The set of features, from the front view and the side profile form a combined feature vector array. Each facial photograph is reduced to this feature vector array which is expected to be unique for each face and is stored in a file with subject index number. To remove the effect of scaling, the distance measure and area measure values of side view have been normalized with respect to the distance between nose point and bridge point. Similarly the measures of front face are normalized with respect to the inter-ocular distance.

12.2.4 Face Identification

The facial images may be acquired by a digital camera in a well illuminated room. The front view is captured by making the subjects look into the camera, while the side view should be taken by keeping the faces at approximately right angle to the camera. A set of optimal features from the front and side face are next extracted and these feature values are stored in a separate file. Let the vectors in the main face data file be called M-vectors. The test-data file contains the 15 dimensional feature vector for the unknown sample.

To reduce the computational complexity of matching the feature vector values along with their subject indices are stored in ascending order. The test feature values are inserted at the appropriate places in the sorted feature list in each feature (column). The two nearest neighbors of the test sample pattern are identified in each column and their similarities are evaluated.

Algorithm for k-nearest neighbor classifier for face matching is as below.

Step 1 Arrange the feature values along with their subject indices in each column in ascending order.

Step 2. Insert each feature vector of the test sample at its appropriate place in each column.

Step 3. Determine the two nearest neighbors in each column and compute the similarities of the test pattern to both the nearest neighbors, each having a subject index.

Step 4. Compute the overall similarity of the test pattern to all the subject indices.

Step 5. Assign the subject index having maximum similarity value to the test face.

If the face in the test-file does not belong to any of those stored in the main-data file, it gives out the class index of the nearest similar face to that of the test face.The algorithm as above yields excellent results on face recognition.

12.3 FACE RECOGNITION USING EIGENFACES

indexEigenface The *eigenface* representation method for face recognition is based on the principal component analysis. The main idea is to decompose face images into a set of eigenfaces (a small set of characteristic feature images), which are the principal components of the original images [6]. These eigenfaces function as the orthogonal basis vectors of a linear subspace called *face space*. The face recognition strategy involves projecting a new face image into the face space and then comparing its position in the face space with those of known faces. In this method the training set of facial image patterns are converted into a vector of $M \times N \times K$ where $M \times N$ is image size and K is number of training samples. This $M \times N$-dimensional space is indeed very large and it is important to reduce the dimensionality of this space before attempting face recognition. *Principal component analysis* (PCA) and *linear discriminant analysis* (LDA) are two common approaches to reduce this large dimensionality.

The basic philosophy in PCA is to map all the $M \times N$-dimensional face samples, x_i, $i = \{1, \cdots, k\}$ to a single vector y_i, $i = \{1, \cdots, n\}$ so that y_i represents x_i, i.e.,

$$y_i = w^T x_i \tag{12.1}$$

where each x_i represents a face and w is a weight vector which represents scaling. The objective function is maximization of variance, i.e.,

$$J = \max \sum_{i=1}^{n} (y_i - \bar{y})^2 \tag{12.2}$$

where

$$\bar{y} = \frac{1}{n} \sum_{i=1}^{n} y_i. \tag{12.3}$$

The set of weight vectors $w_1, \cdots, w_k$ represent the *eigenvectors* of the covariance matrix computed from the set of sample faces and $k < MN$. This implies that we select only a few eigenvectors corresponding to the dominant eigenvalues and thus reduce the dimensionality of facial features.

For an accurate face recognition system, the accuracy should be quite high and at the same time the processing time for a test face image should be low. The recognition system should be invariant to the head rotation and translation, and also to the illumination intensity changes.

12.3.1 Face Recognition Using Fisherfaces

Another interesting technique, known as *Fisherface* also uses the linear projection concept of eigenfaces and *elastic branch graph matching.* Fisherface uses *Fisher linear discriminant analysis* (FLDA) which yields a robust classifier, and enhances the classification accuracies of the face patterns. *Fisher linear discriminant function* (FLDF) creates a linearly separable space which is efficient in discriminating the face patterns. FLDF finds out a number of linear functions which partitions the face space in k distinct classes [7].

While using *Fisherface* approach, the training face patterns includes not only a single face but various expressions of a face pattern. Let there be n_i samples of different expressions of the ith face class and $m^{(i)}$ represent the average feature vector of the n_i samples of the ith class. The objective function to be maximized in this case is

$$J_F = \max \left\{ \frac{W^T S_B W}{W^T S_W W} \right\} \tag{12.4}$$

where S_B is between class scatter matrix given by

$$S_B = \sum_{i=1}^{k} n_i \left(m^{(i)} - m\right)\left(m^{(i)} - m\right)^T, \tag{12.5}$$

and S_W is within class scatter matrix given by

$$S_W = \sum_{i=1}^{k}\left[\sum_{j=1}^{n}\left(x_j^{(i)} - m^{(i)}\right)\left(x_j^{(i)} - m^{(i)}\right)^T\right]. \tag{12.6}$$

In Eqs. 12.4, 12.5, and 12.6, m represents the sample mean vector, n_i is the number of samples in ith class, and $x_j^{(i)}$ represents jth facial expression of the ith face and W is the projection matrix.

The objective function J_F attempts to maximize the Euclidean distance between the face images belonging to different persons, while minimizing the distance between the face images belonging to the same person. The projection direction that maximizes J_F yields the column vectors of the projection matrix W. The objective function J_F is maximized when the column vectors of W are the eigen vectors of $S_W^{-1}S_B$. It may be noted here that if the dimensionality of the sample space is larger than the number of samples in the training set, S_W becomes singular and thus its inverse does not exist. Several strategies have been attempted to take care of such situations.

Both the PCA and LDA preserve the global structure of the face-space, but *Fisherface* method has better discriminating capability compared to the *Eigenface* method.

Another interesting face recognition technique, called *Laplacianface* approach uses locality preserving projection to map into a face subspace [8]. This method attempts to obtain a face subspace that best detects the essential face manifold structure by preserving the local information. The method yields better accuracy compared to *Fisherface* or *Eigenface* methods.

12.4 SIGNATURE VERIFICATION

Signature is an important biometric measure, which are subject to intra-personal variation. An automated system for signature verification is feasible only if the representation of the signature image is insensitive to intra-personal variations, but sensitive to inter-personal variations. The goal is to maximize the distance between signatures of different individuals, the maximizing constraint being that the distance between the signatures of the same person is kept constant or minimized, once the measure of distance is properly defined. A number of interesting techniques on signature verification has been reported [9]–[11].

We now present a four-step methodology for signature verification.

Step 1. Preprocess and binarize the signature image.

Step 2. Extract the medial axis, the thinned version of the signature pattern.

Step 3. Perform water-filling operation. Extract the structural features from the water-filled pattern.

Step 4. Using any matching algorithm, identify the best match from the signature database using the features extracted in step 3.

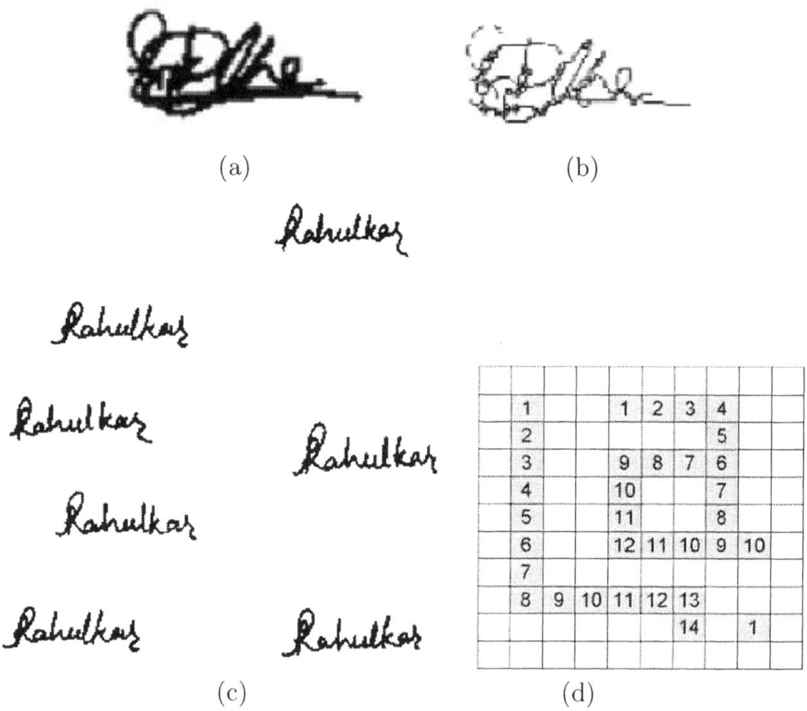

Fig. 12.3 Signature Image (a) Original (b) Thinned version of the original (c) Samples of original and forged signatures (d) example of waterfilling

12.5 PREPROCESSING OF SIGNATURE PATTERNS

The binarized signature image, as in Figure 12.3(a), is thinned using any standard thinning algorithm. The objective is to obtain a trajectory of the pen-tip. It may be observed from Figure 12.3(b) that the thinned image does not accurately capture the pen-tip trajectory, while it adequately preserves the structural information necessary for recognition.

Some artifacts introduced by thinning are:

- loops and holes in the binary image are reduced to single segments.
- strokes intersect on a sequence of points instead of a single point.
- false loops may arise in the thinned version, which can affect the water-filling algorithm.
- there may be a loss in the continuity of the connected components.

Edge map is usually used for extracting object shape, which requires edge linking. This is not a trivial problem, considering the fact that the edge set can be seen as general graphs. A heuristic graph search is not optimal, since vital information may be lost, while optimal methods like dynamic programming may be computationally too expensive. Also the edge detectors are not usually ideal, which may result in creation of false edge points.

To address the above issue, a good solution is to employ the water-filling algorithm, which can extract features directly from the edge map without edge linking or shape representation. The algorithm performs a raster scan on the edge map and fill in water at the first edge pixel that has less than 2 neighbors, i.e., the fill in process starts at an end point. The waterfront then flows along the edges in all paths available (4-connectivity or 8-connectivity). This algorithm can be regarded as a simulation of "flooding of connected canal systems" and hence the name "water-filling algorithm". In the example shown here, 4-connectivity has been depicted. A small part of the signature image (Figure 12.3(a)) has been chosen, whose thinned version is shown in Figure 12.3(b), while the waterfilling process is depicted in Figure 12.3(d). The following quantities are used as primitives and their values refer to the Figure 12.3(d).

1. Filling time: It is the time taken by water to fill a set of connected edges. The filling time is {14, 12, 1}.
2. Group point count: It is the number of pixels, which offer three paths for the waterfront to travel, keeping aside the path from which the front had reached that pixel. Group point count is {0, 0, 0}.
3. Multiple point count: It is the number of pixels, which offer 4 or more paths for the water-front to travel, keeping aside the path from which the front had reached that pixel. It is applicable only for 8-connectivity and the count value is {0, 0, 0}.
4. Edge point count: It is the number of points, which have no paths for the waterfront to flow, keeping aside the path from which the front had reached that pixel. Edge point count is {2, 2, 1}.
5. Loop count: It is the number of simple loops in a set of connected edges. Using the 8-connectivity, there can be many instances of false loops, which need to be removed. Loop count is {0, 1, 0}.

6. Water amount: It is the total amount of water used to fill up the set of edges in terms of the number of pixels. Water count is $\{14, 18, 1\}$.

7. Horizontal (vertical) cover: It is the width (height) of the rectangular bounding box of the set of edges. Horizontal cover is $\{6, 5, 1\}$ and vertical cover is $\{8, 6, 1\}$.

Using these primitives, a set of secondary or structural features can be extracted. Some of these structural features are:

- *Maximum Filling Time* (MFT) and associated *Fork Count* (FC): MFT is defined as the maximum of the individual filling times. MFT & FC are features associated with a salient object in the image.

- *Filling Time Histogram* and associated *Average Fork Count* (FTH & AFC): This is a global feature on all sets of connected edges in the edge map. It represents the edge map by the distribution of the edge "length."

- *Global Loop Count* and *Maximum Loop Count* (GLC & MLC): Global loop count is defined as the summation of all the loop counts. MLC on the other hand is the maximum value of all the loop counts.

12.5.1 Feature Extraction

A multistage matching algorithm is used for signature matching. After the initial pre-processing is performed, the water-filling features are extracted next.

The conventional Hough transform is now applied on the image, and the accumulator values are incremented in the parameter space.

1. Remove all horizontal lines, which contain at least one run of length greater than a parameter, say *maxrun*.

2. Remove all horizontal strips of height less than the parameter say, *strip height*.

The above two steps are performed, since the Hough transform produces all possible lines in the parameter form. If the parts of the image which are not persistent strokes, are not removed, we get unnecessary count in the accumulator cells. The signatures are more or less scribbled in a small patch in a horizontal fashion. A study of scribbled signatures shows that the length of the perfectly horizontal strokes is not persistent.

The top few accumulator values are used as the feature values for matching. A new algorithm for matching is proposed, which directly works in the parameter space of the Hough transform and based on which matching is done.

12.5.1.1 A Matching Strategy The matching strategy should be inherently robust to scaling and translation. It is assumed here that the signatures acquired are more or less horizontal in orientation and that the slight changes in angle due to intra-personal variations is effectively taken care of. Certain assumptions, which are made while designing a signature verifier are listed below.

- The intra-personal signatures may be inherently translated, during the image acquisition stage itself. The algorithm is designed for translation invariance.
- The variations in intra-personal signatures may be a variation in the slant, position or length of the keystroke; which is not too much.
- Same strokes of the same person may be of variable length, in different signatures.

The test signature is matched with those in the database and the signature with minimum distance measure, i.e., which is closest enough to the best match in the database is selected.

The experiment should be carried out on a database of a large number of signatures as shown in Figure 12.3(c) taken from many people and include signatures of all types from very smooth cursive type of signatures to scribbled signatures. The aim of signature verification is to identify the skilled forgeries.

12.6 BIOMEDICAL IMAGE ANALYSIS

Bio medical image processing can broadly be classified into (1) Microscopic image analysis, and (2) Macroscopic image analysis. A brief description of the two are given below.

12.6.1 Microscopic Image Analysis

In microscopic image analysis, we deal with the living organisms, Which are microscopically small objects, and are important in understanding the sciences of biology and medicine. Living organisms are composed of cells and the normal and abnormal cells influence growth, developments, disease or malignancy in human body. Advanced imaging techniques have made it possible to view these hidden aspects of microscopic biological material and helped the understanding of the biomedical sciences over the past several decades.

The digital image processing techniques have been extensively used to improve the clarity of microscope images using several techniques detailed in Chapter 6. Some of the techniques like histogram manipulations, image filtering etc have been employed extensively to get quantitative and morphometric information about the biological organisms.

Image processing techniques have also been used to automate and standardize diagnostic tests performed on blood cells, historical samples cervical smears, chromosome preparations and chromosome analysis and other types of microscopic images.

12.6.2 Macroscopic Image Analysis

In macroscopic image analysis, we deal with the images of human organs such as heart, brain, eye, etc., which are enhanced, segmented and analyzed. Different image processing tasks like image filtering, shape modeling, segmentation, classification and interpretation techniques have been extensively used in diagnostic radiology, cardiology, dentistry, and many other areas.

The preprocessing phase involves enhancement, deblurring, and filtering of the radiographs. The enhancement techniques may be linear or non-linear, and may involve local or global filters. Deblurring techniques consist of inverse or Weiner filters. Edge detection and boundary detections are important steps in many biomedical image analysis applications. Segmentation including thresholding techniques are useful in segmenting objects of interest. Various supervised and unsupervised techniques have been used for analysis of macroscopic images.When adequate prior information is available, matched filters can be used.Shape modeling including 3D representation and graphic manipulations have also been extensively used.

Medical image engineering includes a broad range of image formation modalities such as computer aided conventional radiography, time varying images, stationary non-invasive images, tomographic reconstruction, x-ray computed tomography, positron emission tomography, MRI, volume image reconstruction, nuclear imaging, ultrasonic imaging, diaphonography, electro-encephalography, opthalmography, etc.

12.7 BIOMEDICAL IMAGING MODALITIES

Here we briefly review some important biomedical imaging techniques [12].

12.7.1 Magnetic Resonance Imaging (MRI)

Various constituent organs in human body contains considerable amount of water molecule and fat, and thus there is an abundance of Hydrogen in our body tissues. The MRI signals, emanate from these Hydrogen nuclei, when they are excited by magnetic stimulus and these signals are used for imaging. Noted physicists Bloch and Purcell have first conceived the concept of MRI, an advanced type of imaging technique in the year 1946. The principle involved here is to stimulate the matter magnetically and the imaging signal is obtained by the changes in the fundamental properties of matter, in response to this magnetic stimulus.

MRI (alternatively Nuclear MRI) utilizes a tomographic imaging technique and captures the image in form of slices. Each body slice, abundant in hydrogen, may be viewed as a set of voxels, which are volumetric cells element , where each Hydrogen nuclei represents one voxel. When properly excited, these nuclei represented by a volumetric cell give out NMR signal, and the intensity of the image pixel is proportional to the intensity of NMR signal from the corresponding voxel. Thus from the mapping of the individual tissues, the cumulative mapping of the entire organ is obtained.

The imaging applies one-dimensional magnetic field gradient over a slice. It works on the principle of frequency encoding, where the obtained signal is proportional to the number of spins perpendicular to the applied magnetic vector, and the resonance frequency is a function of the position of the spin. MRI applies 1-dimensional field gradient to a slice and records the response, now the field vector is rotated through a circle in small steps, and the recorded data can be back projected for getting the image.

Using the principles stated above the imaging may employ various techniques such as *Multisliced Imaging*, *Oblique Imaging*, *Spin Echo Imaging*, *Inversion Recovery*, *Gradient Recalled Echo Imaging*, etc. The imaging technique depends upon the pulses that are used for excitation.

MRI is nowadays widely used in visceral imaging for tumor detection and also in other applications in spine, neck, brain, etc. Apart from being an accurate imaging system it enjoys a great advantage of being safe in application. It doesn't employ the conventional belief that the frequency used for imaging should be smaller than the object. It uses phase and frequency variation in the RF range, hence devoid of the hazardous effect of other visceral imaging techniques such as X-Ray.

12.7.2 Computed Axial Tomography

Computed axial tomography, popularly known as CT-scan or CAT scan is another powerful technique for medical imaging. This is employed in imaging of soft tissue system, as well as hard bones, and blood vessel.

This imaging technique employs X-ray photography principle. It sends X-rays with different strength, and depending upon the type of obstruction it faces, the X-ray beams are characterized based on the responses. This employs tomographic imaging techniques, i.e., the imaging is carried on slices.

Structurally a CAT scanner contains a X-ray tube, and a detector. The tube is rotated along a spiral/circular path and the image of the slice is captured by the X-ray detector. During a full rotation, the detector records a large number (nearly 1000 per rotation) of snapshots. The images are further broken into several independent data set and further processed in a number of parallel channels. During this processing the profile is back projected to give the actual image of the tomographic slice.

The CAT scanner is a banana shaped device, that performs the function of X-ray beam firing as well as the photography. It also contains the frame that rotates the X-ray tube for producing a rotating X-ray beam. Apart from the scanner device a CT-scan system requires parallel PC interfacing, in a mutually communicating mode, and also advanced three-dimensional image processing algorithms for image reconstruction by the back projection method. The rotation of the frame, table, control of the X-ray beam is also built on microprocessor based system. Being fully computerized the required zone may be imaged by sliding a window with high precision.

Though originally evolved as a slice imaging, modern CT-scans uses volume imaging for a 3D reconstruction of the imaged viscera. As stacking of slices may lead to misalignment because of voluntary and involuntary movements on the patient's part such as respiration, etc.

Because of the imaging technique, CT-scan provides sufficient perspective views of the object for 3D reconstruction. PC based system organization allows a CT-scan to carry out high quality 3D image processing to provide an excellent opportunity for the diagnosis purpose, CT has got another big advantage in terms of speed. Due to microprocessor based data acquisition and fast parallel processing algorithms, CT-scan has a significant speed advantage over other imaging techniques.

CT uses 3D imaging based on X-ray vision, so it has got the penetration power and hence can be used for virtual endoscopy of colon or bronchial canal, without any physical endoscope. Even in some cases it reaches beyond the scope of the invasive endoscopes.

CT-scan is extensively used in imaging of visceral organs like the brain, lungs, kidneys, liver, pancreas, pelvis, and blood vessels. With the improvement in the imaging techniques gradually it is finding in application in cancer detection as well as diagnosis of heart disease, stroke, etc.

12.7.3 Nuclear and Ultrasound Imaging

In nuclear medicine, radioactive materials are usually given through intravenous (IV), or swallowed or inhaled to obtain images of human organs. The passage of movement of the radioactive substance is tracked by a detector. Radionuclides get tagged with certain substances within the body. It emits gamma radiation which is captured by a sensor in a gamma camera. These images are poor in resolution, yet they visualize the physiological functions, such as metabolism in a clear way.

In ultrasound images, an ultrasonic pulse propagates from a transducer placed on the skin of the patient. The backscattered echosignal is recorded from which an image is constructed. Ultrasound gets transmitted through water. The cysts which are watery fluid structure does not send any echo to the recorder. On the other hand bones, calcification and fats absorb and reflect the ultrasound beam to a small extent and creates acoustic shadowing. Thus cysts can be detected in any organ using ultrasound images.

12.8 X-RAY IMAGING

X-ray images on photographic films are the oldest and most frequently used form of medical imaging. X-ray imaging is the fastest and easiest way for a physician to view and assess broken bones, a cracked skull or in back bone. X-ray is useful in detecting more adverse forms of cancer in bones. Diagnostic X-ray images can be created by passing small highly controlled amounts of radiation through the body, capturing the resulting shadows and reflections on a photographic plate. The X - ray images are caused by various levels of absorption by calcified structures, soft tissues, fatty substances and so on. There are two types of projections (i) Postero-Anterior (PA) when the beam traverses from the back to the front of the patient, and (ii) Antero-Posterior (AP) where radiation traverses in the opposite direction.

12.8.1 X-Ray Images for Lung Disease Identification

An interesting survey of low level image processing including pre-processing, feature extraction and classification techniques for radiographic images may be found in [13]–[17]. Interesting Feature selection and pattern classification techniques for Digital chest radiograph have been highlighted in [18, 19].

Radiographic analysis chest X-ray film (a standard view PA erect 14×17 film) involves number of steps. In radiographic images the regions containing air appears as black, while the regions containing more solid tissues appear light. Bone containing calcium are more radio opaque than soft tissue. The anatomical structures clearly visible on a normal chest X-ray film are the ribs, the throcic spine, the heart, and the diaphragm separating the chest cavity from the abdominal cavity. All the different regions in the chest radiographs may be examined for abnormality, however primary emphasis is on the lungs and heart. There are several disease processes with diagnostic signs that can be identified on a PA chest radiograph. Important indicator of lung diseases:

a) vascular pattern within the lung field

b) increase in tissues adjacent to the small bronchi and blood vessels

c) small tumors, rib fractures and localized pneumonia.

12.8.2 Enhancement of Chest X-Ray

A number of enhancement strategies have been employed by various researchers for enhancement of chest images. Gabor filters have been used for enhancement of the nodule by removing the continuous components. Laplacian of Gaussian filter has also been employed to enhance the isotropic features of the lung nodules. By enhancing the contrast between the region inside the nodule and the background structure, the lung nodules can be made more visible.

12.8.3 CT-scan for Lung Nodule Detection

Lung cancers originate as a small growth, or nodule in the lung. Use of screening CT scans are extremely useful in detecting nodules as small as 2 or 3 mm within the lungs. A CT scanner captures a series of high-speed X-rays, which are used to generate three-dimensional images. These images are interpreted by the radiologists. The scan itself takes less than 20 seconds to perform and no preparation is required on the part of the patient. The CT lung scan uses low-radiation-dose computed tomography to detect cancers of extremely small diameter. Screening CT-scans are capable of detecting lung nodules much smaller than by conventional chest X-ray.

Lung nodules are commonly described as small round like blob object. It is roughly spherical and has a density comparable to water, which is higher than the surrounding lung parenchyma. Radiologist uses circularity and density versus size measurement as criteria for nodule detection.

Some of the features used for the detection and classification of nodule in a chest radiograph are

(a) *Gray level statistics* features are (i) First order and second order statistics like mean, standard deviation computed from the histogram of the gray values of the inner core region and also the outer region of the suspected region, (ii) Contrast between the core part and the outer shell region

(b) *Geometric* features are (i) Circularity of the shape, (ii) Average eccentricity, (iii) Kurtosis, (iv) Directionality of the mass, (v) Shape factor, (vi) Area, perimeter, etc., (vii) Compactness.

(c) Other features are (i) Textures, (ii) Angular second moment, (iii) Entropy, (iv) Inverse Difference Moment.

An appropriate classifier is next designed utilizing a set of features from the above list, for classification of the nodule as benign or malignant. Various classifiers including neural network based classifier have been proposed for the detection of nodules in chest radiographs [20].

12.8.4 X-Ray Images for Heart Disease Identification

The heart size and shape are important features for heart disease detection and classification. These along with other features are extracted from the ordinary PA chest radiograph. Rheumatic heart disease affects the mitral, aortic and the tricuspid valves. The inflammation caused by the disease cause scattering of the valve leaflets. The involved valve or values produce The anatomical changes in the heart and great vessels should be detected in chest radiographs. In mitral stenosis diseases, the changes are characterized by a

a) filling of the region immediately above the heart and lateral to the pulmonary artery and left hilar region.

b) reduction or disappearance of the aortic knob.

c) slight to severe protrusion of the left atrial appendage and a steep descent of the left ventricular arch to the diaphragm.

d) enlargement of the cardiac projection with elongation and greater rounding of the left ventricular arch.

The heart diseases may be classified in five classes – (1) Normal, (2) Mitral stenosis only, (3) Other mitral valve lesions, (4) Aortic and mitral valve involvement, and (5) Aortic involvement.

12.8.5 X-Ray Images for Congenital Heart Disease

Several efforts have been put forward by the researchers to detect congenital heart disease from chest radiographs.Congenital heart disease is associated with a broad range of structural anomalies of the heart. The problems associated with identification of heart diseases is that it is not possible to see all the chambers of the heart on any one projection. To make a diagnosis of conventional disease, the PA chest X-ray is most important because it contains the most diagnostic information. The lateral view is a complimentary view, which shows both the right ventricle and the left ventricle. Radiologists and cardiologists use a right anterior oblique radiograph and a left anterior oblique radiograph. These views are part of a cardiac series but are not routinely taken. The major tasks involved in analyzing the PA chest X-ray are

step 1: Preprocess the chest radiograph

step 2: Extract the heart contour by active contour model.

step 3: Determine the heart size and extract the shape parameters

step 4: Determine the right and left edge of the heart

step 5: Determine the lower right and lower-left corner of the heart.

step 6: obtain the cardiac rectangle from the heart contour.

step 7: Extract a set of features from the cardiac outline.

step 8: Classify the chest X-ray using the shape parameters.

There may be an overlap of the heart boundaries and the right main pulmonary vessels. Using an appropriate thresholding technique the right edge, may be determined.

12.8.6 Enhancement of Chest Radiographs Using Gradient Operators

Robust image enhancement techniques are useful in obtaining good quality medical images, which can efficiently enhance the radiographic images, suppressing the image noise considerably. Gradient operators can be used for the enhancement of images, and are widely used in medical image processing applications. The gradient operators enhance image edges and at the same time enhance noise also and hence the need for a filter before edge enhancement is attempted. Emphasizing the fine details of a radiographic image while suppressing noise is possible by employing a median filter prior to applying a gradient operator. Approximations to the pill box blur and Gaussian low pass filter also yield noise reduction.

Several adaptive image enhancement algorithms are available in the literature. An adaptive algorithm first removes the film artifacts, then computes the gradient images by using the first derivative operators and finally enhances the important features of the radiographic image by adding the adaptively weighted gradient images. The local statistics of the image are utilized for adaptive realization of the enhancement, so that the image details can be enhanced and image noise can be suppressed. The contrast-limited adaptive histogram equalization (CLAHE) produces images in which the noise content of an image is not enhanced, but in which sufficient contrast is provided for the visualization of structures within the image. Images processed with CLAHE have a more natural appearance and facilitate the comparison of different areas of an image. However, the reduced contrast enhancement of CLAHE may hinder the ability of an observer to detect the presence of some significant gray-scale contrast.

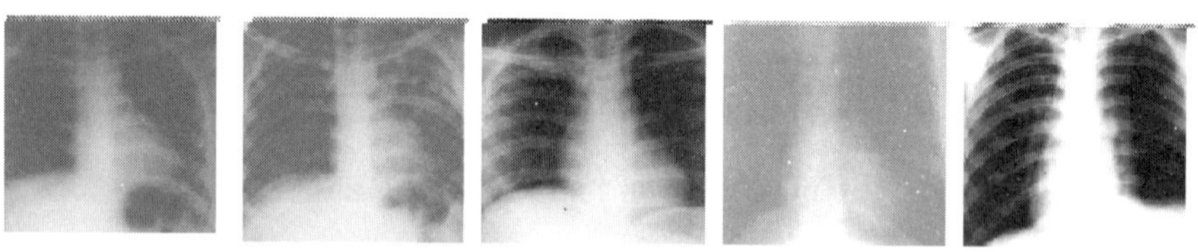

(a) (b) (c) (d) (e)

Fig. 12.4 Chest Image (a) Original, enhanced by (b) histogram equalization, (c) local area histogram equalization, (d) another input original image, (e) enhanced by contrast limited adaptive histogram equalization.

Figure 12.4(a) is the original chest x-ray image. In Figure 12.4(b) and (c), the lung boundaries and edges of the bones are enhanced using histogram equalization and local area histogram equalization respectively. Figure 12.4(d) is another original chest x-ray image and the enhanced image by contrast limited adaptive histogram equalization is shown in Figure 12.4(e).

12.8.7 Bone Disease Identification

The image processing of bones provides important information like the existence of any tumor or growth in the bone and if it is present, it is important to estimate the location, size and shape of the tumor and density of the tumor with reasonable accuracy. Other features of interest like the bone destruction visibility, type of bone destruction, penetration of cortex, fracture sign may also be available by image processing. The common use of bone radiographs is to assist physicians in identifying and treating fractures. Images of skull, spine, joints. Bone X-rays are an essential tool in orthopedic surgery such as spiral repair, joint replacements or fracture reduction. Texture analysis has been observed to detect severe Osteoporosis of bones [21].

12.8.8 Rib-cage Identification

Finding the rib-cage in chest radiograph is an important area of research in X-ray image processing. To find the rib cage is to find and identify the upper and lower contours of each rib. Each of these contours constitutes, description of a rib. The rib cage finder consists of the

Step 1. Preprocessing, which consists of scanning digitizing and filtering the input radiograph.

Step 2. Local edge detection , consisting of gradient operator, a Laplacian and a threshold operator.

Step 3. Global boundary detector which finds candidates for the dorsal and ventral portions of the rib contours by matching straight, parabolic and elliptical curve segments to the

Step 4. Rib linker, which matches and joins the corresponding dorsal and ventral segments from step3 by a fourth degree polynomial. The polynomial represents each rib contour as a separate entity.

In order to enhance the edges, a high pass filtering in the spatial frequency domain with an transformation function is required. The important uses of the Rib cage finder are

(a) Rib image contribute false positives in tumor detection algorithm. Hence detecting the Rib should facilitate removal of these false positives.

(b) The ribs give the radiologists a convenient frame of reference for the description and location of the heart and lung within the thoracic cavity.

(c) The outer envelope of the Rib provides a more accurate basis for computing the boundary of the chest cavity than other available techniques. Earlier work on automatic rib detection has been reported in [22].

12.9 DENTAL X-RAY IMAGE ANALYSIS

Dental caries and periodontal disease are the most common dental diseases in the world. Dental caries has affected human being widely in modern times. Dental caries is an infectious microbiological disease that results in localized dissolution and destruction of the calcified tissues of the teeth. If untreated the caries results in the progressive distraction of the tooth and infection of the dental pulp takes place.

12.10 CLASSIFICATION OF DENTAL CARIES

Classification of dental caries is important for the diagnosis and treatment planning of the dental disease, which has been affecting a very large population throughout the globe. It is also helpful for conducting detailed study and investigations about the nature of the dental disease. Classification of dental diseases is decided on the basis of certain criteria, such as based on whether the lesion is within the enamel, dentin or whether it touches the pulp. Dental caries are, clearly visible in the x-ray changes and it can be detected from the caries lesion present in the radiographs.

Image processing techniques will help check the x-ray and examine the extent to which the caries lesion is present and then classify the type of caries present in the dental radiograph. [23, 24]. A computer aided interpretation and quantification of angular periodontal bone defects on dental radiograph is performed by P.F. Van der stelt and Wil G.M. Geraets. Capabilities of human observers to detect and describe small bone defects objectively are limited. Digital image processing can provide a useful contribution to the diagnostic process. Their procedure was able to rank series of artificial periodontal bone lesions as accurate as experienced clinicians. Comparison of data from clinical inspection of lesions during surgery and quantitative result of the digitized procedure shows that the latter produced reliable information on the lesion size. Dental caries can be classified in a number of ways depending upon the clinical features, which characterize the particular lesion. Dental caries may be classified according to the location of the individual teeth as (a) Pit or Fissure caries and (b) Smooth surface caries.

According to the rapidity of the process dental caries can be classified as (i) Acute dental caries and (ii) Mild dental caries.

Caries may also be classified according to other criteria.

(i) Primary (virgin) caries: Depending on whether the lesion is a new one attacking a previously intact surface,

(ii) Secondary (Recurrent) caries: Depending on whether whether it is localized around the margins of a restoration.

According to the extent of attack dental caries may be classified as

1. ENAMEL CARIES: Caries of the enamel is preceded by the formation of a microbial (dental) plaque. The processes vary slightly, depending upon the occurrence of the lesion on smooth surfaces or in pit or fissures. When the caries have affected the outer enamel portion alone and if the inner dentine and pulp regions are healthy then the type of caries is called enamel caries. Enamel caries can further be classified as (a) Smooth surface caries and (b) Pit and fissures caries.

2. DENTINAL CARIES: Caries of the dentin begins with the natural spread of the process along the natural spread of great numbers of the dentinal tubules each which acts as a tract leading to the dentinal pulp along which the micro-organism may travel at a variable rate of speeds depending upon some factors. In some instances carious invasion appears to occur through an enamel lamella so that little if any visible alteration in the enamel occurs. Thus when lateral spread at the dentino-enamel junction occurs with involvement of under lying dentin, a cavity of considerable size may actually from with only slight clinically evident changes in the overlying enamel except for its undermining.

3. PULPAL CARIES: The carious lesion discussed prior to this point has been limited chiefly to coronal caries, and process has involved basically the enamel and dentin of that portion of the tooth. Another form of disease does exit which is known as root caries or root surface caries. At one time it was also referred to as caries of cementum. It is generally recognized that the longer life span of persons today, with the retention of teeth into the later decades of life has increased the no: of population exhibiting gingival recession with clinical exposure of cemental surfaces and there by probably increasing the prevalence of root caries. Root surface must be exposed to the oral environment before caries can develop here. In this section dental caries classification is performed based on the edge detection: The caries affected tooth is selected , which is subsequently used for caries classification.

12.10.1 Classification of Dental Caries

The dental caries classification algorithm is briefly explained below. The radiographic image is first captured using an appropriate X Ray imaging device, connected to the image analysis system. The image is captured along a line parallel to the long axis of the tooth. The dental x ray image is initially segmented into individual tooth, which is followed by binarization of the tooth pattern. The edge detection of the segmented tooth yields the outline of the dental cavity. The classification may be achieved using a simple rule based system. By determining the number of carries affected pixels, the region area may be extracted. If there exists only one black region and there is an adjacent white border, i.e., black carries region is adjacent to the white border

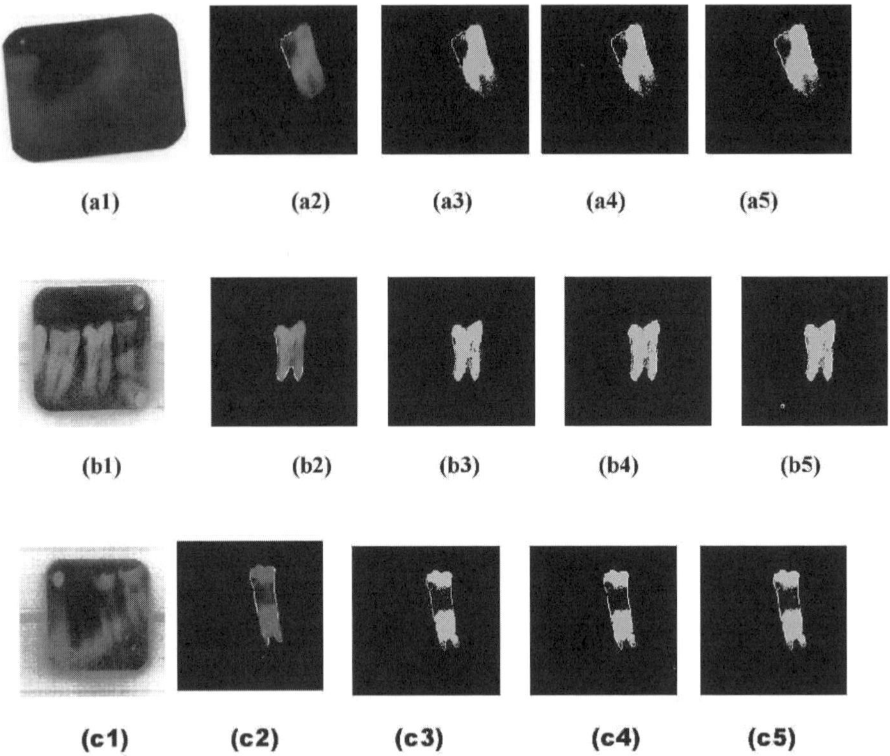

Fig. 12.5 Dental Image

enclosing the tooth, then the carries is classified as palpal. If on the other hand there exists two or more number of black regions and the width of the black region is less than 2 mm then it is Enamel carry. It may be pointed out here that the thickness of enamel around the tooth is approximately 2 mm. Alternately if it is more than 2 mm that means it is dentinal carries. The result of the above procedure implemented on two different sets of enamal caries, dentinal caries and pulpal caries are shown in Figure 12.5.

12.11 MAMMOGRAM IMAGE ANALYSIS

The detection and classification of various types of tumors in digital mammograms using mammogram image analysis system has found paramount importance in recent times [25]–[28]. Breast masses, both non-cancerous and cancerous lesions, appear as white regions in mammogram films. The fatty tissues appear as black regions, while other components of the breast, such as glands, connective tissue, tumors, calcium deposits, etc appear as shades of gray, more towards the brighter intensity on a mammogram. Some types of breast cancer show *signs of invasion* or *spread* - they are spreading types, while in some others there is no evidence of invasion, and these are called non-spreading types. The breast cancer starts as the non-spreading type, and later starts spreading and invading as the cancer cells become more abnormal

Ductal Carcinoma In Situ (DCIS) is a non-spreading cancers, which is detected during routine mammography.

12.11.1 Breast Ultrasound

Mammography continues to be the most widely used imaging device, because of its cost-effectiveness and accuracy. There are, however, other methods of imaging the breast.

Ultrasound uses pulses of high-frequency sound waves to detect abnormalities in the breast. One of the major advantages of ultrasound is that, it can distinguish a benign fluid-filled cyst from a solid lesion that may be cancerous. Cystic areas are rarely cancerous. The ultrasound does not always detect micro calcifications, microscopic deposits of calcium which are often indicative of breast cancer.

12.11.2 Steps in Mammogram Image Analysis

The steps in a mammogram image analysis system are as follows.

Step 1. Enhancement of mammogram image

Step 2. Suspicious area segmentation

Step 3. Feature extraction:

Step 4. Classification of masses

Step 5. Analysis of classifier performance.

12.11.3 Enhancement of Mammograms

The first step in the analysis of mammogram images involves enhancement or denoising of the image. The denoising technique should not deteriorate or destroy the information content in the image. The objective is to remove background noise while preserving the edge information of suspicious areas in the images.

Various enhancement techniques, viz, histogram equalization, neighborhood based contrast enhancement algorithm, selective Median filtering enhancement method based on multiscale analysis etc. may be used for enhancing the contrast of mammogram images.

A mammographic image utilizes a narrow range of gray levels, without a well-defined histogram structure. Thus conventional histogram equalization techniques may not be suitable in enhancing the mammogram images.

One may attempt fuzzy spatial filtering for enhancement of mammogram images. The conventional spatial filtering utilizes an averaging procedure to generate the smooth image. The weights used for averaging are crisp and invariant to image data. Thus all the pixels inside the window region of the image, which can be brought under an arbitrary neighborhood W, are equally affected. The conventional spatial filtering method by masking or averaging does not take into account the effect of the difference of gray levels between the central pixel and a neighboring pixel and it does not always take into account the diminishing influence of the pixels that are situated in increasing distance from the central pixel.

These two limitations of the conventional spatial filtering (averaging) are taken care of while designing the fuzzy spatial filter.

12.11.4 Suspicious Area Detection

The efficiency of a mammogram image analyzer depends on the effective segmentation of the lesion region in an image. The mammogram images are complicated images where segmentation of not only the uniform smooth regions of images is required but also the discrete cluttered portions. So a novel approach towards development of such an adaptive thresholding technique using the concepts of wavelets and Multi-resolution analysis may possibly give better results.

A suspicious area is a tumor area – benign or otherwise. A tumor like template may be defined based on several characteristics of benign tumor areas, namely brightness contrast, uniform density, approx circular shape etc.

The detection accuracy will be evaluated in terms of the number of true positives (TP) for a given number of false positives (FP) detections. TP is an object whose area overlaps the centroid of the biopsy proven areas. FP comprises of all other objects classified as potential masses.

Petrick et al. [29] employed density weighted contrast enhancement (DWCE) segmentation algorithm to extract lesions from their surrounding tissues. Markov random elds have been used to classify the different regions in a mammogram based on texture[30]. Segmenting lesions in digital mammograms using a radial gradient index (RGI) based algorithm and a probabilistic algorithm were reported in [31]. These techniques are seeded segmentation algorithms. In their work, RGI is a measure of the average proportion of the gradients directed radially outward. Because of inherent nature of RGI, it gives more emphasis to round shaped lesions. Other interesting techniques of automated seeded Lesion Segmentation in Digital Mammogram and detection of circumscribed masses in mammograms, may be found in [32, 33]

The goal of seeded lesion segmentation is to separate suspected masses from surrounding tissues as effectively as possible. Segmentation of lesion regions in a mammographic image is not easy due to the low contrast and the fuzzy nature of the transition of a malignant lesion from its core region to surrounding tissues.

Here we present a method for lesion segmentation based on mean shift algorithm, a simple nonparametric procedure for estimating density gradients [32]. Iteration of mean shift gives rise to natural clustering algorithm. An objective measure has been developed to differentiate a target T (lesion) from its background B (Tissue). The performance of the proposed method is compared against RGI method. It results in more meaningful features being extracted from potential lesion regions, and, ultimately, in better classification of malignant lesions from normal tissue regions.

12.11.5 LESION SEGMENTATION

The lesion tissues are usually brighter than its surrounding tissues and possess uniform density inside the area. The malignant lesion area possesses an approximately uniform and more or less compact circular shape of varying size [?]. Quite often they are observed to possess fuzzy edges. Given a sub image or region-of-interest (ROI) of dimension n by m containing the suspect lesion, the set of coordinates in the ROI is represented by

$$I = \{(i,j) : i = 1, 2, \cdots, n \text{ and } j = 1, 2, \cdots, m\}.$$

The image function describing the pixel gray levels of this sub image is given by $f(i,j)$. The pixel values of $f(i,j)$ are normalized between 0 (black) and 1 (white).

Lesions tend to be compact, meaning that their shapes are typically convex. To incorporate this knowledge into the creation of the partitions, the original image is multiplied by a function, called the constraint function that suppresses distant pixel values. One may use an isotropic Gaussian function centered on the seed point location (μ_i, μ_j) with a fixed variance σ^2 as the constraint function. The function $h(i,j)$ resulting from the multiplication of the original ROI with the constraint function is given by

$$h(i,j) = f(i,j)N(i,j;\mu_i,\mu_j,\sigma^2), \tag{12.7}$$

where $N(i,j;\mu_i,\mu_j,\sigma^2)$ is circular normal distribution centered at (μ_i, μ_j) with a variance σ^2.

12.11.5.1 Clustering based on non-parametric density-estimation There are several nonparametric methods available for probability density estimation: histogram, naive method, the nearest neighbor method, and kernel estimation. The kernel estimation method is one of the most popular techniques used in estimating density. Given a set of n data points x_i, $i = 1, 2, \cdots n$ in a d-dimensional Euclidian space, the multivariate kernel density estimator with kernel ϕ and window radius (bandwidth) h_n is defined as

$$p_n(x) = \frac{1}{n} \sum_{i=1}^{n} \frac{1}{V_n} \phi \left(\frac{x - x_i}{h_n} \right) \tag{12.8}$$

After density estimation we identify candidate-clusters by using gradient ascent (hill-climbing) to pinpoint local maxima of the density $p_n(x)$. Specifically, the k-nearest neighbors of every point is determined, whereupon each point is linked to the point of highest density among these neighbors (possibly itself). Upon iteration, this procedure ends up assigning each point to a nearby density-maximum, thus carving up the data set in compact and dense clumps.

12.11.5.2 Density Gradient Estimation and Mean Shift Method Application of the mean shift leads to the steepest ascent with a varying step size according to the magnitude of the gradient [34, 35] Assuming that the probability density function $p(x)$ of the p-dimensional feature vectors x is multimodal and also assuming that a small sphere S_x of radius r centered on x contains the feature vectors y such that $|y - x| \leq r$. The expected value of the vector $z = y - x$, given x and S_x, may be derived as

$$E\left(x \,|\, x \in S_x\right) - x = \frac{r^2}{p+2} . \frac{\nabla p(x)}{p(x)} \tag{12.9}$$

The mean shift vector, which is the vector of difference between the local mean and the center of the window, is proportional to the gradient of the probability density at x. The proportionality factor is reciprocal to $p(x)$. This is beneficial when the highest density region of the probability density function is sought. Such region corresponds to large $p(x)$, i.e. small mean shifts. On the other hand, low-density regions correspond to large mean shifts. The shifts are always in the direction of probability density maximum, which describes the mode. At the mode the mean shift is close to zero. The mean shift algorithm is tool needed for feature space analysis. The unimodality condition, assumed during derivation of Eq. 12.9, can be relaxed and extended to multimodal conditions by randomly choosing the initial location of the search window. The algorithm then converges to the closest high-density region.

The width of the constraint function in Eq. 12.8 should be chosen based on the knowledge of lesions. Within the same segmentation class an image containing large homogeneous regions should be analyzed at higher resolution than an image with many textured areas. The simplest measure of the visual activity can be derived from the global covariance matrix.

12.11.5.3 Objective Measure for Segmentation The basic idea involves developing an objective measure to differentiate a target T (lesion) from its background B (Tissue) within the mammographic image. The target T has a greater density within the mammogram, thus having higher mean gray level intensity component compared to the surrounding background B. A good objective measure should aim to yield high value at the point where the contrast between target T and background B is high. The measure is initially computed by determining the difference between mean gray values in the target and background areas as

$$\delta_\mu = \mu_T - \mu_B \tag{12.10}$$

In addition, the objective measure should at the same time yield low values with the spread of gray scales in the target area compared with the background area. This reduction can be achieved by the ratio of the gray level variances as

$$\delta_\sigma = \frac{\sigma_T}{\sigma_B} \tag{12.11}$$

The resulting target to background contrast ratio R can be computed as

$$R = \frac{\delta_\mu}{\delta_\sigma} \tag{12.12}$$

This effective segmentation measure leads to a large value of R when the contrast between target T and background B is high. In order to quantify the performance differences between the different segmentation methods, an overlap measure (O) may be used as follows:

$$O = \frac{Area(L \bigcap T)}{Area(L \bigcup T)} \tag{12.13}$$

where L and T are the segmented region obtained from the mean shift algorithm based segmentation and that obtained by the expert radiologist respectively. The value of O lies between between 0 (no overlap) and 1 (exact overlap).

Figure 12.6(a) shows a set of noisy mammogram images. Figure 12.6(b) shows the corresponding segmented regions of the lesions in the mammogram images in Figure 12.6(a) using mean shift algorithm and Figure 12.6(c) shows the same using RGI algorithm.

12.11.6 Feature Selection and Extraction

The methods for classification of non-tumor, benign, malignant masses utilize the area measurement, signature of boundary shape, edge distance and edge intensity variation, etc. Contour modelling generates boundary model and measures the similarity between the generated model and extracted region. The local mean intensity, tumor circularity, normalized radial length, mean and standard deviation of the normalized radial length along with the Fourier descriptor based shape factor, moment based shape factors etc are some of the prominent features. The selection of optimal set of features is an important decision in the final classification of the mammogram. Also features like, area, shape descriptor, edge intensity variation features, etc., are used for

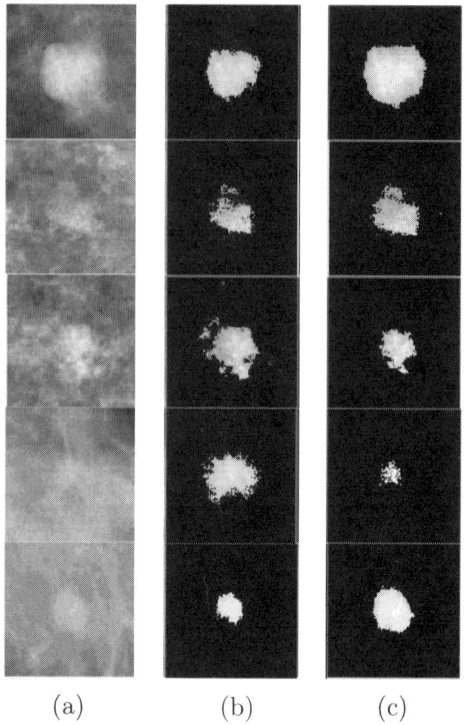

Fig. 12.6 (a) Original Mammogram Images, (b) segmented regions using mean shift, and (c) segmented regions using RGI.

tumor and non-tumor classification. An expert system provides the user with a facility to classify various test mammogram images for diagnosis. The design of the expert system uses a set of fuzzy rules capable of handling the pervasive fuzziness of information in the knowledge base for mammogram interpretation.

12.11.7 Wavelet Analysis of Mammogram Image

The given mammogram image is decomposed into multi-resolution hierarchy of localized information at different spatial frequencies. Multi-scale wavelet representation suggests a mathematical coherent basis not only for existing multi-grid techniques but also for exploiting non-linear systems. Multi-resolution wavelet analysis provides a natural hierarchy in which to embed an interactive paradigm for accomplishing scale-space feature analysis. Similar to tradition coarse to fine matching strategies, the radiologist may first choose to look for coarse features (e.g. Dominant masses) within low frequency levels of wavelet transform and later examine finer features (e.g. Micro calcifications) at higher frequency levels. Choosing wavelets that are simultaneously localized in both space and frequency results in powerful methodology for image analysis. The following are the key features of wavelet analysis of mammogram image:

1. By modifying wavelet coefficients associated with multi-scale edges from level 2 alone improves the local contrast of both micro and macro-calcifications clusters not visible in original low-contrast mammogram.

2. Radiologists have observed that the subtle features including calcifications and the penetration of fibro glandular structure into the obvious mass tissue are made clearer. In addition, the geometric shapes of calcifications (important for diagnosis) are made more visible and improved definition is seen in the ductules (intra and extra lobular units) as well as in arterial structure within the less dense tissue of breast.

3. A mammogram is generally dominated by low frequency information while micro-calcifications appear in High frequency bands. This makes wavelet methods suitable for extraction of micro-calcifications since wavelets allow separation of image into frequency bands without affecting spatial locality.

4. Calcifications in general are small and they appear in certain levels of wavelet decomposition of the image.

Also, from the biological mechanisms of HVS, multi-resolution and multi-orientation are known features of HVS. There exist cortical neurons, which respond specifically to stimuli within certain orientations and frequencies.

12.12 SUMMARY

Biometric and biomedical imaging are important areas of image processing applications today. In this chapter, we have reviewed some of the techniques on automatic identification of biometric features such as human faces and signatures. A number of feature extraction techniques based on front and side facial images have been described. Similarly, a novel waterfall based algorithm for extraction of signature features has been discussed. Chest radiographs and dental imageries are two most important application areas in biomedical imaging. We have discussed techniques of enhancement, segmentation and classification of diseases related to chest and dental abnormalities in this chapter. Digital mammogram is another important medical imaging technique. It is widely being used by medical practitioners to detect breast cancers and other applications. We presented a scheme for automated identification of lesions in digital mammogram.

REFERENCES

1. L. D. Harmon, M. K. Khan, R. Lasch, and P. F. Ramig, "Machine identification of human faces," *Pattern Recognition*, 13(2), 1981, 97–110.

2. R. Chellapa, C. Wilson, and S. Sirohey, "Human and machine recognition of faces: A survey," *Proc. IEEE*, 83(5), 1995, 705–740.

3. A. Samal and P. A. Iyengar, "Automatic recognition and analysis of human faces and facial expressions: A survey," *Pattern Recognition*, 25, 1992, 65–77.

4. O. Nakamura, S. Mathur, and T. Minami, "Identification of human face based on isodensity maps," *Patterrn Recognition*, 24(3), 1991, 263–272.

5. S. Y. Lee, Y. K. Ham, and R. H. Park, "Recognition of human front faces using knowledge based feature extraction and neuro fuzzy algorithm," *Pattern Recognition*, 29(11), 1996, 1863–1876.

6. M. Turk and A. Pentland, "Eigenfaces for recognition," *The Journal of Cognitive Neuroscience*, 3(1), 1991, 71–86.

7. P. N. Belhumeur, J. P. Hespanha, and D. J. Kriegman, "Eigenfaces vs. Fisherfaces: Recognition Using Class Specific Linear Projection," *IEEE Tran. on Pattern Analysis and Machine Intelligence*, 19(7), 1997, 711–720.

8. X. He, S. Yan, Y. Hu, P. Niyogi, and H. J. Zhang, "Face Recognition Using Laplacianfaces," *IEEE Tran. on Pattern Analysis and Machine Intelligence*, 27(3), 2005, 328–340.

9. W. Nemcek and W. Lin, "Experimental investigation of automatic signature verification," *IEEE Transactions on System, Man, and Cybernetics*, 4, 1974, 121–126.

10. Y. Qi and B. R. Hunt, "Signature verification using global and grid features," *Pattern Recognition*, 27(12), 1994, 1621–1629.

11. K. Han and I. K. Sethi, "Handwritten signature retrieval and identification," *Pattern Recognition Letters*, 17, 1996, 83–90.

12. A. K. Ray and T. Acharya, "Information Technology: Principles and Applications", *Prentice Hall of India*, New Delhi, 2004.

13. E. L. Hall, R. P. Kruger, S. J. Dwyer, D. L. Hall, G. W. Mclaren, and G. S. Lodwick, "A survey of pre-processing and feature extraction techniques for radio- graphic images," *IEEE Transactions on Computers*, Vol-C-20, September 1971, 1032–1044.

14. R. N. Sutton and E. L. Hall, "Texture measures for automatic classification of pulmonary diseases," *IEEE Tran. on Computers*, 21, July 1972, 667–676.

15. J. S. DaPonte and M. D. Fox, "Enhancement of chest radiographs with gradient operators," *IEEE Tran. on Medical imaging*, 7(2), June 1988, 109–117.

16. J. Rogowaska, K. Preston Jr., and D. Sashin, "Evaluation of digital unsharp masking and local contrast stretching as applied to chest radiographs," *IEEE Tran. on Biomedical Engineering*, 35(10), October 1988, 817–827.

17. T. L. Ji, M. K. Sunderasan, and H. Roehrig, "Adaptive image contrast enhancement based on human visual properties," *IEEE Tran. on Medical imaging*, 13(4), December 1994, 573–586.

18. M. F. McNitt-Gray, H. K. Huang, and J. W. Sayre, "Feature selection in the pattern classification problem of digital chest radiograph segmentation," *IEEE Tran. on Medical imaging*, 14(3), October 1995, 537–547.

19. M. I. Sezan, A. M. Tekalp, and R. Schatzing, "Automatic anatomically selective image enhancement in digital chest radiography," *IEEE Trans. On Medical Imaging*, 8, 1989, 154–162.

20. M. Penedo, M. Carreira, A. Mosquera, and D. Cabello, "Computer-aided diagnosis: A neural-network-based approach to lung nodule detection," *IEEE Trans. Medical Imaging*, 17(6), 1998, 872–880.

21. D. Terzopoulos and S. W.Zucker, "Detection of Osteoporosis imperfecta by automated Texture analysis," *Computer graphics and image processing*, 20, 1982, 229–243.

22. Z. Yue, A. Goshtasby, and L. V. Ackerman, "Automatic detection of Rib boarders in Chest Radiograph," *IEEE Transactions on Medical imaging*, 14(3), September 1995, 525-536.

23. P. F. Van Der stelt and Q. G. M. Geraets, "Computer-aided interpretation and quantification of angular periodontal bone defects on dental radiographs," *IEEE Transactions on Biomedical engineering*, 38(4), 1991, 334–338.

24. G. C. Burdea, S. M. Dunn, C. H. Immendorf, and M. Mallick, "Real time sensing of tooth position for dental digital subtraction radiography," *IEEE Transactions on Biomedical Engineering*, 38(4), 1991, 366–378.

25. A. F. Laine, S. Schuler, J. Fan, and W. Huda, "Mammographic feature enhancement by multiscale analysis," *IEEE Trans. Medical Imaging*, 13(4), 1994, 725–752.

26. H. Kobatake, M. Murakami, H. Takeo, and S. Nawano, "Computerized detection of malignant tumors on digital mammograms," *IEEE Trans. Medical Imaging*, 18(5), May 1999, 369–378.

27. S.C. B. Lo, H. Li, Y. Wang, L. Kinnard, and M. T. Freedman, "A multiple circular path convolution neural network system for detection of mammographic masses," *IEEE Trans. Medical Imaging*, 21(2), 2002, 150–158.

28. J. Kildaus, F. Palmieri, and M. D. Fox, "Classifying mammographic lesions using computerized image analysis," *IEEE Trans. Medical Imaging*, 12(4), 1993, 664–669.

29. N. Petrick, H. P. Chan, D. Wei, B. Sahiner, M. A. Helvie, and D. D. Adler, "Automated detection of breast masses on mammograms using adaptive contrast enhancement and texture classification," *Int Journ. Med. Phys.*, 23(10), 1996, 1685–1696.

30. H. D. Li, M. Kallergi, L. P. Clarke, and V. K. Jain, "Markov random field for tumor detection in digital mammography," *IEEE Trans. Med. Imag.*, 14, 1995, 565–576.

31. M. A. Kupinski and M. Giger, "Automated Seeded Lesion Segmentation on Digital Mammograms," *IEEE Trans. Med. Imag.*, 17, 1998, 510–517.

32. O. Nagaraju and A. K. Ray, "Automated seeded Lesion Segmentation in Digital Mammogram," *Proc. Indian Conference on Medical Informatics and Telemedicine*, Indian Institute of Technology, Kharagpur, India, 193–196, 2005.

33. S. M. Lai, X. Li, and W. Bischof, "On techniques for detecting circumscribed masses in mammograms," *IEEE Trans. Med. Imaging* 8, 1989, 377–386.

34. K. Fukunaga, Introduction to Statistical Pattern recognition, *Academic Press*, 1990.

35. Y. Cheng, "Mean Shift, Mode Seeking, and Clustering," *IEEE Trans. Pattern Analysis and Machine Intelligence*, 17(8), August 1995, 790–799.

13

Remotely Sensed Multispectral Scene Analysis

13.1 INTRODUCTION

Information regarding the natural resources of a country is extremely useful for planning purposes. Resources include agricultural resources (e.g., food stock like rice, wheat, vegetables, etc.), hydrological resources (e.g., water bodies like rivers, canals, ponds, etc.), mineral resources (e.g., metal mines, coal, etc), forest resources, geological and strategic resources, etc. Such information can be conveniently extracted from remote sensing images. Remote sensing techniques utilize satellite and aerial image data to estimate the available resources in a country. For remotely sensed scene analysis, the images of objects on the earth's surface are captured by sensors in remote sensing satellites or by multispectral scanners housed in an aircraft. These images are then transmitted to the earth station for further processing.

Digital processing of remotely sensed image data has been of great importance in recent times [1]–[4]. The availability of digital pictures of earth's surface from various sources, such as LANDSAT multispectral imageries, SPOT high-resolution multiband and panchromatic data, IRS (Indian Remote Sensing) Satellite, and MODIS from Terra (EOS AM) and Aqua (EOS PM) satellites, etc., have enabled researchers to pursue extensive low-level and high-level processing tasks on these satellite borne images of the earth's terrain.

The classification of remotely sensed data involves procedures for assigning the multiband image pixels obtained from remotely sensed satellite imageries to an appropriate set of meaningful object classes. From the classified images, useful information can be made available for the proper utilization of natural resources.

Various approaches, such as statistical, knowledge-based, and neural network based methods have been proposed for the classification of the remotely sensed scenes. With an increase in the spatial, spectral, and temporal resolution of the sensing instruments, more advanced and efficient techniques are required for the classification and interpretation of remotely sensed data. In this chapter, we present brief description of the various satellite data along with some of the techniques of satellite based image analysis.

13.2 SATELLITE SENSORS AND IMAGERIES

Remotely sensed images of the earth are captured from satellites or aircrafts for agricultural, hydrological, geological, military and many other applications. Prominent among satellite based systems are LANDSAT *Multi Spectral Scanner* (MSS), *Thematic Mapper* (TM), *SPOT*, *Indian Remote Sensing Satellite* (IRS), *Moderate Resolution Imaging Spectroradiometer* (MODIS), *Multiangle Imaging Spectro Radiometer* (MISR), and *Synthetic Aperture Radar* (SAR).

13.2.1 LANDSAT Satellite Images

The first remote sensing satellite LANDSAT was launched by United States in 1972. It contained onboard image vidicon cameras operating in three separate visible wave lengths bands. The multispectral scanning system of LANDSAT provided four-band images of the earth's surface. These four-band images were captured in four wave-lengths; three in the visible and one in the near-infrared portion of the electromagnetic spectrum. The satellite located at about 912 kilometer away from the earth completes one revolution around the earth in around 1 hour 43 minutes. It completes fourteen orbits a day with repetitive coverage of eighteen days. Four more LANDSAT satellite were subsequently launched which were located at an orbital altitude of 705 kilometer. The LANDSAT-MSS sensors having cross track scanning were incorporated in all the five LANDSAT satellites. The spatial resolution of LANDSAT MSS data is 56 m $\times$ 79 m.

LANDSAT-TM (Thematic Mapper) having cross-track scanner with bidirectional oscillating scan mirror and arrays of detectors has seven spectral bands, ranging from the visible to infra red region with 30 m spatial resolution, improved detector sensitivity and radiometric resolution. The satellite also had sensor in the thermal IR band with 120m resolution. The seven band Thematic Mapper (TM) sensor provides information with greater radiometric precision. It yields much better boundaries and accentuates textural appearance of the different categories of objects.

SPOT, a French satellite, provides three-band data of 20 m resolution, and also a single band 10 m Panchromatic data.

13.2.2 Indian Remote Sensing Satellite Imageries

The *Indian Remote Sensing Satellite* (IRS) has a scanner known as *Linear Image scanning System* (LISS). Various combination of bands (which were adopted in LISS-I and LISS-II cameras of IRS series of satellites) provide information for specific purposes. LISS-I Camera has spatial resolution of 72.5 m having a viewing width of 148.48 km. Two cameras each of LISS-II, LISS-IIA & LISS-IIB, with spatial resolution of 36.25 m and a strip width of 74.24 km, provide a combined strip width of 145.45 km allowing a 3 km overlap. The utilities of each band are as follows:

- Band 1 (0.45–0.52 μm): In this band, plant pigments exhibit a strong spectral reflectance. There is a comparatively higher penetration in water in this band, which is useful for mapping suspended sediments/water quality and various under water studies in the coastal region.
- Band 2 (0.52–0.59 μm): The discrimination of vegetation cover, and also the identification of iron oxides may be best achieved in this band.
- Band 3 (0.62–0.68 μm): This band is centered near the chlorophyll absorption band of vegetation, and is useful for the identification of plant species.
- Band 4 (0.77–0.86 μm):Healthy vegetation exhibits high reflectance in this band and hence this band is useful for green biomass estimation, crop vigor studies, etc. Water absorption in this region clearly demarcates land and water boundary.

IRS-IB was launched as a backup for IRS-IA, The second generation remote sensing satellite IRS-IC has a polar sun-synchronous orbit at an altitude of 817 km and IRS-ID provides continuity of data flow for earth monitoring. These two satellites have been provided with improved versions of LISS- III, WIFS (Wide Field Sensor) and PAN cameras for multiple purposes.

13.2.3 Moderate Resolution Imaging Spectroradiometer (MODIS)

MODIS is a key instrument onboard the *Terra* (EOS AM) and *Aqua* (EOS PM) satellites, collecting data in 36 spectral bands. It provides high radiometric sensitivity (12 bit)in 36 spectral bands ranging from 0.4μm to 14.4μm. Two bands are imaged at a nominal resolution of 250 m at nadir, with five-bands at 500 m, and the remaining 29 bands at 1 km resolution. The images captured from MODIS of Terra and Aqua satellites are shown in Figure 13.1. Their color version of Figure 13.1 is shown in color pages section.

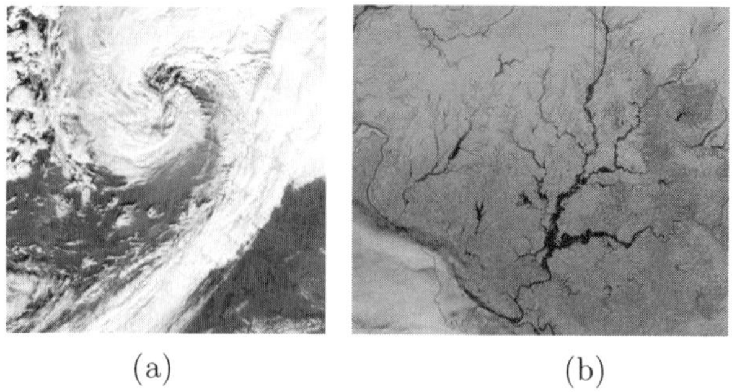

Fig. 13.1 Remotely sensed images from MODIS

The information of the various bands of MODIS are as follows:

- Bands 1–2 (620–876 nm): Land, cloud and aerosols boundary.
- Bands 3–7 (459–2155 nm): Land, cloud and aerosols properties.
- Bands 8–16 (405–877 nm): Ocean color, phytoplankton, and biogeochemistry.
- Bands 17–19 (890–965 nm): Atmospheric water vapor.
- Bands 20–23 (3.660–4.080 μm): Surface/cloud temperature.
- Bands 24–25 (4.433–4.549 μm): Atmospheric temperature.
- Bands 26–28 (1.360–7.475 μm): Cirrus clouds, water vapor.
- Band 29 (8.400–8.700 μm): Cloud properties.
- Band 30 (9.580–9.880 μm): Ozone.
- Bands 31–32 (10.780–12.270 μm): Surface/cloud temperature.
- Bands 33–36 (13.285–14.385 μm): Cloud top altitude.

13.2.4 Synthetic Aperture Radar (SAR)

It is a coherent microwave imaging method, which plays an important role in remote sensing. Many physical and geometric parameters of the earth objects in the scene contribute to the gray values of a SAR image pixel [2]. SAR is able to reliably map the Earth's surface and acquire information about its physical properties, such as topography, morphology and roughness. SAR can be most beneficially used over land, ice, and space surfaces. As the space borne SAR systems operate in the microwave region of the spectrum and provide their own illumination, they can acquire information globally and almost independently of meteorological conditions and the sun illumination. They are, therefore, most suitable for operational monitoring tasks. The side looking imaging geometry, pulse compression techniques as well as the synthetic aperture concept are employed to achieve geometric resolutions in the order of tens of meters with physical antennas of modest size. The use of space-borne SAR as interferometers also serves as an extremely powerful tool for mapping the Earth's land, ice, and even the sea surface topography.

13.3 FEATURES OF MULTISPECTRAL IMAGES

Remote sensing images are available in two forms–photographic film form and digital form. Variations in the scene characteristics are represented as variations in brightness on photographic films. A particular part of a scene reflecting more energy appears bright, while a different part of the same scene reflecting lesser energy appears relatively dark. The pixel intensity depicts the average radiance of a relatively area within a remotely sensed scene. The size of this area affects the reproduction of details within the scene. With reduction in the pixel size greater scene detail is preserved in the digital representation.

13.3.1 Data Formats for Digital Satellite Imagery

Although there is no fixed standard for the storage and transfer of remotely sensed data, the CEOS (Committee on Earth Observation Satellites) format is a widely accepted standard. An image consisting of four spectral channels can be visualized as four superimposed images, with corresponding pixels in each band registering exactly to those in the other bands. Remote sensing data are organized using one of the following three common formats:

- Band Interleaved by Pixel (BIP)
- Band Interleaved by Line (BIL)
- Band Sequential (BQ)

13.3.2 Distortions and Corrections

The satellite images transmitted to the earth stations are associated with various types of distortions and each distortion has specific corrective strategy, such as, radiometric corrections and geometric corrections.

Radiometric Distortions: The radiation from the sun is incident on the ground pixel and then gets reflected to the sensor. The molecules of oxygen, carbon-di-oxide, ozone, and water present in the atmosphere attenuate the radiation very strongly in certain wavelengths. Scattering by these atmospheric particles is the dominant mechanism that leads to radiometric distortion in image data. Radiometric Corrections are carried out when an image is recorded by sensors which contain errors in the measured brightness values of pixels. These errors are referred to as radiometric errors and can result from

- The instruments used to record the data

- The effects of the atmosphere

Radiometric processing influences the brightness values of an image by correcting for sensor malfunctions or by adjusting the values to compensate for atmospheric degradation. Radiometric distortion may result in a situation where the relative distribution of brightness over an image in a given band can be different from that in the ground scene. Also sometimes the relative brightness of a single pixel from one band to another can be distorted compared to the spectral reflectance characteristics of the corresponding region on the ground.

The following procedures are used for the removal of the above defects:

1. **Duplicating correction:** Sometimes, due to malfunctional detectors, a set of adjacent pixels may contain spurious intensity values. In such cases the defective lines may be replaced by either a duplicate of the previous line or the next line. If the value of the spurious pixel at position (x, y) is $f(x, y)$, then it is cleaned by the rule $F(x, y) = f(x, y-1)$ or $f(x, y+1)$. Even the average of the two lines also yields good results, i.e., $F(x, y) = [(f(x, y-1) + f(x, y+1)]/2$.

2. **Destripping correction:** Sometimes a set of detectors in a certain spectral band may go out of adjustment. This may result in an image, where patterns of lines with consistently high or low intensity values occur repeatedly. Such horizontal strips of patterns from satellite images need correction. The destripping correction enhances the visual quality of the image. It also enhances the objective information content of the image.

3. **Geometric correction:** While capturing images of the earth's surface, one needs to consider the curvature of the earth, platform motion, and nonlinearities in the scanning motion, which produce geometrical distortions in satellite images. These distortions are corrected to produce rectified images.

The resultant images, after correction, may still lack in contrast and image enhancement techniques are useful for getting better quality images for further processing.

13.4 SPECTRAL REFLECTANCE OF VARIOUS EARTH OBJECTS

Different categories of earth objects have different sensitivities at a particular spectral band. For example, plant biomass is sensitive at 0.77 μm to 0.86 μm while chlorophyll has more sensitivity in 0.62 μm to 0.68 μm. Green reflectance of vegetation is sensitive at 0.52 μm to 0.59 μm spectral band. The spectral reflectance of various features on the earth's surface are discussed briefly below.

13.4.1 Water Regions

In multispectral satellite images, water is the unique land cover type for which the band reflectance values decrease as the band number increases, except in band 6 which is the thermal band. In other words, if pixel (i, j) lies entirely within a water region, then $b_{ij}^1 > b_{ij}^2 > b_{ij}^3 > b_{ij}^4 > b_{ij}^5 > b_{ij}^7$, where the intensity vector b_{ij} at each pixel (i, j) contains the seven band reflectances: $b_{ij} = [b_{ij}^1, b_{ij}^2, \cdots, b_{ij}^7]$.

It is easy to locate water bodies in remote sensing images using this property. However, the various types of water bodies, like clear water and turbid water, manifest different properties in the visible wavelengths. Clear water absorbs relatively little energy in the wavelength, less than 0.6 μm. However, as the turbidity of the water changes (because of the presence of organic and inorganic materials), the reflectance changes dramatically. For example, water containing large quantities of suspended materials derived from the erosion rocks, soil, etc. normally has much higher visible reflectance than clear water. Moreover, water absorbs energy in the near- and far-infrared wavelengths.

13.4.2 Vegetation Regions

Green vegetation usually has a high reflectance at near-infrared wavelengths (band 4 in LANDSAT TM data) and a low-reflectance at the red wavelengths (band 3 in TM data). Thus it is possible to discriminate between two different plant species using a feature which is a ratio of the reflectances at these two wavelengths. This index, known as *vegetation index* at pixel (i, j) of the scene, is defined as

$$\Gamma_{veg} = \frac{r_{ij}^{(4)}}{r_{ij}^{(3)}} \tag{13.1}$$

where $r_{ij}^{(k)}$ represents the reflectance at the kth band for vegetation class.

Depending on the value of the vegetation index, a pixel is classified as either belonging to a vegetation class or nonvegetation class. The spectral reflectance curve for green (healthy) vegetation, almost always manifests the *peak and valley* configuration. The pigments in the plant leaves are responsible for the valleys in the visible portion of the spectrum.

Chlorophyll present in green plants strongly absorbs energy in the wavelength bands centered at around 0.45 μm and 0.65 μm. The plant leaves in general absorb the red and blue energies very heavily and strongly reflect green energy. Plant reflectance in the 0.7 μm to 1.3 μm range is largely due to the internal structure of plant leaves, which varies greatly from one plant species to another. Hence, the measurement of reflectance values in this range often permits us to discriminate one plant species from another. One may observe a decrease in the reflectance of plant leaves at 1.4 μm, 1.9 μm, and 2.7 μm, because water molecules in the leaves strongly absorbs the radiation at these wavelengths.

13.4.3 Soil

Soil usually does not exhibit significant variation in reflectance in the entire visible band. However two interrelated factors affecting soil reflectance are the moisture content and the texture of soil. For example, the presence of moisture in the soil decreases the reflectance. The moisture content of soil again strongly depends upon its texture; coarse sandy soils are usually well drained resulting in low moisture content and relatively high spectral reflectance. Poorly drained (high textured) soils, on the other hand, generally have lower reflectance. Two other factors that reduce soil reflectance are surface roughness and organic matter content. The presence of iron oxide in soil also significantly decrease reflectance in the visible wavelength.

13.4.4 Man-made/Artificial Objects

Interestingly, there does not exist unique spectral signature of some man-made materials like urban/suburban features. Numerous man-made materials have roughly the same spectral reflectance properties. Thus the spectral reflectance of urban scenes obtained from remote sensing satellites is often not unique, and accurate identification require a domain knowledge of the scene, in addition to simple spectral response. The features like size, shape, and texture information of the remotely sensed scene are essential for accurate identification and classification of urban features in the scene. Often shadows provide useful information for the interpretation of artificial objects. Concrete and asphalt pavements have spectral reflectances typical of man-made materials, i.e., generally increasing reflectance with increasing wavelengths. The brightness of asphalt varies with the age of the pavement as a result of weathering, i.e., new asphalt is relatively dark and old asphalt is relatively bright. The shape of the spectral profile is highly variable due to the influence of vehicles, paint markings, oil marks, etc. Asphalt can be discriminated from the roofs of building structures on the basis of differences in contrast, i.e., asphalt generally has a lower brightness and temperature compared to the roofs (of the buildings, because the ground base below the pavement acts as a heat sink conducting heat away from the surface). Roofs generally have a higher reflectance than asphalt because of the lack of such a heat sink.

13.5 SCENE CLASSIFICATION STRATEGIES

One of the most important aspects of remotely sensed digital image processing involves segmentation and pixel classification of the image data. Usually the maximum likelihood classifier based on Bayes' decision theory is employed to classify the pixels.

It has been observed that Bayes decision theory works well when the probability density function of each class of earth objects is estimated properly. In such cases one achieves very high recognition accuracy while classifying the pixels in a multispectral remotely sensed image. We present two neural network architectures, namely backpropagation and counterpropagation networks, for remotely sensed satellite image classification.

13.5.1 Neural Network-Based Classifier Using Error Backpropagation

Neural networks are promising techniques for the classification of remotely sensed data [5]–[8].

To classify multispectral data using a neural network, the spectral values in different bands are mapped onto a set of input neurons. Different schemes can be used for input data representation. Some researchers used one input node (neuron) for each spectral band of the satellite image [7]. One may use other options for feeding the multispectral data to the network. For making the neural network classification invariant to the changes in the gray values of the image due to seasonal variations, normalization may be performed in each spectral band of the image. The normalized gray value of a pixel in a spectral band may be chosen as

$$f_{norm} = \frac{f_p - f_{min}}{f_{max} - f_{min}}. \tag{13.2}$$

Here f_{min} and f_{max} are the minimum and maximum gray values in a spectral band and f_p is the gray value of the pixel under consideration. Each class is represented by one output neuron in the output layer, i.e., the number of output nodes is equal to the number of categories of the object classes. An input pattern is assigned to class i, if the output node i has the highest activation of all output nodes. If either the activation of all output nodes is below a specified threshold, or the activations of two or more output nodes are above threshold and are very close, then the pixel class cannot be ascertained.

The *multilayer perceptron* (MLP) network is trained with finitely large number of input samples, i.e., pixels belonging to each class of regions present in the scene input. Once the network training is over, i.e., it has converged, the network is tested using test pixel samples. During the training phase, experiments should be carried out by varying the parameters like learning constant, momentum, etc., and the performance of the network is evaluated for each parameter setting. The network configuration, such as number of hidden layers or the number of neurons in a hidden layer, should also be varied as training progresses.

13.5.2 Counterpropagation network

The counterpropagation network combines the hidden layer, which is a Kohonen layer with competitive units that undergoes unsupervised learning [8]. The top layer is a non competitive Grossberg layer and the hidden layer are fully interconnected. The Grossberg layer is trained by the Widrow-Hoff or Grossberg rule. The architecture used for the classification of multispectral image data consists of a three layered CPN network, which has been discussed in Chapter 8. Here the Kohonen (competitive) layer has been chosen as a rectangular grid of nodes. The weight updation for the nodes in the hidden layer is done not only for the winner node but also for the nodes in its neighborhood. The output layer (Grossberg layer) consists of a number of nodes equivalent to the number of earth surface categories present in the scene to be classified.

Adaptation of the output layer weights is achieved using the Widrow-Hoff rule for the winning unit $Z_c = 1$; all other units are set to zero. Thus the only weight adjusted for the output unit j is the weight connected to the winning unit c. For the two IRS image data sets used for training and testing the neural network, we have used fourteen and eight nodes respectively as the output nodes.

The network uses four nodes in the input layer, where each node represent the normalized pixel gray value $[0, 1]$ in each spectral band.

13.5.3 Experiments and Results

The aim of remotely sensed scene classification is to create a map covering a set of major categories. For example, in rural areas the major land cover categories may be: (1) agricultural land, (2) built-up land, (3) forest, and (4) water.

The two aforesaid neural network architectures may be used for the classification of remotely sensed satellite data for land-cover analysis. The classification accuracy estimation of the results obtained by neural networks may be performed by comparing the ground truth pixels (samples) and the classified pixels through the confusion or error matrix.

The performance of backpropagation network is affected by many factors, such as number of hidden nodes, learning rate and appropriate selection of training samples. Therefore, several experiments may be conducted to gain insight into the network operation. The parameters of the network may be chosen as (1) nodes in the hidden layer, (2) learning rate, (3) momentum factor. The peak performance may be achieved with a particular set of parameters. The effect of variations of these parameters on the system performance should be observed. While for smaller learning rates the neural network training is slow, for larger values of the learning rate the learning algorithm oscillates.

The neural network should be trained using a sufficiently large number of training samples generated from the ground truth data from different classes within the data sets. The method of selecting the training sets also affects the classification accuracy. All the training data sets were processed to delete the occurrence of any repeated patterns and also to remove the confusing patterns, which are located on the boundary demarcating two pattern classes. After the training phase is over, the network should be tested using the whole scene.

Performance of the counterpropagation network is affected by many factors such as the learning rate of the Kohonen layer and the output (Grossberg) layer, the size and the arrangement of nodes in the Kohonen layer (linear array or rectangular grid), the strategy to update weights (only for winner node or for winner node and the neighboring nodes) in the Kohonen layer, and the activation function used for the nodes in the Kohonen layer. Several experiments are required to be conducted to get insight into the network operations. The effect of variations in the grid size of the Kohonen layer on the performance of the network should also be observed. In many applications it is found that the classification accuracy increases with an increase in the grid size of the Kohonen layer.

13.5.4 Classification Accuracy

The classification accuracy is the most important aspect of assessing the performance and reliability of a classifier. The best way of representing the classification accuracy is by comparing the ground truth pixels and the classified pixels through the confusion/error matrix.

The confusion/error matrix is in the form of a $m \times m$ matrix, where m is the number of classes. It shows the number of samples (pixels) classified correctly and, for the misclassified samples, it shows how they are distributed among the other classes. The samples which do not belong to any of the specified classes are termed as unknown class samples. The sum of the pattern samples lying on the main diagonal of the confusion/error matrix gives the total number of correctly classified samples. Dividing this sum by the total number of labeled samples gives the classification accuracy of the network. One of the most important characteristics of the error matrix is its ability to summarize the errors of omission and commission. The error of omission for each class is computed by summing the number of samples assigned to incorrect classes along each class row and dividing this by the total number of samples in that class. The error of commission of a class is computed by summing the number of samples assigned to incorrect classes along each column and dividing this by the total number of samples in that class.

Analysis of the two network architectures indicates that both the backpropagation network and the counterpropagation network yield very good results. The problem of slow training of a MLP network can be reduced to a reasonable level by the use of adaptive backpropagation algorithms. The time taken for classification varies linearly with the size of the image. Each pixel is classified independently of its neighbors. Therefore, large images can be split into smaller parts to speedup the classification process using coarse grain multiprocessor machines. The counterpropagation network is a faster alternative to the backpropagation network. The improvement in training time with counterpropagation network is substantial. The performance in terms of classification accuracy is also comparable to the backpropagation network.

The classification results of both the backpropagation and the counterpropagation neural networks are comparable to the maximum likelihood classifier. Moreover, both the networks are insensitive to the form of the underlying probability density function.

13.6 SPECTRAL CLASSIFICATION–A KNOWLEDGE-BASED APPROACH

In case of satellite images, the domain knowledge of the image can be exploited to improve the quality of image segmentation techniques [3, 4, 9, 10].

The absence of such knowledge usually makes the choice of a correct number of region types a difficult problem. The segmentation techniques can generally be divided into two categories *partial* (or general-purpose) segmentation and *complete* (or knowledge-guided) segmentation.

There are some segmentation techniques which segment an image into homogeneous regions without any apriori knowledge about the image context or the image-generation processes. Many popular segmentation techniques like thresholding, split and merge, and region growing techniques fall into this category.

The knowledge used in the segmentation process can be divided into two types: *spectral* and *spatial*. In remote sensing, spectral properties associated with different land cover types have been extensively studied. Spectral knowledge plays an important role in image segmentation. Spatial knowledge deals with the spatial relationships like proximity, connectivity and associativity among various objects in an image. Such knowledge has been widely used in the interpretation of aerial photographs. The generation of spatial rules has been automated for aerial photographs but not for LANDSAT images, possibly, due to the complexity of LANDSAT images and the difficulty associated with the acquisition of spatial knowledge from the domain experts.

13.6.1 Spectral Information of Natural/Man-Made Objects

The human expert first identifies some regions in the image and provides an initial interpretation of the image. The automated image interpretation information that can be reliably extracted from image features is called spectral knowledge. Spectral knowledge is usually extracted from an image by selecting a representative site for each class and extracting the statistics of that class.

13.6.2 Training Site Selection and Feature Extraction

The selected training sites representing the land cover classes of interest should not be typical ones but should represent the norms for each class. The image coordinates for these sites are then identified and used to extract features from the data of each of these sites. Training data should be selected in such a way that the features should be of the same value if the environment from which they are obtained is relatively homogeneous. However, if conditions change dramatically, as the study progresses across the study area, it is likely that the features extracted from the training sites acquired at one end of the study area will not be true representatives of the spectral conditions found in the other end of the same study area. Hence, the selection of training sites plays a major role in the acquisition of spectral knowledge related to the interpretation of remotely sensed images.

13.6.3 System Implementation

The knowledge-based classification of satellite images is the most important step in the development of an image understanding system. From the map of the area of interest, some of the regions are selected as the representative target classes. Analysis is then done for finding the features of the classes from the representative training sites.

The mean feature vector and variances of the different categories of the study area are used for framing the knowledge rules used for the classification. By using this methodology, different knowledge rules can be framed for the classification of satellite scenes. These knowledge rules almost completely describe the study area.

Spectral knowledge organization: In the development of the spectral knowledge base, two types of relations are used for describing the features: (1) band-to-band (BB) and (2) category-to-background (CB). The band-to-band feature represents the relationship between different spectral bands i.e. the shape of the spectral reflectance curve. The band-to-band constraints are used to characterize the shape of the reflectance curve in terms of inequalities of combinations of spectral bands to avoid the use of absolute thresholds.

The category-to-background feature is based on the observation that land cover categories are often consistently brighter or darker than the surrounding pixels that belong to other categories. This feature is defined in terms of the contrast gradient between the category pixels and the normal background response. The sign of the constraint gradient is defined for each spectral band.

13.6.4 Rule Creation

All rules do not necessarily support binary decisions, i.e., accept if true and reject if false. In most of the cases the antecedents may be satisfied only partially. Also even if the antecedents are fully satisfied, the consequent may be fuzzy.

For the development of a knowledge based segmentation system, following three types of knowledge rules or constraints are used for representing the domain knowledge of the scene under consideration.

1. *Mandatory constraints*: These rules describe the spectral properties that are expected to be satisfied by all the pixels belonging to a category. In other words, these rules represent the general features of target categories.

2. *Optional constraints*: These rules describe the spectral properties to be satisfied by a group of pixels belonging to a category.

3. *Contradictory constraints*: These are the rules that describe spectral properties that are known to contradict a category.

An interesting approach is to represent these knowledge rules in terms of three parameters that are used to direct the accumulation of evidence, i.e., a constraint, a supporting weight and an opposing weight. The supporting and opposing weights are associated with each of the target classes. The constraint is used for rule validation and is evaluated as a function of the local distribution of reflectance values of each pixel. The weights determine the relative significance of the evidence provided by a given constraint in supporting or opposing a category.

If the constraint is validated, then the supporting evidence is incremented by incrementing the supporting weight, otherwise the opposing weight is incremented depending on one of the three types of constraints used for classification. If the constraint is mandatory, then on validation the supporting weight will be increased, otherwise the opposing weight will be increased. On the other hand, if the constraint under consideration is an optional one, then on validation we increment the supporting weight. If, however, the constraint under consideration is contradictory, then if it is invalid the opposing weight is incremented, but no change is done to the supporting weight. The step size for incrementing the supporting and opposing weights is selected on the basis of the level to which a rule can describe the category of interest.

13.6.5 Rule-Base Development

A number of rules may be created for the pixel classification of TM imagery. These rules should embody the spectral knowledge of the scene under consideration. The consequent of each rule is associated with an action to be taken to update weights (incrementation or decrementation of the weights). Some of the rules for the classification of general classes like water, and structure class are given below:

- **RULE 1** (mandatory):

  IF (the mean value of pixel follows the band decreasing property) *THEN* {the pixel belong to water class and hence increment the supporting weight for the water class} *ELSE* {the pixel does not belong to water class and hence increment the opposing weight for the water class}.

  The extent to which the weights are incremented is user specified and is data dependent.

- **RULE 2** (optional):

  IF (the average of bands 4 and 5 is greater than the average of bands 1 and 2) ***THEN*** {pixel may belong to the structure class and hence increment the supporting weight for the structure class} ***ELSE*** {pixel may not belong to the structure class and increment the opposing weight for the structure class}.

  The optional rule need not be followed strictly. These rules will partially enhance the truth values of an evidence. The optional rules may be checked only in cases where confusion exists.

- **RULE 3** (contradictory):

  IF (if intensity of a segment in band 7 is less than its average intensity in band 1 and band 2) ***THEN*** {the segment is not *concrete* and hence increase the opposing weight for the *concrete class*}.

  For a contradictory rule, the consequence leads to the enhancement of the possibility of nonbelongingness to a class.

The rules presented above use band-to-band type of knowledge. Likewise the rules for differentiating the different land cover categories present in the image under consideration may be used for the classification of the image. Classification decisions are made on the basis of convergent evidence, i.e., the category having the highest ratio of supporting to opposing evidence at the disposal of the final rule is selected as the representative class of that pixel. The selection of a representative class is usually done on the basis of the scores for different target classes. The scores for each of the classes are found by using the scores for the different classes. The score for a class may be computed by using the following formulaes:

$$SCORE = 100 \times \left[1 - \frac{E_{opp.i}}{E_{sup.i}}\right] \text{if} E_{sup.i} > E_{opp.i}$$

$$SCORE = 100 \times \left[1 - \frac{E_{sup.i}}{E_{opp.i}}\right] \text{if} E_{opp.i} > E_{sup.i}$$

where $E_{sup.i}$ and $E_{opp.i}$ are the supporting and opposing weights for class i.

The maximum possible score of 100 indicates that there are no opposing evidences and only supporting evidence, while the minimum score represents the presence of only opposing evidence and no supporting evidence. In case there are equal amount of supporting and opposing evidences, the score is zero and the decision in this case is hard.

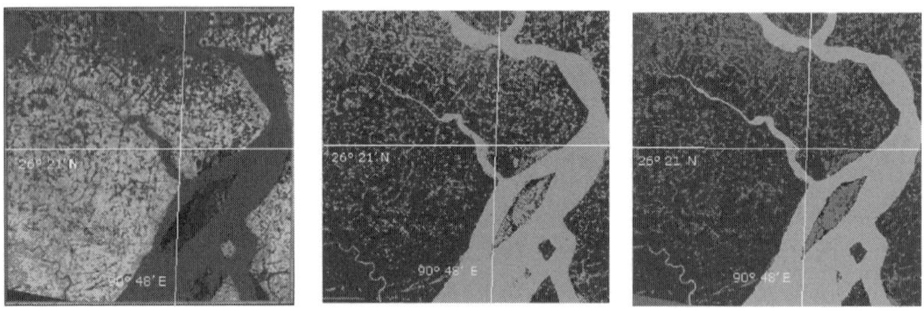

(a) (b) (c)

Fig. 13.2 Satellite Image (a) original, (b) segmented by multilayered perceptron (error backpropagation), (c) segmented by Fuzzy self-organizing feature map.

Figure 13.2(a) shows a satellite image of a river surrounded by vegetation and forestry. Figures 13.2(b) and 13.2(c) are the results of segmentation of the satellite image into different regions by *multilayer perceptron* and *fuzzy self-organizing feature map* network approaches respectively. The color version of Figure 13.2 is provided in color pages section. The major land cover categories in the scene has been seen appropriately classified by both the techniques. The errors of omission and commission are quite low in both the segmentation results.

13.7 SPATIAL REASONING

Since geometry is an important aspect of defining the world, spatial reasoning plays a useful role in the interpretation process of an image understanding system. Usually, after the segmentation and classification stage we get the information regarding the class of a segment, but not the structural and contextual information of a segment, which is very important for the interpretation of any scene. This can be explained by using the fact that, though two segments may belong to the water class one of them may be a river and the other a water pond. From the spectral classification, we can only infer that both belong to the water class. In spatial reasoning that utilizes the spatial parameters of the segmented objects in a scene yields good recognition of the objects.

In general the image understanding system utilizes two different types of spatial information to recognize an object: (1) intrinsic properties, i.e., attributes such as size, shape, length, etc. and (2) semantic knowledge of the scene, i.e., spatial relations among the different objects present in the scene. In a scene understanding system both the intrinsic and the semantic knowledge should be incorporated for better interpretation results.

13.7.1 Evidence Accumulation

The first step in the design of a spatial data analysis system involves the accumulation of evidence for spatial reasoning. The spatial evidence is obtained by the spatial reasoning expert and the spatial knowledge rule base is developed from the apriori structural and contextual information about the different objects in the scene. These sets of rules describe the structural properties of different objects present in the scene and may be derived by using the map information. For the final inference, forward or backward chaining inference mechanism may be used, which gives a final interpretation of the segment. Here we discuss about the accumulation of evidence, spatial knowledge base, and the inference mechanism.

After extracting a segment by any appropriate segmentation strategy, the parameters like area, location, perimeter, compactness, shape, length, breadth, end points, aspect ratio, and the adjacent information are extracted to interpret the segment.

The information of adjacent pixels is extremely important for distinguishing between objects closely resembling one another. The following seven features are useful:

1. **Length of a segment:** For identifying linear segments like rivers, canals, roads, railway line, airport runways, the length of the segment is an important parameter for spatial reasoning. The length of any segment is usually the length of its thinned segment, which is almost always the number of pixels in the skeleton. After the process of segment extraction, the object of interest is skeletonized and the number of pixels in the thinned segment is considered as the length of the segment. For skeletonizing a segment any good thinning algorithm, like the *Safe Point Thinning Algorithm* (SPTA), may be used. It is possible that the skeleton of a segment may consist of many branches and only the main branch with the maximum length is considered as the length of the segment. For linear segments, the end points of a segment provide a useful guide for the interpretation of a segment.

2. **Area, breadth, and perimeter:** The area of any given segment or the connected component is calculated as the number of pixels in the segment. The average breadth of a given segment is calculated as the ratio of its area to its length. The perimeter of the segment is computed from the number of border points of the segment.

3. **Aspect ratio:** The aspect ratio of a segment is defined as the ratio of length to breadth of a segment, and is mainly used for the representation of the regular-shaped objects present in the scene.

4. **Compactness:** The compactness information is used for finding the shape of a given segment. It is calculated as the ratio of the square of the perimeter to the area of a given segment. It is known that the compactness is maximum for circle in the continuous case, and in the case of digital domain it is more for square and hexagonal objects and less for linear segments. By using this property the shape of the segment can be inferred.

5. **Location of a Segment:** The centers of gravityies may be chosen as the location of a regular-shaped segment. The X-coordinate of the center of gravity is calculated as the average of the X-coordinates of all the pixels and the Y-coordinate of center of gravity as the average of the Y-coordinates of all the pixels of the segment respectively.

6. **Moment of inertia:** The moment of inertia is also useful for finding the shape of a given segment, which is calculated by using the following formula.

$$M.I._x = \frac{\sum (X_i - CG_{x-d})^2}{n}, \quad M.I._y = \frac{\sum (Y_i - CG_{y-d})^2}{n}$$

   and hence

$$M.I. = \sqrt{M.I._x^2 + M.I._y^2}$$

   and X_i and Y_i are the coordinates of the ith pixel and n the number of pixels in the segment. $M.I._x$ and $M.I._y$ represent moment of inertia about the axes parallel to the x-axis and y-axis passing through CG of the segment.

7. **The presence of CG within the segment:** Location of the CG of a segment provides an important information of the segment. To find out whether the CG of a segment lies inside the segment or not, the measure of relative distances of the CG point with all pixels of the segment can be used. All the above spatial parameters may be used for spatial reasoning.

13.7.2 Spatial Rule Generation

A set of spatial rules are framed by using the domain knowledge of the scene. These rules are developed to form an interface between the forward-chaining inference mechanism and the facts obtained from the above evidences for the spatial reasoning. The final interpretation of the scene is obtained at this stage. A few of the rules in the rule base used for spatial reasoning are discussed here.

In this reasoning we first divide all the segments into either linear or regular shaped categories. Some of the rules used for this purpose are as below.

- **Rule 1:** ***IF*** (moment of inertia/ area of a segment is *high*) AND (average breadth of the segment is *not high*) ***THEN*** {the segment is *linear*}.
- **Rule 2:** ***IF*** (moment of inertia/ area of a segment is low) AND (center of gravity of a segment is inside another segment) ***THEN*** {the segment is regular-shaped}.

  Depending on the spatial parameters, the segments have been interpreted as follows.

- **Rule 3:** ***IF*** (segment is *linear* AND average breadth of the segment is *low*) And the segment class is *water*), ***THEN*** { the segment is a *canal* or a *river*}.

  Rule 4: ***IF*** (segment is regular-shaped AND belongs to the *water* class) ***THEN*** {the segment is *lake*}.

The rules are generated from the domain knowledge of the problem. The objective of the analysis is to classify the different land cover regions in various categories of interest and then locate the rivers, water ponds, parks, residential areas, etc., present in the scene.

13.8 OTHER APPLICATIONS OF REMOTE SENSING

Interesting applications for the generation of the Cover Map of a Remotely Sensed Image for GIS Applications has been reported in [11]. Other important applications include change detection, which is briefly discussed below.

13.8.1 Change Detection using SAR Imageries

Repeat pass change detection incorporates the basic principle of imaging the same area after an elapsed time and is particularly useful in detecting changes in the that area.

Change detection techniques for SAR data can be divided into several categories, each corresponding to different image quality requirements. In a first category, changes are detected based on the temporal tracking of objects or stable image features of recognizable geometrical shape. Absolute calibration of the data is not required, but the data must be rectified from geometric distortions due to differences in imaging geometry or SAR processing parameters, and the accurate spatial registration of the multi-date data is essential. Applications include sea-ice monitoring and motion tracking, monitoring of glaciers, landslides, and oceanic features. In the second category, changes are detected based on temporal differences in radar backscatter. The requirements are a stable calibration accuracy of the data, and an accurate spatial registration of the multi-date data. Typical applications include monitoring of crops, volcanic activity, snow extent and conditions, glacial melt, soil moisture, and vegetation water content.

The methods of change detection consist of image ratioing, differencing techniques as well as techniques based on wavelet decompostion and singular value decomposition. Application of 2D Gabor filter has been shown to detect the changes in SAR images [12]. We have introduced the Gabor filter and its characteristics in Chapter 5. Gabor filter is particular suitable for characterizing textures in an image. The orientation parameters of the Gabor filter are useful in delineating the texture orientations in satellite (SAR) image.

SAR images are characterized by more textural information than gray tone variations. The features associated with texture can be effectively, extracted. The texture based segmentation algorithm is applied on the SAR repeat pass images followed by image differencing and ratioing techniques in order to achieve change detection in the images.

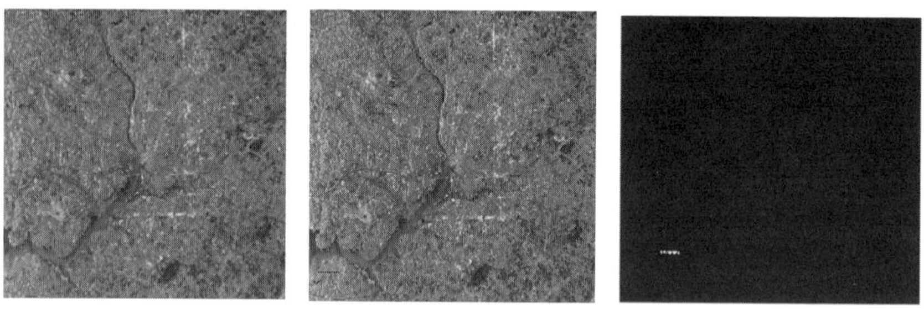

(a) (b) (c)

Fig. 13.3 Change detection from repeat pass SAR image: (a) original, (b) repeat pass of the original after a time interval, (c) changes in them.

The test image Figures 13.3(a) and 13.3(b) show two SAR repeat pass images of an urban area of size 31.2 miles by 27.9 miles. The radar illumination is from the left side of the image. The image shows a single channel of SIR-C radar data: L-band, horizontally transmitted and received. Image differencing technique have been applied to the original Figure 13.3(a) and changed images Figure 13.3(b) and the result of changes that has taken place during the time interval of the repeat pass is shown in Figure 13.3(c).

13.9 SUMMARY

In this chapter we have presented brief description of the different satellite Imageries, their utilities and applications in diverse areas of resource monitoring and management using satellite image data. The techniques of remotely sensed scene segmentation and pixel classification have been discussed here. Pixel classification using neural network based techniques, viz, multilayered perceptron with error backpropagation and counterpropagation based networks have been discussed. Knowledge based systems play an important role in remote sensing. Design of a system using spectral rules has been detailed here. Different aspects of spatial reasoning have been encoded for finer classification and recognition of regions. An interesting application of image processing is on change detection from remotely sensed images which has been discussed in this chapter.

REFERENCES

1. T. M. Lillesand, R. W. Keifer, and J. W. Chipman, " Remote Sensing and Image Interpretation," *Wiley*, 5th Ed., 2003.

2. G. Franceschetti and R. Lanari, "Synthetic Aperture RADAR Processing," *CRC Press*, Boca Raton, 1999.

3. J. Ton, J. Sticklen, and A. K.Jain, "Knowledge Based Segmentation of Landsat Images," *IEEE Trans. on Geo Science & Remote Sensing*, 29(2), March 1991, 222–231.

4. S. W. Wharton, "A Spectral-Knowledge Based approach for urban land Cover discrimination," *IEEE Trans. on Geo Science & Remote Sensing*, Vol. GE-25, May 1987, 272–282.

5. J. A. Benediktsson, P. H. Swain, and O. K. Ersoy, "Neural network approaches versus statistical methods in classification of multi source remote sensing data," *IEEE Trans. GeoSci. Remote Sensing*, 28(4), July 1990, 540–554.

6. P. D. Heermann and N. Khazenie, "Classification of multispectral remote sensing data using a backpropagation neural networks," *IEEE Trans. GeoSci. Remote Sensing*, 30(1), January 1992, 81–88.

7. H. Bischof, W. Schneider, and A. J. Pinz, "Multispectral classification of landsat-images using neural networks," *IEEE Trans. GeoSci. Remote Sensing*, 30(3), May 1992, 482–490.

8. R. Hecht-Nielsen, "Applications of counterpropagation networks," *IEEE Trans. Neural Networks*, 1(2), 1988, 131–140.

9. B. Nicolin and R. Gabler, "A Knowledge Based System for the analysis of aerial images," *IEEE Trans. on Geo Science & Remote Sensing*, Vol. GE-25(3), May 1987, 317–328, .

10. T. Matsuyama, "Knowledge based aerial image understanding systems and Expert systems for image processing," *IEEE Trans. on Geo Science and Remote Sensing*, Vol. GE-25(3), May 1987, 305–316.

11. A. Chakrabarti, A. Jain, and A. K. Ray, "A Novel Algorithm to Generate the Cover Map of a Remotely Sensed Image for GIS Applications", *Indian Conference on Computer Vision, Graphics and Image Processing*, Bangalore, India, December 2000, 118-123.

12. A. Samanta, S. Ganguly, A. K. Ray, S. K. Ghosh, and D. Gray, "Application of 2-D Fuzzy Gabor Filter in SAR Change Detection," *International Conference on Microwaves, Antenna, Propagation and Remote Sensing*, 2004.

14

Dynamic Scene Analysis: Moving Object Detection and Tracking

14.1 INTRODUCTION

The detection and classification of moving objects is an important area of research in computer vision. The problem assumes immense importance because of the fact that our visual world is dynamic and we constantly come across video scenes that contain a finitely large number of moving objects. To segment, detect, and track these objects from a video sequence of images is possibly the most important challenge that the vision experts confront today. In this chapter, we will present the problems and the possible solutions of each subtask in *dynamic image analysis.* These systems find applications in human surveillance, security systems, traffic monitoring, industrial vision, defense surveillance, etc. [1]–[3]

14.2 PROBLEM DEFINITION

The first step for moving object detection and tracking involves foreground-background separation by subtracting the background from each frame of a video sequence. This difference shows the moving objects, e.g., a moving man or a moving car in the scene and is known as *foreground* detection. The problem complexity in extracting the moving objects in a dynamic scene, however, increases with the increasing presence of motion due to various other phenomena. For example factors such as the change in illumination, variation of sunlight as the day progresses, change of appearance of color due to white-balancing (and/or color correction) in the color digital camera itself, change of pixel values due to flickering and other phenomena, create false motions in the dynamic scene. Also motions due to shadow movement, tree movement, and many such other events in the scene need to be considered and detected as false motion. In view of the above, we need to model the background continuously so that only moving objects of interest may be detected and tracked accurately.

14.3 ADAPTIVE BACKGROUND MODELING

The key issues in the understanding of background modeling are:

- Why background modeling is at all necessary
- What are the conventional approaches of background modeling and
- The new strategies based on some of the concepts put forward by several researchers over the last decade

By the term *background* we mean the static part of the scene that does not change with time. It is only the moving objects that change their positions in the scene. Interestingly, however, there are several factors resulting in a change of the background. It is thus important that while dealing with continuous monitoring of the scene, as in the case of video surveillance, we should model this background continuously.

Background modeling is necessary due to the following reasons:

1. There is a possibility of frequent change in illumination in the outdoor scenes, resulting from clouds, rains, fogs, etc.
2. Natural phenomena such as storms and winds also cause changes in background due to the movement of tree branches, bushes, etc.
3. Even on a normal day, the overall outdoor illumination changes from dawn to dusk as the brightness of the sunlight changes continuously.
4. Changes are observed due to glints and glare of the sun, which are quite prominent as the sun angle changes.
5. Manmade artificial lights like street lights also cause a change in the background.
6. There are other moving objects, including shadows, which obscure the visual field and produces clutter and changes in the background.

Each of the above factors causes change in the background, which is not a static entity. Modeling of the background is thus an essential component in dynamic scene analysis. The background modeling methods fail if the algorithm does not model the background continuously, since the background undergoes continuous changes because of the factors outlined above. Without continuous and adaptive background modeling, the errors in the modelled background accumulate continuously. As a consequence, the detection and tracking of moving objects may be erroneous.

14.3.1 Basic Background Modeling Strategy

In a continuously changing scene, it is important to note that each pixel may have a constant motion. This means that a particular pixel may undergo constant change in its brightness value with the change in environmental conditions. Another interesting phenomenon occurs quite often. A pixel may belong to the class of tree leaf at a particular frame and the same pixel may assume another class, say a human being, moving in front of the tree, in the next frame. It is thus important that each pixel variation should be modeled in a statistical sense. Continuous estimation of this statistical model, which embodies all such variations, is the main philosophy of the background modeling.

In cases where we have very few or no moving objects in the background, we may simply average the background frames to create an approximated static background. Simple background modeling methods work when we get large number of background frames without any moving objects in the scenes. In most of the practical cases, however, there are considerable number of moving objects. This necessitates the adoption of robust background modeling [7, 8].

14.3.2 A Robust Method of Background Modeling

A pixel in a scene may be conceived as belonging to one of a multiple number of pattern classes, such as, tree leaves, grass, etc. Each class assumes a specific probability distribution. Each specific class of patterns will be modeled by a specific histogram of intensity values which provides a measure of its distribution. When the pixel belongs to a single monolithic class, the ideal distribution of values should be a compact, Gaussian-like cluster around some mean point.

However, since each pixel is in a constant state of motion, a particular pixel in the scene may be considered as a combination of many Gaussian distributions and a single Gaussian would not be sufficient to model the pixel value.

In practice, multiple objects may appear in the view of a particular pixel along with a change in the lighting condition. Thus multiple, adaptive Gaussians are required for modeling a pixel. We use an adaptive mixture of Gaussians to approximate this process.

Let us consider the past history of a pixel over time, which is essentially a time series of scalars for gray values (or vectors for color pixel values). At a given time instant t, the history of a particular pixel $\{X_1, X_2, \cdots, X_{t-1}\}$, is given by a weighted combination of a number of Gaussian distributions, (say K) as below. The higher is the value of K, the better is the approximation.

$$P(X_t) = \sum_{i=1}^{K} \omega_{i,t}\, \eta_i(x_{i,t}, \mu_{i,t}, \sigma_{i,t}),$$

where $\omega_{i,t}$ is the estimated weight at tth frame of the ith Gaussian distribution, $\eta_i(x_{i,t}, \mu_{i,t}, \sigma_{i,t})$ with $\mu_{i,t}$ as the mean value and $\sigma_{i,t}$ as the variance representing the ith pattern class. The Gaussian for $i$the class is

$$\eta_i(x_{i,t}, \mu_{i,t}, \sigma_{i,t}) = \exp\left[-\frac{1}{2}\frac{(x_{i,t}-\mu_{i,t})^T|\Sigma_{i,t}|^{-1}(x_{i,t}-\mu_{i,t})}{(2\pi)^{d/2}|\Sigma_{i,t}|^{1/2}}\right]$$

The parameters that we need to estimate and update are (1) mean feature vector $\mu_{i,t}$, and (2) covariance matrix $\Sigma_{i,t}$. If the features are independent, then the covariance elements of $\Sigma_{i,t}$ tend to zero. Assuming that the correlation among the red (R), green (G) and blue (B) components of a pixel is very small and their variances are identical, the covariance matrix $\Sigma_{i,t}$ can be expressed as

$$\Sigma_{i,t} = {\sigma_{i,t}}^2\, I$$

where I is an identity matrix.

A new pixel value, in general, is represented by one of the K number of major components of the mixture model and this is used to update the model. When the intensity value of a pixel falls within a predefined range of distribution, we say that the pixel matches with the distribution. If none of the K distributions match the current pixel value, the least probable distribution is replaced with a new distribution with the current value as its mean value.

The prior weights $\omega_{i,k,t+1}$ of the K distributions at time $t+1$ are adjusted as follows:

$$\omega_{i,k,t+1} = \alpha\omega_{i,k,t} + \beta M_{i,t},$$

where α and β are two learning constants and $M_{i,t}$ is 1 for the model matched and 0 for the unmatched models. The mean and covariance parameters are updated as follows.

$$\mu_{i,t+1} = \rho_1\mu_{i,t} + \rho_2 x_{i,t+1}$$

and

$$\sigma^2_{i,t+1} = \rho_1{\sigma_{i,t}}^2 + \rho_2(x_t - \mu_{i,t})^T(x_{i,t} - \mu_{i,t}).$$

The learning constants ρ_1 and ρ_2 in the above updation equations controls the speed of learning and $\rho_1 + \rho_2 = 1$.

The basic purpose of background model estimation is to determine which of the Gaussians of the mixture are most likely produced by background process.

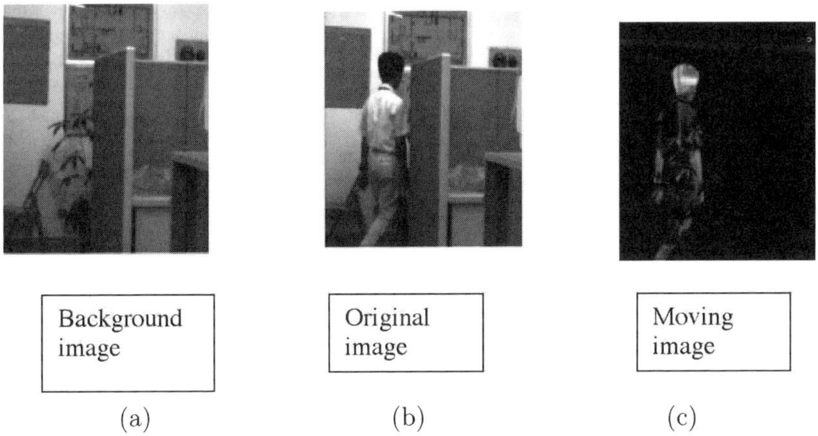

Fig. 14.1 Results of background modeling and segmentation of a moving human figure: (a) modelled background, (b) an input frame, (c) segmented human object.

After ranking the distributions by an appropriate strategy, the most likely background distribution remain on top of ranked list and the less probable transient background distributions gravitate toward the bottom, which may eventually be replaced by new distributions.

One of the significant advantages of this background modeling is that, when a pixel in the foreground is allowed to become part of the background, it doesn't destroy the existing model of the background.

14.4 CONNECTED COMPONENT LABELING

Once we get a set of foreground pixels in each frame, after background modeling and differencing, it is important to threshold them so as to yield a set of binary points. These labeled foreground binary pixels representing the object and background are next segmented into regions by *connected component labeling* algorithm, which yields a set of smooth blobs corresponding to each of the moving objects. The objective of connected component labeling is to determine all the connected set of components in an image an assign a distinct label to each pixel in the same connected component.

In a binary image scanned from left to right, an unlabeled object pixel is assigned a label, say X and each of its neighboring object pixels is assigned the same label till there is no more unlabelled object pixel in the image. Such a recursive algorithm for connected component labelling may be efficiently computed on a mesh-connected parallel machine. An alternate strategy of connected component labelling can be defined in two passes. In the first pass, each object pixel is assigned a label according to the following criteria.

(a) If both the upper neighbor $P(i-1, j)$ and left neighbor $P(i, j-1)$ of the object pixel $P(i, j)$ in a 4-connected neighborhood have the same label X, then assign label X to $P(i, j)$.

(b) If either the upper neighbor $P(i-1, j)$ or the left neighbor $P(i, j-1)$ of the object pixel $P(i, j)$ in a 4-connected neighborhood has the label X, then assign label X to $P(i, j)$.

(c) If the upper neighbor $P(i-1, j)$ has label X and left neighbor $P(i, j-1)$ of the object pixel $P(i, j)$ has a different label Y (i.e., $x \neq Y$) in a 4-connected neighborhood, then assign label X to $P(i, j)$. Enter X and Y in an equivalence table, say E.

(d) If the upper neighbor $P(i-1, j)$ and left neighbor $P(i, j-1)$ of the object pixel $P(i, j)$ in a 4-connected neighborhood are both nonobject pixels, then assign new label Z to the pixel $P(i, j)$. Enter label Z in the equivalence table E.

Usually we assign numeric labels to each component. The equivalence table E contains a set of equivalent labels.

In the second pass, the equivalent labels are merged to create unique labels for each connected component in the image. If X and Y are two equivalent labels in E, then reassign Y by X when $X < Y$ or vice versa. As a result, each connected component is assigned a unique label.

Hence the above connected component labeling algorithm is a two pass algorithm. In the first pass the object pixels belonging to the same connected component are assigned different labels and are recorded as equivalent in the equivalence table E. In the second pass where the equivalent labels are merged together so that each connected component is assigned a unique label.

14.5 SHADOW DETECTION

Shadow detection and suppression is possibly one of the most important tasks in video surveillance [4]–[8]. If we can suppress the shadows the task of object recognition and scene interpretation becomes easy. One of the most important properties of shadow is that the pixels forming the moving object and its shadow are detected simultaneously from the inter-frame differences. The shadows and the moving object which causes it are always located adjacent to each other. Thus they may be enclosed in the same blob. Also quite often the shadows of two moving objects may get merged into a single blob, which may result in false object detection.

Thus we have to understand some of the important properties of shadows which may be useful in differentiating shadows from the moving objects.

Shadows may be of different types. Sometimes a portion of the object may be dark because this part was not illuminated at all. These shadows are termed *self-shadows*. On the contrary, shadows produced by an object on a surface are called *cast-shadows*. As the object, say a human figure moves, the shadow also moves. There are three different properties of the shadows:

1. Shadows are better distinguished in HSV color space, which resembles more of human perception. This implies that the shadows produced by the objects, change the H, S, and V more significantly.

2. Also the shadow points cast by the object reduce the intensity of the background pixels, i.e., shadows have a darkening effect on the background.

3. It has been observed that shadows lower the saturation of points, which means that the difference between the background pixel saturation values and the shadow pixel saturation values will be positive.

Shadow suppression algorithm using both intensity and color information yields good results [5].

14.6 PRINCIPLES OF OBJECT TRACKING

Once appropriate background modeling and shadow detection and elimination have been performed, the next task is tracking the object in the midst of clutter. Many interesting approaches have been reported on different tracking algoritms [10]–[12] The following eight design considerations may be incorporated in a target tracking system.

1. **Stationary Background**: When the scene contains multiple objects, the background is stationary while all or part of the objects in the foreground may be in motion.

2. **Target size variation**: The target size reduces as the target moves further away from the camera. Thus a scaling mechanism needs to be incorporated during the process of tracking.

3. **Occlusion or Temporary loss of target**: During the tracking phase the target may be temporarily lost as it goes behind another object. This is known as occlusion. In such cases the system will recover the target automatically.

4. **Target Model**: The model of the target needs to be incorporated. In case of human tracking, for example, a human figure may be modeled as an ensemble of several ellipses, where each ellipse represents the individual body parts like head, torso, hands, and legs, etc. The color, shape, intensity, and other attributes of the object may vary while the object is in motion, and yet the tracker should be able to track correctly.

5. **Automatic Target detection**: The tracker should be able to detect all the new targets automatically and start tracking the them.
6. **Real time**: The tracking algorithm should be computationally simple and optimum so that the tracking can be implemented in real time.
7. **Target trajectory**: The target may or may not follow a particular trajectory. There may be abrupt changes in the target path.
8. **Target speed**: Speed of the target can change abruptly; it may be constant, increasing, or decreasing.

14.7 MODEL OF TRACKER SYSTEM

A tracking system may be modeled as a three-state sequential machine as shown in the figure. The sequential machine has three states, i.e., *locking*, *tracking* and *recovery*. Functions of each state are as follows:

1. **Locking state:** Initially the system is in locking state, when the camera is in search mode, i.e., searching for targets. During this state the processing is carried out on the whole image frame. The system will partition the image frame captured by camera into a number of moving objects. The history of these objects is extracted by checking the trajectory followed by the objects, and confirmation of the moving object is carried out in automatic mode. Once the target is confirmed the control of the system is transferred to tracking state.
2. **Tracking state:** This stage should use computationally inexpensive techniques. Current location extracted by locking state is used for processing. Next position of the target is identified, and that positional information is stored in history database. If the target does not exist in the predicted window area, then the system control is transferred to recovery state.
3. **Recovery state:** Quite often the moving object of interest may be lost temporarily or permanently. In this state if the target is lost, the system will try to recover the target from low-resolution image. If the target is recovered in a few frames, then the system will transfer control to tracking state; otherwise it remains in recovery state till its predefined time expires. After the time is elapsed, control transfers to locking state.

14.8 DISCRETE KALMAN FILTERING

A thorough and precise approach for dynamic state estimation is formulated under Bayesian methods [9, 12]. The methodology is to construct the probability distribution function (PDF) of the state vector based on all available information. This PDF summerises the current state of knowledge about the state vector and then the optimal computational approach can be formulated based on this. The PDF will be Gaussian in the case of linear/Gaussian problem and the Kalman filter is found to yield optimal solution to discrete data filtering. A distinctive feature of Kalman filter is that its mathematical formulation is described in state space concepts and its solution is computed recursively. Each updated estimate is computed from previous estimate and from the new input data. This offers the advantage that only the previous estimate need to be stored in memory. Kalman filter has been applied widely in the field of tracking and navigation.

Kalman filter addresses the problem of estimation of the state $x \in \Re^n$ of a discrete-time controlled process which is governed by linear stochastic difference equation

$$x_k = Ax_{k-1} + Bu_k + w_{k-1}$$

with a measurement $z \in \Re^m$ i.e

$$z_k = Hx_k + v_k$$

The random variables w_k and v_k represent the process and measurement noise respectively. These are assumed to be independent white noise,with normal probability distributions.

$$P(w) \sim N(0, Q), P(v) \sim N(0, R)$$

where Q and R are the process and measurement noise covariances respectively.The matrix A of dimension nxn relates the state at the previous time step $k-1$ to the state at the current time step k. The matrix B is of dimension nxl and it relates to the control input to the state $u \in \Re^l$. The matrix H is of size mxn relating the state of measurement.

Let $\hat{x}_k^- = x_k/Z_{(k-1)}$ denote the *apriori* state estimate at step k, which is based on the knowledge of the process up to $k-1$. Let $\hat{x_k} = x_k/z(k)$ be the *aposteriori* state estimate at step k given the measurement $z(k)$.

We define *apriori* and *aposteriori* error estimate as

$$e_k^- = x_k - x_k/Z_{k-1}$$

$$e_k = x_k - x_k/z_k$$

The *a priori* estimate of error covariance is defined as

$$P_k^- = [e_k^- e_k^{-T}]$$

and *a posteriori* estimate of error covariance is

$$P_k = [e_k e_k^T]$$

In formulating the Kalman filter equation, the objective is to determine the equation relating *a posteriori* state estimate $\hat{x_k}$ as a linear combination of *a priori* estimate $\hat{x_k^-}$ as

$$\hat{x_k} = x_k/Z_{k-1} + K\left[z_k - Hx_k/Z_{k-1}\right]$$

The term $[z_k - Hx_k/Z_{k-1}]$ is called measurement innovation or residual. The residual reflects the discrepancy between predicted measurement Hx_k/Z_{k-1} and the actual measurement z_k.

The nxm size matrix K is defined as Kalman Gain.One of the popular form of expressing Kalman gain is

$$K_k = \bar{P_k}H^T(H\bar{P_k}H^T + R)^{-1} = \frac{\bar{P_k}H^T}{H\bar{P_k}H^T + R}$$

The Kalman filter equations can be described in various forms and the equations formulated above is one of them.

14.8.1 Discrete Kalman Filter Algorithm

Kalman filter estimates a process by using a form of feedback control. The filter estimates the process at some time and then obtain feedback in the form of noisy measurements. The equations for Kalman filter fall in two groups: *Time update equations* and *measurement equations.* The time update equations are responsible for projecting forward the current state and the error co variance estimates to obtain the *a priori* estimates for the next step. The measurement update equations cater for incorporating a new measurement into *a priori* estimate to obtain an improved *a posteriori* estimate. The discrete Kalman time update equations are

$$\hat{x}_k^- = A\hat{x}_{k-1} + Bu_k$$

$$P_k^- = AP_{k-1}^-A^T + Q$$

where P_k^- is $\varepsilon[\bar{e_k}, \bar{e_k}^T]$, *apriori* estimate error covariance and P_{k-1} is $\varepsilon[e_{k-1}, e_{k-1}^T]$, *aposteriori* estimate error covariance.

The measurement update equations are

$$K_k = P_k^- H^T(HP_k^- H^T + R)^{-1}$$

$$\hat{x_k} = \hat{x_k^-} + K_k(z_k - H\hat{x_k^-}$$

$$P_k = (I - K_kH)P_k^-.$$

The algorithm can be formulated as

- Compute the Kalman Gain K_k using $K_k = \bar{P_k}H^T(H\bar{P_k}H^T + R)^{-1}$

- Obtain the measurement data$z(k)$.
- Compute and update the *a posteriori* state estimate

$$\hat{x_k} = \hat{x_k}^- + K_k(z_k - H\hat{x_k}^-)$$

- Compute and update the error covariance

$$P_k = (I - K_k H)P_k^-$$

- Compute and project the new state estimate

$$\hat{x_{k+1}^-} = A\hat{x_k} + Bu_k$$

- Compute and project the error covariance

$$P_{k+1}^- = A\bar{P_k}A^T + Q$$

After each time and measurement update pair, the process is repeated with previous *aposteriori* estimates which are used to predict new *apriori* estimates. This recursive nature makes the practical implementation computationally feasible to solve real time tracking problems.

14.9 EXTENDED KALMAN FILTERING

As described earlier, the Kalman filter addresses the general problem of estimating the state $x \varepsilon \Re^n$ of a discrete time controlled stochastic difference equation. However, if the process to be estimated and/or the measurement relationship to the process is non-linear, a method known as *Extended Kalman Filtering* wherein the current mean and covariance is linearized is used. *Extended Kalman* filter is the most popular approach for implementing recursive non-linear filters. This is a linearization technique based on the first order Taylor series expansion of the non-linear system and measurement functions about the current estimate of the state.

In the non-linear case, the process equation is defined as

$$x_k = f(x_{k-1}, u_k, w_{k-1})$$

with a measurement $z = h(x_k, v_k)$, wherein the random variables w_k and v_k represent the process and measurement noise. In this case, the non-linear function f in the above difference equation relates the state at previous time instant $k-1$ to the state at current time instant k. The non-linear function h in the measurement equation relates the state x_k to the measurement z_k.

As it may not be feasible to obtain individual values of noise w_k and v_k at each time step periodically, the approach is to approximate the state and measurement vector without them as

$$\tilde{x_k} = f(\hat{x_{k-1}}, u_k, 0)$$

and

$$\tilde{z_k} = (\tilde{x_k}, 0)$$

where $\tilde{x_k}$ is the *aposteriori* estimate of the state computed data up to time instant $k-1$.

For estimating the process with non-linear difference and measurement relationships, the equations can be expressed as

$$x_k \approx \tilde{x_k} + A(x_{k-1} - \hat{x_{k-1}}) + W w_{k-1}$$

$$z_k \approx \tilde{z_k} + H(x_k - \tilde{x_k}) + V v_k$$

where x_k and z_k are actual state and measurement vectors, $\tilde{x_k}$ and $\tilde{z_k}$ are approximate state and measurement vectors. $\hat{x_k}$ is the *aposteriori* estimate at time instant k.

A is Jacobian matrix of partial derivatives of f with respect to x, i.e.,

$$A_{[i,j]} = \frac{\delta f_{[i]}}{\delta x_{[j]}}(\hat{x_k}, u_k, 0)$$

H is Jacobian matrix of partial derivatives of h with respect to x, i.e.

$$H_{[i,j]} = \frac{\delta h_{[i]}}{\delta x_{[j]}}(\tilde{x_k}, 0)$$

W is Jacobian matrix of partial derivatives of f with respect to w, i.e.,

$$W_{[i,j]} = \frac{\delta f_{[i]}}{\delta w_{[j]}}(\hat{x_k}, u_k, 0)$$

V is Jacobian matrix of partial derivatives of h with respect to v, i.e.,

$$V_{[i,j]} = \frac{\delta_{h[i]}}{\delta v_{[j]}}(\tilde{x_k}, 0)$$

We define the prediction error and measurement residual error as

$$e_{\tilde{x_k}} \equiv x_k - \tilde{x_k}$$

and

$$e_{\tilde{z_k}} \equiv z_k - \tilde{z_k}$$

With the above definition, the governing equations for the error process can now be approximated as

$$\tilde{e}_{x_k} \approx A(x_{k-1} - \hat{x}_{k-1}) + \varepsilon_k$$

and

$$\tilde{e}_{z_k} \approx H\tilde{e}_{x_k} + \eta_k$$

where ε_k and η_k denote the new independent random variables having zero mean and covariance matrices WQW^T and VRV^T

The above equations closely resemble the difference and measurement equations of *Discrete Kalman Filter* and are linear. Hence actual measurement residual $e_{\tilde{z}_k}$ and a hypothetical second Kalman filter can be used to estimate the prediction error $\tilde{e}_{x_k} \approx A(x_{k-1} - \hat{x}_{k-1}) + \varepsilon_k$ defined as $\hat{x_k}$ so as to obtain the *a posteriori* state estimate for the non-linear process as

$$\hat{x_k} = \tilde{x_k} + e_{\tilde{x}_k}$$

and the random variables ε_k, η_k and $e_{\tilde{x}_k}$ will have approximately the following probability distributions:

$$P(e_{\tilde{x}_k}) \sim N(0, \varepsilon[e_{\tilde{x}_k}, e_{\tilde{x}_k}{}^T])$$

$$P(\tilde{\epsilon_k}) \sim N(0, WQ_kW^T)$$

$$P(\tilde{\eta_k}) \sim N(0, VR_kV^T)$$

Now considering the predicted value $\hat{e_k}$ to be be zero, the Kalman equation can be expressed as

$$\hat{e_k} = K_k e_{\tilde{z}_k}$$

and the state equations can be expressed as

$$\hat{x_k} = \tilde{x_k} + K_k e_{\tilde{z}_k} = \tilde{x_k} + K_k(z_k - \tilde{z_k})$$

The above equations can be used for the measurement update in the *Extended Kalman Filter.* To summerize, the *Extended Kalman filter* time update equations are

$$\hat{x_k^-} = f(\hat{x_{k-1}}, u_k, 0)$$

$$\bar{P_k} = A_k P_{k-1} A_k^T + W_k Q_{k-1} W_k^T$$

and the measurement equations are

$$K_k = \bar{P_k} H_k^T (H_k \bar{P_k} H_k^T + V_k R_k V_k^T)^{-1}$$

$$\hat{x_k} = \hat{x_k}^- + K_k(z_k - h(\hat{x}_k^-, 0)$$

$$P_k = (I - K_k H_k) P_k^-$$

An important feature of *Extended Kalman filter* is that the Jacobian H_k in the equation for Kalman gain K_k serves to correctly propagate only the relevant component of measurement information.

For a particular problem, if the assumptions of the Kalman filter hold, then no other algorithm can out perform it. However, for a variety of real world computer vision applications, these assumptions are often unrealistic. So alternate techniques which approximate the function is employed. Particle filtering has emerged as a powerful tool in solution to these problems. A number of different types of particle filter exists which are optimal for certain class of applications.

14.10 PARTICLE FILTER BASED OBJECT TRACKING

A number of tracking strategies have been suggested and important ones are based on Kalman filters and extended Kalman filtering. In tracking problem, if the assumptions of the Kalman filter hold, then no other algorithm can outperform. However, in real life applications, these assumptions may not hold at all the time and approximate techniques must be employed [9]. Extended Kalman Filter approximates the models used for the dynamics and measurement process, in order to be able to approximate the probability density of Gaussian. Particle filtering approximates the density directly as a finite number of samples. Different types of particle filters exists and some form of implementation outperform others when used for specific applications. In this section, we will present the tracking algorithm using Particle filters.

With advancements in the computing field, particle filtering has emerged as a powerful methodology for sequential signal processing. Broadly stating, Particle filtering is the recursive implementation of Monte-Carlo based statistical signal processing. A number of excellent particle filtering based tracking may be found in [9]–[16] and many others where this method has been applied to solve a wide range of application areas.

Particle filter techniques are formulated on the concepts of *Baysien theory* and *Sequential Importance Sampling*. This approach has been found to be very effective in dealing with non-gaussian and non-linear problems. The basic approach is to approximate the relevant distributions with random measures composed of properly weighted particles.

The process equation can be defined as

$$x_k = f_k(x_{k-1}, w_k)$$

with a measurement vector $z_k = h_k(x_k, v_k)$, where h_k is the measurement function, f_k is the system transition matrix at time instant k. The random variables w_k and v_k represent the process and measurement noise, which are assumed to be independent.

In particle filtering based tracking, we use discrete random measures to approximate the continuous probability distribution functions [3, 11, 12, 17]. These discrete random measures are formed out of particles which are suitably weighted, where the weights are probability masses, computed by using Bayes theory. The particles are samples of unknown states from state space.

One of the features of particle filtering is that the approximation of desired probability distribution function is attempted against the approach of linearization around the current estimate. In this technique, the concept of importance sampling is important. Since, this is a sequential process, it is known as sequential importance sampling.

Assuming the noise samples are independent, the joint *a posteriori* distribution of $(x_0, x_1, ...x_k)$ can be expressed as

$$P(\hat{x_k}/Z_k) \propto P(x_0/z_0)\prod_{t=1}^{K} P(z_t/x_t).P(x_t/\hat{x_{t-1}})$$

The recursive formula for obtaining $P(\hat{x_k}/Z_k)$ from $(\hat{x_{k-1}}/Z_{k-1})$ is

$$P(\hat{x_k}/Z_k) = \frac{P(z_k/x_k).P(x_k/x_{k-1})}{P(z_k/Z_{k-1})}.P(\hat{x_{k-1}}/Z_{k-1})$$

As mentioned earlier, we use the methods based on approximations. These approximations can be expressed as,

$$P(x) \approx \Sigma_{m-1}^{M} w^{(m)}.\delta(x - x^{(m)}),$$

where $x^{(m)}$ are the particles and $w^{(m)}$ are the corresponding weights and M is the number of particles. $\delta(.)$ is a Dirac delta function. With approximation, the expectation operations become summations as

$$\varepsilon[f(x)] \approx \sum_{m=1}^{M} w^m.f(x^m)$$

In particle filtering, another factor is importance sampling. The probability distribution function is approximated by particles, each of them assigned with equal weights $1/M$. When the direct sampling from the distribution function is not feasible, then the particles $x^{(m)}$ are generated from a distribution $\prod(x)$, known as importance function and then assign weights as

$$w^{*(m)} = \frac{P(x)}{\prod(x)}.$$

On normalization of the weights, this is expressed as

$$w^m = \frac{w^{*(m)}}{\sum_{i=1}^{M}}.$$

If the importance function can be factored as

$$\prod(\hat{x_k}/Z_k) = \prod(x_k/\hat{x_{k-1}}, Z_k).\prod(\hat{x_k}/Z_{k-1})$$

and if

$$\hat{x_{k-1}}^{(m)} \approx \prod(\hat{x_k}/Z_{k-1})$$

and the weights:

$$\hat{x_{k-1}}^{(m)} \propto \frac{P(\hat{x_{k-1}}^{(m)}/Z_{k-1})}{\prod(\hat{x_{k-1}}^{(m)}/Z_{k-1})}$$

then, we can determine $x_k^{(m)}$ as

$$x_k^{(m)} \approx \prod(x_k/\hat{x_{k-1}}^{(m)}, z_k)$$

with weight $w_k^{(m)}$ as

$$w_k^{(m)} \approx \frac{P(z_k/x_k^{(m)}).P(x_k^{(m)}/\hat{x_{k-1}}^{(m)})}{\prod(x_k^{(m)}/\hat{x_{k-1}}^{(m)}, z_k)} .w_{k-1}^{(m)}$$

and the posterior density function

$$P(x_k/Z_k) \approx \sum_{m=1}^{M} w_k^m \delta\left(x_k - x_k^{(m)}\right)$$

As $M \to \infty$, the approximation approaches the true *a posteriori* density $P(x_k/Z_k)$. It is important to choose a proper importance function. The importance function must have the same support as the PDF which is being approximated. The closer the selected importance function to the PDF, the better the approximation.

One of the problem in implementation of the particle filtering is that the discrete random measure degenerates quickly, implying the deterioration of performance of particle filter. This effect can be reduced by careful selection of importance sampling function and *resampling*. Resampling is the process by which, the particles which have been assigned with small and negligible weights are resampled with particles with large weights.

Broadly, when the particle filter techniques are used for tracking a human object, the major steps involved in the algorithm can be stated as :

1. Resampling
2. Dynamics and
3. Weight updation using observation

In the first phase, initially we choose a particle and n number of copies of the same.

14.10.1 Particle Attributes

For each new particle generated, the particle generating system may determine a set of attributes that are assigned initially to the particle. Some of these attributes may be *position*, *velocity*, *size*, *chromatic* and *achromatic* values, *transparency*, *shape*, and *lifetime*. Several parameters of a particle system control the initial position of its particles. A particle system has a position in three-dimensional space that defines its origin. Two angles of rotation about a coordinate system through this origin give it an orientation. We may choose some shape which defines a region around its origin into which newly born particles are randomly placed. These regions may be any regular or even arbitrary shaped regions, e.g., a rectangle of length l and width w, or a sphere of radius r, or a circle of radius r in the x-y plane.

14.10.2 Particle Filter Algorithm

The algorithm is as below.

Step 0 (initialization phase): Choose the centroids of the human blobs as obtained from the output of the morphological transforms as the seed particles.

Step 1 (resampling phase): Estimate the scale $s (0 < s < 1)$ from the blob location and size of the blob using a lookup table. Compute the radius of circle r, in which the particles are located as also the number of resampled particles N corresponding to each blob. Please note that both r and N are also functions of scale.

Step 2 (dynamic estimation): In this phase we estimate the dynamics of motion of the particles. Here we have chosen a random walk strategy to estimate the next location, i.e., the dynamics is based on random walk.

$$x_{k+1} = x_k + \eta_1(\mu_x, \sigma_x)$$
$$y_{k+1} = y_k + \eta_2(\mu_y, \sigma_y)$$

Select $\mu_x = \mu_y = 0$ and $\sigma_x = (a_1 + b_1\Delta_x)$ which is variance around each blob I, a_1 is the minimum variance that the particle search space will cover even when there is very little or no movement of the particle, and b_1 is a constant which is multiplied with the directivity constancy factor Δ_x along x-direction $\sigma_y = (a_1 + b_1\Delta_y)s$, where the parameters are defined in a similar way.

Select a circle with radius r around from look up table. Here r is a function of scale.

Step 3 (weight assignment): Assign weight to each particle

$$w = \frac{1}{1 + d(p_i, q_i)},$$

where $d(p_i, q_i)$ yields a matching measure between the ith particle and the actual blob in the image frame. Normalize the weights if necessary and rank the particles according to the weights. Go to step 1 (i.e., resampling phase).

14.10.3 Results of Object Tracking

Figure 14.2 shows a sequence of video frames, where there is a moving human figure in a static background. The color version of the figure is provided in the color page section. The tracked human figure in each color frame is shown in a blob, enclosed in a red rectangular bounding box. Here a set of hundred particles have been chosen at each frame representing the moving target. A simple random walk guides the motion of the human and the search region shape has been chosen as a square window of 50 × 50 pixels. The moving object could be continuously tracked in every frame in a sequence of 10,000 frames.

Fig. 14.2 Results of particle filter based tracking - moving human figure encapsulated in rectangular bounding boxes at an interval of 10 frames.

14.11 CONDENSATION ALGORITHM

The condensation algorithm is a form of particle filtering. This is based on factored sampling and extended to apply iteratively to successive images in a sequence. The tracking mechanism based on Particle filters essentially involves tracking a variable as it evolves over time with a non gaussian multi-modal probability distribution function [11]. The objective here is to track features and outlines of foreground objects modeled as curves as they move in cluttered environment. The variable we are talking about may be the position of a target or the pose of a moving robot. At a particular instant our variable, say position of a target, may be represented as an ensemble of a set of samples. where each sample is known as a particle. Thus each particle is a copy of the variable. Also each particle has a weight, which specifies the contribution of the particle to the overall estimate of the variable. How do we estimate our variable of interest (i.e., position of the target)? This may be obtained from the weighted sum of all the particles.

Discrete-time propagation of state density: The state of the modeled object at time t is denoted x_t and the its history is $X_t = \{x_1, \ldots, , x_t\}$. The set of image features at time t is z_t with history $Z_t = \{z_1, \ldots, z_t\}$.

Stochastic dynamics: The new state is conditioned directly only on the immediately preceding state, independent of the earlier history. This still allows quite general dynamics.

Condensation Algorithm
From the "old" sample-set $\left\{s_{t-1}^{(n)}, \phi_{t-1}^{(n)}, c_{t-1}^{(n)}, n = 1, \ldots, N\right\}$ at time-step $t-1$, construct the "new" sample-set $\left\{s_t^{(n)}, \phi_t^{(n)}, c_t^{(n)}\right\}$, $n = 1, \cdots, N$ for time t.

Construct the n_{th} of N new samples as follows:

1. **Select** a sample s_t as follows:

   (a) Generate a random number $r \in [0, 1]$, uniformly distributed.

   (b) Find, by binary subdivision, the smallest j for which $c_{t-1}^{(j)} \geq r$

   (c) Set $s_t = s_{t-1}^j$

2. **Predict** by sampling from

$$p(x_t, x_{t-1}) = s_t$$

   to choose each s_t. For instance, in the case that the dynamics are governed by a linear stochastic differential equation, the new sample value may be generated from random walk - see item 1.

3. **Measure** and assign weight to the new position in terms of the measured features z_t:

$$\phi_t = p(z_t | z_t = s_t)$$

4. **Normalize** ϕ_t so that $\sum_n \phi_t^{(n)} = 1$ and store together with cumulative probability as (s_t, ϕ_t, c_t).

5. Repeat Steps 1 to 4 to construct the N new samples.

6. **Estimate** the moments of the tracked position at time-step t as

$$E[f(x_t)] = \sum_{n-1}^{N} \phi_t f(s_t).$$

Thus the variable of interest (the mean position of the tracked object in this example) may be tracked as it propagates through a cluttered environment.

14.12 SUMMARY

Dynamic estimation and Tracking of objects in a scene from an image sequence has a wide range of application, like video surveillance, Defence, Bio medical and weather modeling etc. The methods described above has been extensively used in real world applications. Adaptive background modeling is important so as to cater for the frequent change in illumination of the observed scene. A robust background model is essential to eliminate unwanted changes in the scenes. If background model is not updated, this could result in wrong moving object estimate.Once the pixels have been classified as foreground and background, a thresholding is to be applied to filter out noise and isolated points, which are falsely detected. Connected component labeling yields a a set of smooth blobs corresponding to each moving objects. If there is only one object, which need to be tracked in a clear non varying background, then the problem is relatively simple. In case of multiple object tracking, and with varying background, which need to be eliminated, complex approaches are needed. For this, formulation of suitable tracking model is important. Typically, this is modeled as a three state sequential machine. i.e. Locking, Tracking and Recovery. For a linear Gaussian process, the Kalman filter provides an optimal recursive estimate. The mathematical formulation is in state space concepts. Each updated estimate is computed from previous estimate and the new input data. If the process to be estimated/tracked and/or the measurement relationship to the process is non-linear, then the approach known as Extended Kalman Filtering is used.This has been the popular approach for implementation of recursive state estimate for a non-linear process. Particle filtering is a powerful technique and is the recursive implementation of Monte-Carlo based statistical signal processing. These are formulated on the concepts of bayesian theory and sequential importance sampling. Multiple object tracking under varying background in real time / near real time is a very challenging area of research. Development of a robust adaptive background model will compliment the effectiveness of object tracking. Research on effective management of shadow detection and object occlusion problem can lead to building a robust tracking solution.

REFERENCES

1. D. Reid, "An algorithm for tracking multiple targets," *IEEE Trans. on Automation and Control*, December 1979, vol. AC-24, 84–90.

2. I. Haritaoglu, D. Harwood, and L. S. Davis, "W4: real-time surveillance of people and their activities," *IEEE Transactions on Pattern Analysis and Machine Intelligence*, 22(8), August 2000, 809–830.

3. R. Cucchiara, C. Grana, G. Neri, M. Piccardi, and A. Prati, "The Sakbot System for Moving Object Detection and Tracking," *Video-Based Surveillance Systems-Computer Vision and Distributed Processing*, 2001, 145-157.

4. C. Jiang and M. O. Ward, "Shadow Identification," *Proc. IEEE Int'l Conf. Computer Vision and Pattern Recognition*, 1992, 606–612.

5. R. Cucchiara, C. Grana, M. Piccardi, A. Prati, and S. Sirotti, "Improving shadow suppression in moving object detection with hsv color information," *Proceedings of IEEE Int'l Conference on Intelligent Transportation Systems*, August 2001, 334-339.

6. I. Mikic, P. Cosman, G. Kogut, and M. M. Trivedi, "Moving Shadow and Object Detection in Traffic Scenes," *Proc. Int'l Conf. Pattern Recognition*, 1, September 2000, 321-324.

7. T. Horprasert, D. Harwood, and L. S. Davis, "A Statistical Approach for Real-Time Robust Background Subtraction and Shadow Detection," *Proc. IEEE Int'l Conf. Computer Vision*, 1999.

8. G. Medioni, "Detecting and Tracking Moving Objects for Video Surveillance," *Proc. IEEE Int'l Conf. Computer Vision and Pattern Recognition*, 2, 1999, 319–325.

9. S. Arulampalam, S. Maskell, N. Gordon, and T. Clapp, "A tutorial on particle filters for on-line non-linear/non-Gaussian Bayesian tracking," *IEEE Transactions on Signal Processing*, 50, February 2002, 174–188.

10. M. Isard and A. Blake, "Contour tracking by stochastic propagation of conditional density," *Proc. Eur. Conf. on Computer Vision* (ECCV), 1996, 343-356.

11. M. Isard and A. Blake, CONDENSATION - Conditional Density Propagation for Visual Tracking, International Journal on Computer Vision, 1(29), 1998, 5-28.

12. G. Welch and G. Bishop, "An introduction to the Kalman filter", *Tech Rep. TR95041*, University of North Carolina, Dept. of Computer Science, 2000.

13. K. Nummiaro, E. Koller-Meier and L. Van Gool, An Adaptive Color-Based Particle Filter, Journal of Image and Vision Computing, 21(1), 2003, 99-110.

14. D. Comaniciu, V. Ramesh, and P. Meer, "Real-Time Tracking of Non-Rigid Objects using Mean Shift," *Computer Vision and Pattern Recognition*, 2, 2000, 142–149.

15. S. J. McKenna, S. Jabri, Z. Duric, A. Rosenfeld, and H. Wechsler, "Tracking groups of people," *Computer Vision and Image Understanding*, 80(1), October 2000, 42-56.

16. G. Funka-Lea and R. Bajcsy, "Combining Color and Geometry for the Active, Visual Recognition of Shadows," *Proc. IEEE Int'l Conf. Computer Vision*, 1995, 203-209.

17. P. M. Djuric, J. H. Kotecha, J. Zhang, Y. Huang, T. Ghirmai, M. F. Bugallo, and J. Migues, "Particle Filtering," *IEEE Signal Processing Magazine*, 2003, 19-38.

15

Introduction to Image Compression

15.1 INTRODUCTION

Demand for communication of multimedia data through the telecommunications network and accessing the multimedia data through the Internet is growing explosively. In order to handle this pervasive multimedia data usage, it is essential that the data representation and encoding of multimedia data be standard across different platforms and applications. Image data comprise a significant portion of the multimedia data and they occupy the lion's share of the communication bandwidth for multimedia communication. As a result, development of efficient image compression techniques continues to be an important challenge to us, both in academia and in industry.

Despite the many advantages of digital representation of signals compared to the analog counterpart, they need a very large number of bits for storage and transmission. For example, a television-quality low-resolution color video of 30 frames per second with each frame containing 640 × 480 pixels (24 bits per color pixel) needs more than 210 megabits per second of storage. As a result, a digitized one-hour color movie would require approximately 95 gigabytes of storage. The storage requirement for upcoming high-definition television (HDTV) of resolution 1280 × 720 at 60 frames per second is far greater. A digitized one-hour color movie of HDTV-quality video will require approximately 560 gigabytes of storage. A digitized 14 × 17 square inch radiograph scanned at 70 μm occupies nearly 45 megabytes of storage. Transmission of these digital signals through limited-bandwidth communication channels is even a greater challenge and sometimes impossible in its raw form. Although the cost of storage has decreased drastically over the past decade due to significant advancement in microelectronics and storage technology, the requirement of data storage and data processing applications is growing explosively to outpace this achievement.

Interestingly enough, most of the sensory signals such as still image, video, and voice generally contain significant amounts of superfluous and redundant information in their canonical representation as far as the human perceptual system is concerned. By human perceptual system, we mean our eyes and ears. For example, the neighboring pixels in the smooth region of a natural image are very similar and small variation in the values of the neighboring pixels are not noticeable to the human eye. The consecutive frames in a stationary or slowly changing scene in a video are very similar and redundant. Some audio data beyond the human audible frequency range are useless for all practical purposes. This fact tells us that there are data in audio-visual signals that cannot be perceived by the human perceptual system. We call this *perceptual redundancy*. There are many such examples of redundancy in digital representation in all sorts of data.

Data compression is the technique to reduce the redundancies in data representation in order to decrease data storage requirements and hence communication costs. Reducing the storage requirement is equivalent to increasing the capacity of the storage medium and hence communication bandwidth. Thus the development of efficient compression techniques will continue to be a design challenge for future communication systems and advanced multimedia applications.

15.2 INFORMATION THEORY CONCEPTS

The *Mathematical Theory of Communication*, which we also call *Information Theory* here, pioneered by Claude E. Shannon in 1948 [1]–[4] is considered to be the theoretical foundation of data compression research. Since then many data compression techniques have been proposed and applied in practice.

Representation of data is a combination of *information* and *redundancy* [1]. Information is the portion of data that must be preserved permanently in its original form in order to correctly interpret the meaning or purpose of the data. However, redundancy is that portion of data that can be removed when it is not needed or can be reinserted to interpret the data when needed. Most often, the redundancy is reinserted in order to regenerate the original data in its original form. Data *compression* is essentially a redundancy reduction technique. The redundancy in data representation is reduced such a way that it can be subsequently reinserted to recover the original data, which is called *decompression* of the data. In the literature, sometimes data compression is referred to as *coding* and similarly decompression is referred to as *decoding*.

Usually development of a data compression scheme can be broadly divided into two phases–*modeling* and *coding*. In the modeling phase, information about redundancy that exists in the data is extracted and described in a model. Once we have the description of the model, we can determine how the actual data differs from the model and encode the difference in the *coding* phase. Obviously, a data compression algorithm becomes more effective if the model is closer to the characteristics of the data generating process, which we often call the *source.* The model can be obtained by empirical observation of the statistics of the data generated by the process or the source. In an empirical sense, any information-generating process can be described as a source that emits a sequence of symbols chosen from a finite alphabet. Alphabet is the set of all possible symbols generated by the source. For example, we can think of this text as being generated by a source with an alphabet containing all the ASCII characters.

15.2.1 Discrete Memoryless Model and Entropy

If the symbols produced by the information source are statistically independent to each other, the source is called a *discrete memoryless source.* A discrete memoryless source is described by its source alphabet

$$A = \{a_1, a_2, \cdots, a_N\}$$

and the associated probabilities of occurrence

$$P = \{p(a_1), p(a_2), \cdots, p(a_N)\}$$

of the symbols $a_1, a_2, \cdots, a_N$ in the alphabet A.

The definition of the *discrete memoryless source* model provides us a very powerful concept of quantification of *average information content per symbol* of the source, or *entropy* of the data. The concept of "entropy" was first used by physicists as a thermodynamic parameter to measure the degree of "disorder" or "chaos" in a thermodynamic or molecular system. In a statistical sense, we can view this as a measure of degree of "surprise" or "uncertainty." In an intuitive sense, it is reasonable to assume that the appearance of a less probable event (symbol) gives us more surprise, and hence we expect that it might carry more information. On the contrary, the more probable event (symbol) will carry less information because it was more expected.

With the above intuitive explanation, we can comprehend Shannon's definition of the relation between the source symbol probabilities and corresponding codes. The amount of *information* content, $I(a_i)$, for a source symbol a_i, in terms of its associated probability of occurrence $p(a_i)$ is

$$I(a_i) = \log_2 \frac{1}{p(a_i)} = -\log_2 p(a_i). \tag{15.1}$$

The base 2 in the logarithm indicates that the information is expressed in binary form, or bits. In terms of binary representation of the codes, a symbol

a_i that is expected to occur with probability $p(a_i)$ is best represented in approximately $-\log_2 p(a_i)$ bits. As a result, a symbol with higher probability of occurrence in a message is coded using a fewer number of bits.

If we average the amount of information content over all the possible symbols of the discrete memoryless source, we find the average amount of information content per source symbol from the discrete memoryless source. This is expressed as

$$E = \sum_{i=1}^{N} p(a_i) I(a_i) = -\sum_{i=1}^{N} p(a_i) \log_2 p(a_i). \tag{15.2}$$

This is popularly known as *entropy* in information theory. Hence entropy is the expected length of a binary code over all possible symbols in a discrete memoryless source.

The concept of entropy is very powerful. In "stationary" systems, where the probabilities of occurrence of the source symbols are fixed, it provides a bound for the compression that can be achieved. This is a very convenient measure of the performance of a coding system. Without any knowledge of the physical source of data, it is not possible to know the entropy, and the entropy is estimated based on the outcome of the source by observing the structure of the data as source output. Hence estimation of the entropy depends on observation and assumptions about the structure of the source data sequence. These assumptions are called the *model* of the sequence.

15.2.2 Noiseless Source Coding Theorem

The *Noiseless Source Coding Theorem* by Shannon [1] establishes the minimum average code word length per source symbol that can be achieved, which in turn provides the upper bound on the achievable compression losslessly. The *Noiseless Source Coding Theorem* is also known as *Shannon's first theorem*. This is one of the major source coding results in information theory [1, 2, 3].

If the data generated from a discrete memoryless source A are considered as grouped together in blocks on n symbols, to form an *n-extended source*, then the new source A^n has N^n possible symbols $\{a_i\}$, with probability $P(a_i) = P(a_{i_1})P(a_{i_2})\cdots P(a_{i_n}), i = 1, 2, \cdots, N^n$. By deriving the entropy of the new n-extended source, it can be proven that $E(A^n) = nE(A)$, where $E(A)$ is the entropy of the original source A. Let us now consider encoding blocks of n source symbols at a time into binary codewords. For any $\epsilon > 0$, it is possible to construct a codeword for the block in such a way that the average number of bits per original source symbol, $\bar{L}$, satisfies

$$E(A) \leq \bar{L} < E(A) + \epsilon.$$

The left-hand inequality must be satisfied for any uniquely decodable code for the block of n source symbols.

The *Noiseless Source Coding Theorem* states that any source can be losslessly encoded with a code whose average number of bits per source symbol is arbitrarily close to, but not less than, the source entropy E in bits by coding infinitely long extensions of the source. Hence, the noiseless source coding theorem provides us the intuitive (statistical) yardstick to measure the information emerging from a source.

Example: We consider a *discrete memoryless source* with alphabet $A_1 = \{\alpha, \beta, \gamma, \delta\}$ and the associated probabilities are $p(\alpha) = 0.65$, $p(\beta) = 0.20$, $p(\gamma) = 0.10$, $p(\delta) = 0.05$ respectively. The entropy of this source is $E = -(0.65 \log_2 0.65 + 0.20 \log_2 0.20 + 0.10 \log_2 0.10 + 0.05 \log_2 0.05)$, which is approximately 1.42 bits/symbol. As a result, a data sequence of length 2000 symbols can be represented using approximately 2820 bits.

Knowing something about the structure of the data sequence often helps to reduce the entropy estimation of the source. Let us consider that the numeric data sequence generated by a source of alphabet $A_2 = \{0, 1, 2, 3\}$ is $D =$ 0 1 1 2 3 3 3 3 3 3 3 3 3 2 2 2 3 3 3 3, as an example. The probability of appearance of the symbols in alphabet A_2 are $p(0) = 0.05$, $p(1) = 0.10$, $p(2) = 0.20$, and $p(3) = 0.65$ respectively. Hence the estimated entropy of the sequence D is $E = 1.42$ bits per symbol. If we assume that correlation exists between two consecutive samples in this data sequence, we can reduce this correlation by simply subtracting a sample by its previous sample to generate the residual values $r_i = s_i - s_{i-1}$ for each sample s_i. Based on this assumption of the model, the sequence of residuals of the original data sequence is $\bar{D} =$ 0 1 0 1 1 0 0 0 0 0 0 0 0 -1 0 0 1 0 0 0, consisting of three symbols in a modified alphabet $\bar{A}_2 = \{-1, 1, 0\}$. The probability of occurrence of the symbols in the new alphabet $\bar{A}$ are $P(-1) = 0.05$, $p(1) = 0.2$, and $p(0) = 0.75$ respectively as computed by the number of occurrence in the residual sequence. The estimated entropy of the transformed sequence is $\bar{E} = -(0.05 \log_2 0.05 + 0.2 \log_2 0.2 + 0.75 \log_2 0.75) = 0.992$ (i.e., 0.992 bits/symbol).

This is a simple example to demonstrate that the data sequence can be represented with fewer numbers of bits if encoded with a suitable entropy encoding technique and hence resulting in data compression.

15.2.3 Unique Decipherability

Digital representation of data in binary code form allows us to store it in computer memories and to transmit it through communication networks. In terms of length of the binary codes, they can be *fixed-length* as shown in column A of Table 15.1 with alphabet $\{\alpha, \beta, \gamma, \delta\}$, as an example, where all the symbols have been coded using the same number of bits. The binary codes could also be *variable-length* codes as shown in columns B or C of Table 15.1 in which the symbols have different code lengths.

Table 15.1 Examples of variable-length codes

Symbol	A	B	C
α	00	0	0
β	01	10	1
γ	10	110	00
δ	11	111	01

Consider the string $S = \alpha\alpha\gamma\alpha\beta\alpha\delta$. The binary construction of the string S using variable-length codes A, B, and C is as follows:

$$C_A(S) = 00001000010011$$

$$C_B(S) = 001100100111$$

$$C_C(S) = 000001001.$$

Given the binary code $C_A(S) = 00001000010011$, it is easy to recognize or uniquely decode the string $S = \alpha\alpha\gamma\alpha\beta\alpha\delta$ because we can divide the binary string into nonoverlapping blocks of 2 bits each and we know that two consecutive bits form a symbol as shown in column A. Hence the first two bits "00" form the binary code for the symbol α, the next two bits "00" is similarly mapped to the symbol α, the following two bits "10" can be mapped to symbol γ, and so on. We can also uniquely decipher or decode the binary code $C_B(S) = 001100100111$ because the first bit (0) represents the symbol α; similarly the next bit (0) also represents the symbol α according to the code in column B. The following three consecutive bits "110" uniquely represent the symbol γ. Following this procedure, we can uniquely reconstruct the string $S = \alpha\alpha\gamma\alpha\beta\alpha\delta$ without any ambiguity.

But deciphering the binary code $C_C(S) = 000001001$ is ambiguous because it has many possibilities–$\alpha\gamma\gamma\beta\gamma\beta$, $\alpha\gamma\alpha\delta\gamma\beta$, or $\alpha\alpha\alpha\alpha\alpha\beta\gamma\beta$ to name a few. Hence the code $C_C(S) = 000001001$ is not uniquely decipherable using the code in column C in Table 15.1.

It is obvious that the *fixed-length* codes are always uniquely decipherable. But not all the *variable-length* codes are uniquely decipherable. The uniquely decipherable codes maintain a particular property called the *prefix property.* According to the prefix property, no codeword in the code-set forms the *prefix* of another distinct codeword [5]. A codeword $C = c_0c_1c_2 \cdots c_{k-1}$ of length k is said to be the prefix of another codeword $D = d_0d_1 \cdots d_{m-1}$ of length m if $c_i = d_i$ for all $i = 0, 1, \cdots, k-1$ and $k \leq m$.

Note that none of the codes in column A or in column B is a prefix of any other code in the corresponding column. The codes formed using either column A or column B are uniquely decipherable. On the other hand, binary code of α in column C is a prefix of both the binary codes of γ and δ.

Some of the popular variable-length coding techniques are Shannon-Fano Coding [6], Huffman Coding [7], Elias Coding [8], Arithmetic Coding [9], etc. It should be noted that the *fixed-length* codes can be treated as a special case of uniquely decipherable *variable-length* code.

15.3 CLASSIFICATION OF COMPRESSION ALGORITHMS

In an abstract sense, we can describe *data compression* as a method that takes an input data D and generates a shorter representation of the data $c(D)$ with a fewer number of bits compared to that of D. The reverse process is called *decompression*, which takes the compressed data $c(D)$ and generates or reconstructs the data D' as shown in Figure 15.1. Sometimes the *compression* (coding) and *decompression* (decoding) systems together are called a *CODEC*, as shown in the broken box in Figure 15.1.

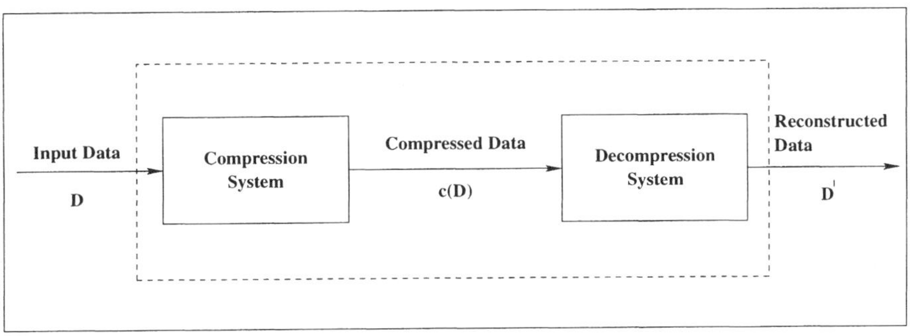

Fig. 15.1 CODEC.

The reconstructed data D' could be identical to the original data D or it could be an approximation of the original data D, depending on the reconstruction requirements. If the reconstructed data D' is an exact replica of the original data D, we call the algorithm applied to compress D and decompress $c(D)$ to be *lossless*. On the other hand, we say the algorithms are *lossy* when D' is not an exact replica of D. Hence as far as the reversibility of the original data is concerned, the data compression algorithms can be broadly classified in two categories–*lossless* and *lossy* . Usually we need to apply lossless data compression techniques on text data or scientific data. For example, we cannot afford to compress the electronic copy of this textbook using a lossy compression technique. It is expected that we shall reconstruct the same text after the decompression process. A small error in the reconstructed text can have a completely different meaning. We do not expect the sentence "You should *not* delete this file" in a text to change to "You should *now* delete this file" as a result of an error introduced by a lossy compression or decompression algorithm. Similarly, if we compress a huge ASCII file containing a program written in C language, for example, we expect to get back the same C code after decompression because of obvious reasons. The lossy compression techniques are usually applicable to data where high fidelity of reconstructed data is not required for perception by the human perceptual system. Examples of such types of data are image, video, graphics, speech, audio, etc. Some image compression applications may require the compression scheme to be lossless (i.e., each pixel of the decompressed image should be exactly identical to the original one). Medical imaging is an example of such an application where compressing digital radiographs with a lossy scheme could be a disaster if it has to make any compromises with the diagnostic accuracy. Similar observations are true for astronomical images for galaxies and stars.

Sometimes we talk about *perceptual lossless* compression schemes when we can compromise with introducing some amount of loss into the reconstructed image as long as there is no perceptual difference between the reconstructed data and the original data, if the human perceptual system is the ultimate judge of the fidelity of the reconstructed data. For example, it is hardly noticeable by human eyes if there is any small relative change among the neighboring pixel values in a smooth non-edge region in a natural image.

In this context, we need to mention that sometimes *data compression* is referred to as *coding* in the literature. The terms *noiseless* and *noisy coding*, in the literature, usually refer to *lossless* and *lossy compression* techniques respectively. The term "noise" here is the "error of reconstruction" in the lossy compression techniques because the reconstructed data item is not identical to the original one. Throughout this book we shall use *lossless* and *lossy compression* in place of *noiseless* and *noisy coding* respectively.

Data compression schemes could be *static* or *dynamic*. In *static* methods, the mapping from a set of messages (data or signal) to the corresponding set of compressed codes is always fixed. In *dynamic* methods, the mapping from the set of messages to the set of compressed codes changes over time. A dynamic method is called *adaptive* if the codes adapt to changes in ensemble characteristics over time. For example, if the probabilities of occurrences of the symbols from the source are not fixed over time, we can adaptively formulate the binary codewords of the symbols, so that the compressed file size can adaptively change for better compression efficiency.

15.4 SOURCE CODING ALGORITHMS

In this section, we present some of the popular source coding algorithms used for data compression. From an information theoretic perspective, *source coding* can mean both lossless and lossy compression. However, it is often reserved by researchers to indicate lossless coding only. In the signal processing community, the source coding is used to mean source model-based coding. We adopt this convention here and by source coding we mean lossless coding only. These algorithms can be used directly to compress any data losslessly. Depending on the characteristics of the data, each algorithm may give different compression performance. So selection of the particular algorithm will depend on characteristics of the data themselves. In lossy image compression mode, the source coding algorithms are usually applied in the entropy encoding step after transformation and quantization.

15.4.1 Run-Length Coding

The neighboring pixels in a typical image are highly correlated to each other. Often it is observed that the consecutive pixels in a smooth region of an image are identical or the variation among the neighboring pixels is very small. Appearance of runs of identical values is particularly true for binary images where usually the image consists of runs of 0's or 1's. Even if the consecutive pixels in gray scale or color images are not exactly identical but slowly varying, it can often be preprocessed and the consecutive processed pixel values become identical. If there is a long run of identical pixels, it is more economical to transmit the length of the run associated with the particular pixel value instead of encoding individual pixel values.

Run-length coding is a simple approach to source coding when there exists a long run of the same data, in a consecutive manner, in a data set. As an example, the data $d =$ 5 5 5 5 5 5 5 19 19 19 19 19 19 19 19 19 19 19 19 0 0 0 0 0 0 0 0 8 23 23 23 23 23 23 contains long runs of 5's, 19's, 0's, 23's, etc. Rather than coding each sample in the run individually, the data can be represented compactly by simply indicating the value of the sample and the length of its run when it appears. In this manner, the data d can be run-length encoded as (5 7) (19 12) (0 8) (8 1) (23 6). For ease of understanding, we have shown a pair in each parentheses. Here the first value represents the pixel, while the second indicates the length of its run.

In some cases, the appearance of runs of symbols may not be very apparent. But the data can possibly be preprocessed in order to aid run-length coding. Consider the data $d =$ 26 29 32 35 38 41 44 50 56 62 68 78 88 98 108 118 116 114 112 110 108 106 104 102 100 98 96. We can simply preprocess this data, by taking the sample difference $e(i) = d(i) - d(i-1)$, to produce the processed data $\bar{e}=$ 26 3 3 3 3 3 3 6 6 6 6 10 10 10 10 10 -2 -2 -2 -2 -2 -2 -2 -2 -2 -2 -2. This preprocessed data can now be easily run-length encoded as (26 1) (3 6) (6 4) (10 5) (-2 11). A variation of this technique is applied in the baseline JPEG standard for still-picture compression [14]. The same technique can be applied to numeric databases as well.

On the other hand, binary (black-and-white) images, such as facsimile, usually consist of runs of 0's or 1's. As an example, if a segment of a binary image is represented as

$$d = 0000000001111111111100000000000000011100000000000001001111111111$$

and it can be compactly represented as $c(d) = (9, 11, 15, 3, 13, 1, 2, 10)$ by simply listing the lengths of alternate runs of 0's and 1's. While the original binary data d requires 65 bits for storage, its compact representation $c(d)$ requires 32 bits only under the assumption that each length of run is being represented by 4 bits. The early facsimile compression standard (CCITT Group 3, CCITT Group 4) algorithms were developed based on this principle [11].

15.5 HUFFMAN CODING

From Shannon's *Source Coding Theory*, we know that a source can be coded with an average code length close to the entropy of the source. In 1952, D. A. Huffman [7] invented a coding technique to produce the shortest possible average code length given the source symbol set and the associated probability of occurrence of the symbols. Codes generated using this coding technique are popularly known as *Huffman codes*. Huffman coding technique is based on the following two observations regarding optimum prefix codes.

1. The more frequently occurring symbols can be allocated with shorter codewords than the less frequently occurring symbols.

2. The two least frequently occurring symbols will have codewords of the same length, and they differ only in the least significant bit.

Average length of these codes is close to entropy of the source.

Let us assume that there are m source symbols $\{s_1, s_2, \cdots, s_m\}$ with associated probabilities of occurrence $\{p_1, p_2, \cdots, p_m\}$. Using these probability values, we can generate a set of Huffman codes of the source symbols. The Huffman codes can be mapped into a binary tree, popularly known as the Huffman tree. We describe the algorithm to generate the Huffman tree and hence the Huffman codes of the source symbols below. We show a Huffman tree in Figure 15.2.

1. Produce a set N=$\{N_1, N_2, \cdots, N_m\}$ of m nodes as leaves of a binary tree. Assign a node N_i with the source symbol s_i, $i = 1, 2, \cdots, m$ and label the node with the associated probability p_i.
(**Example:** As shown in Figure 15.2, we start with eight nodes N_0, N_1, N_2, N_3, N_4, N_5, N_6, N_7 corresponding to the eight source symbols a, b, c, d, e, f, g, h, respectively. Probability of occurrence of each symbol is indicated in the associated parentheses.)

2. Find the two nodes with the two lowest probability symbols from the current node set, and produce a new node as a parent of these two nodes.
(**Example:** From Figure 15.2 we find that the two lowest probability symbols g and d are associated with nodes N_6 and N_3 respectively. The new node N_8 becomes the parent of N_3 and N_6.)

3. Label the probability of this new parent node as the sum of the probabilities of its two child nodes.

(**Example:** The new node N_8 is now labeled by probability 0.09, which is the sum of the probabilities 0.06 and 0.03 of the symbols d and g associated with the nodes N_3 and N_6 respectively.)

4. Label the branch of one child node of the new parent node as 1 and the branch of the other child node as 0.
(**Example:** The branch N_3 to N_8 is labeled by 1 and the branch N_6 to N_8 is labeled by 0.)

5. Update the node set by replacing the two child nodes with smallest probabilities by the newly generated parent node. If the number of nodes remaining in the node set is greater than 1, go to Step 2.
(**Example:** The new node set now contains the nodes N_0, N_1, N_2, N_4, N_5, N_7, N_8 and the associated probabilities are 0.30, 0.10, 0.20, 0.09, 0.07, 0.15, 0.09, respectively. Since there are more than one node in the node set, Steps 2 to 5 are repeated and the nodes N_9, N_{10}, N_{11}, N_{12}, N_{13}, N_{14} are generated in the next six iterations, until the node set consists only of N_{14}.)

6. Traverse the generated binary tree from the root node to each leaf node N_i, $i = 1, 2, \cdots, m$, to produce the codeword of the corresponding symbol s_i, which is a concatenation of the binary labels (0 or 1) of the branches from the root to the leaf node.
(**Example:** The Huffman code of symbol h is 110, formed by concatenating the binary labels of the branches N_{14} to N_{13}, N_{13} to N_{11} and N_{11} to N_7.)

It is needless to mention that any ensemble of binary codes, which can be mapped into a binary tree, consists of prefix codes. Hence Huffman code is also a prefix code. The Huffman code generation process described above is a bottom-up approach, since we perform the code construction process on the two symbols with least probabilities.

Example: Assume the alphabet $S = \{a, b, c, d, e, f, g, h\}$ with 8 source symbols and their corresponding probabilities are $p(a) = 0.30$, $p(b) = 0.10$, $p(c) = 0.20$, $p(d) = 0.06$, $p(e) = 0.09$, $p(f) = 0.07$, $p(g) = 0.03$, and $p(h) = 0.15$ respectively. The Huffman tree generated by the Huffman Coding algorithm is shown in Figure 15.2 and the corresponding Huffman code table is shown in Table 15.2.

Let us consider a string M of 200 symbols generated from the above source, where the numbers of occurrences of a, b, c, d, e, f, g and h in M are 60, 20, 40, 12, 18, 14, 6 and 30 respectively. Size of the encoded message M using the Huffman codes in Table 15.2 will be 550 bits. Here it requires 2.75 bits per symbol on the average. On the other hand, the length of the encoded message M will be 600 bits if it is encoded by a fixed-length code of length 3 for each of the symbols. This simple example demonstrates how we can achieve compression using variable-length coding or source coding techniques.

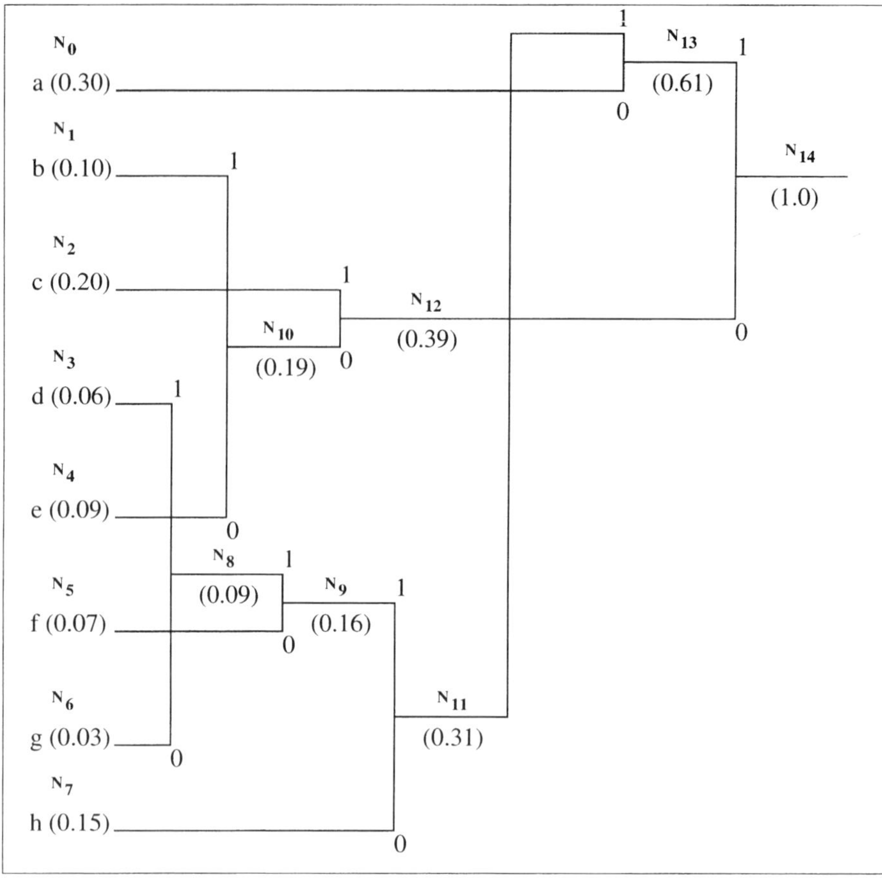

Fig. 15.2 Huffman tree construction for Example 1.

15.6 ARITHMETIC CODING

Arithmetic coding is a variable-length source encoding technique [12]. In traditional entropy encoding techniques such as Huffman coding, each input symbol in a message is substituted by a specific code specified by an integer number of bits. Arithmetic coding deviates from this paradigm. In arithmetic coding, a sequence of input symbols is represented by an interval of real numbers between 0.0 and 1.0. The longer the message, the smaller the interval to represent the message becomes, as will be evident in the following discussions. More probable symbols reduce the interval less than the less probable symbols and hence add fewer bits in the encoded message. As a result, the coding result can reach to Shannon's entropy limit for a sufficiently large sequence of input symbols as long as the statistics are accurate.

Table 15.2 Huffman Code Table

Symbol	Probability	Huffman Code
a	0.30	1 0
b	0.10	0 0 1
c	0.20	0 1
d	0.06	1 1 1 1 1
e	0.09	0 0 0
f	0.07	1 1 1 0
g	0.03	1 1 1 1 0
h	0.15	1 1 0

Arithmetic coding offers superior efficiency and more flexibility compared to the popular Huffman coding. It is particularly useful when dealing with sources with small alphabets such as binary alphabets and alphabets with highly skewed probabilities. Huffman coding cannot achieve any compression for a source of binary alphabets. As a result arithmetic coding is highly efficient for coding bilevel images. However, arithmetic coding is more complicated and is intrinsically less error resilient compared to the Huffman coding.

15.6.1 Encoding Algorithm

The arithmetic coding algorithm is explained here with an example. We consider a four-symbol alphabet $A = \{a, b, c, d\}$ with the fixed symbol probabilities $p(a) = 0.3$, $p(b) = 0.2$, $p(c) = 0.4$, and $p(d) = 0.1$ respectively. The symbol probabilities can be expressed in terms of partition of the half-open range [0.0, 1.0) as shown in Table 15.3.

Table 15.3 Probability model

Index	Symbol	Probability	Cumulative Probability	Range
1	a	0.3	0.3	[0.0, 0.3)
2	b	0.2	0.5	[0.3, 0.5)
3	c	0.4	0.9	[0.5, 0.9)
4	d	0.1	1.0	[0.9, 1.0)

The algorithm for arithmetic coding is presented below. In this algorithm, we consider N is the length of the message (i.e., total number of symbols in the message); $F(i)$ is the cumulative probability of ith source symbol as shown in Table 15.3.

Algorithm: Arithmetic Coding
begin

$L = 0.0;$
$H = 1.0;$
$F(0) = 0;$
for $(j = 1 \text{ to } N)$ {
  $i =$ index of Symbol(j);
  $L = L + (H - L) * F(i - 1);$
  $H = L + (H - L) * F(i);$
}
Output $(\frac{L+H}{2})$;

end

Example: We would like to encode a message "$c\,a\,c\,b\,a\,d$" using the above fixed model of probability estimates. At the beginning of both encoding and decoding processes, the range for the message is the entire half-open interval [0.0, 1.0), which can be partitioned into disjoint subintervals or ranges [0.0, 0.3), [0.3, 0.5), [0.5, 0.9), and [0.9, 1.0) corresponding to the symbols a, b, c, and d respectively, as shown by the range $R(start)$ in Figure 15.3 in terms of the vertical bar with ticks representing the symbol probabilities stipulated by the probability model. As each symbol in the message is processed, the range is narrowed down by the encoder as explained in the algorithm.

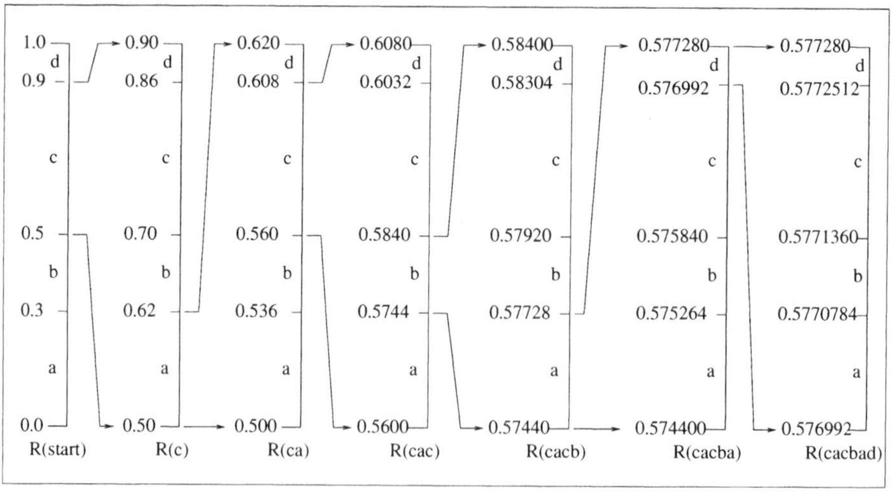

Fig. 15.3 Arithmetic coding technique: an example.

Since the first symbol of the message is c, the range is first narrowed down to the half-open interval $R(c)$=[0.5, 0.9). This range is further partitioned into exactly the same proportions as the original one, yielding the four half-open disjoint intervals [0.5, 0.62), [0.62, 0.70), [0.70, 0.86), and [0.86, 0.9) corresponding to a, b, c and d respectively as shown in Figure 15.3. As a result, the range is narrowed down to $R(ca)$=[0.5, 0.62) when the second symbol a in the message is processed. This new range [0.5, 0.62) is now partitioned into four disjoint intervals [0.5, 0.536), [0.536, 0.560), [0.560, 0.608), and [0.608, 0.62). After processing the third symbol, c, the range is accordingly narrowed down to $R(c\,a\,c)$=[0.560, 0.608). This is again partitioned into [0.560, 0.5744), [0.5744, 0.5840), [0.5840, 0.6032), and [0.6032, 0.608) in order to process the next symbol in the message. After processing the fourth symbol, b, the range is now narrowed down to $R(c\,a\,c\,b)$=[0.5744, 0.5840). This is again partitioned into four intervals [0.5744, 0.57728), [0.57728. 0.57920), [0.57920, 0.58304), and [0.58304, 0.584) corresponding to the symbols a, b, c, and d respectively. After processing the fifth symbol, a, the range is now narrowed down to $R(c\,a\,c\,b\,a)$=[0.5744, 0.57728). This is further partitioned into the disjoint intervals [0.5744, 0.575264), [0.575264, 0.575840), [0.575840, 0.576992), and [0.576992, 0.57728). The last symbol in the message is d and hence the final range for the message becomes $R(c\,a\,c\,b\,a\,d)$=[0.576992, 0.57728). As a result, the message "$c\,a\,c\,b\,a\,d$" can be encoded by any number in the range [0.576992, 0.57728) because it is not necessary for the decoder to know both ends of the range produced by the encoder. If we use the midpoint of the interval, the encoded value will be 0.577136. Assuming that we choose 0.577, the decoding processing is explained below using the same probability model in Figure 15.3 used in the encoding process.

15.6.2 Decoding Algorithm

Both the encoder and the decoder have the same probability model. Initially the decoder starts with the range [0.0, 1.0), which is partitioned into four intervals [0.0, 0.3), [0.3, 0.5), [0.5, 0.9), and [0.9, 1.0) corresponding to the symbols a, b, c, and d in the alphabet. As soon as the decoder receives an encoded number 0.577, it can immediately decode that the first symbol of the message is c because the number 0.577 belongs to the range [0.5, 0.9) and the range is narrowed down to [0.5, 0.9) and partitioned into [0.5, 0.62), [0.62, 0.70), [0.70, 0.86), and [0.86, 0.9) in a similar fashion as the encoder. Since the number 0.577 belongs to the range [0.5, 0.62), it can immediately decode the second symbol to be a. The range is now narrowed down to [0.5, 0.62) and partitioned into [0.5, 0.536), [0.536, 0.560), [0.560, 0.608), and [0.608, 0.62). Since the number 0.577 belongs to the range [0.560, 0.608), the decoder can decode the third symbol to be c. The range is now narrowed down to [0.560, 0.608) and partitioned into the four subintervals [0.560, 0.5744), [0.5744, 0.584), [0.584, 0.6032), and [0.6032, 0.608). Since the number 0.577 belongs in the range [0.5744, 0.584), the decoder deduces that the next symbol is b and narrows the range down to be [0.5744, 0.584). The range is now subdivided into [0.5744, 0.57728), [0.57728. 0.5792), [0.5792, 0.58304), and [0.58304, 0.584). Since the number 0.577 belongs within the range [0.5744, 0.57728), the next symbol decoded is a and the range is narrowed down to

[0.5744, 0.57728) and partitioned into four subintervals [0.5744, 0.575264), [0.575264, 0.575840), [0.575840, 0.576992), and [0.576992, 0.57728). Since 0.577 belongs to the range [0.576992, 0.57728), it is very natural that the decoder decodes the next symbol to be d and narrows the range down to [0.576992, 0.5770784), [0.5770784, 0.577136), [0.577136, 0.5772512), and [0.5772512, 0.57728) respectively. Hence the decoder could uniquely decode the message "$c\,a\,c\,b\,a\,d$" until this step. If the decoder is aware of the length of the message, it can stop decoding here. Otherwise, it can continue decoding the next symbol to be a because 0.577 belongs to the range [0.576992, 0.5770784) and so on indefinitely. Hence the decoder faces the problem of detecting the end of the message in order to stop. To resolve the ambiguity, we can ensure that each message ends with a special terminating symbol known to both encoder and decoder. In this example, if we assume that d is the special terminating symbol, the decoder will effectively stop after decoding the message "$c\,a\,c\,b\,a\,d$." Otherwise the length of the original message needs to be known to the decoder in order to stop decoding effectively.

15.6.3 The QM-Coder

The QM-coder [14] is an enhancement of the Q-coder [13]. The QM-coder is the adaptive binary arithmetic coding algorithm used in the JBIG (Joint Bilevel Image Processing Group) standard for bilevel image compression. Although it follows the same principle of arithmetic coding, QM-coder is designed for simplicity and speed. The input symbols to the QM-coder are single bits of the bilevel image and it is free from multiplications by approximating the computation of intervals by fixed-precision integer arithmetic operations (addition, subtraction, and shift operations only).

The main idea behind the QM-coder is to map the input bits into *more probable symbol* (MPS) and *less probable symbol* (LPS). This can be explained in terms of a black-and-white image. If bits 0 and 1 represent the black and white pixels respectively, then in a mostly black region 0 will be mapped to MPS and 1 will be mapped to LPS, whereas in a mostly white region 1 will be mapped to MPS and 0 will be mapped to LPS. Before the next bit is input, the QM-coder determines which bit is MPS (the other bit is LPS) and compresses this information instead of the input bit directly. During the decoding process, the QM-decoder decodes whether the bit just decoded is MPS or LPS and then converts this information to actual binary pixel value. Hence the QM-coder assigns the intervals to the MPS and LPS symbols instead of the 0 or 1 input bit. If the probability estimate of LPS is Q, then the probability estimate of MPS is $(1 - Q)$ because there are only two symbols in the alphabet. For interval A, the QM-coder divides the interval into two subintervals according to the value of Q. The sizes of the subintervals assigned to LPS and MPS are AQ and $A(1 - Q)$ respectively as shown in Figure 15.4.

In QM-coder, the value of A is always assumed to maintain close to 1. As a result, the subintervals of LPS and MPS can be approximated to $AQ \approx Q$ and

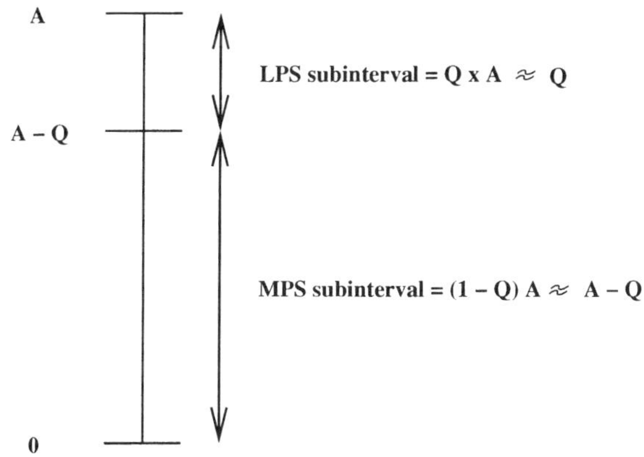

Fig. 15.4 Subinterval assignment in QM-coder.

$A(1-Q) \approx A-Q$ respectively and hence the multiplication is avoided. The subinterval of LPS is placed above the subinterval of MPS as shown in Figure 15.4. Accordingly, the MPS and LPS are assigned the subintervals $[0, A-Q)$ and $[A-Q, A)$ respectively. Actually the value of A is always maintained within the range $1.5 > A \geq 0.75$. Whenever the value of A drops below 0.75 during the encoding process, the *renormalization* is done by repeated doubling (shifting left) A until it is greater than or equal to 0.75. Whenever we renormalize A, we need to apply the same renormalization to C as well to keep these two parameters in sync.

We denote the output code stream of the QM-coder by C. Ideally C can be any value within the current interval as we explained in the arithmetic coding algorithm in the previous section. However, for simplicity of implementation, the QM-coder points C at the bottom of the current interval. If the current input is MPS (or LPS), C is updated by adding the bottom of the MPS (or LPS) subinterval to the current value of C. Since the bottom of MPS subinterval is 0, C actually remains unchanged when MPS is encoded. During encoding of LPS, C is updated by adding $A-Q$ to the current value of C since $A-Q$ is the bottom of the LPS subinterval as shown in Figure 15.4. It should be noted that the encoder is initialized with the $A = 1$ at the beginning of the encoding process. Hence the encoding algorithm can be described as follows.

When MPS is encoded:

begin
  C is unchanged;
  $A = A - Q$;
  if $(C < 0.75)$ **then**
    Renormalize A and C;

```
        endif;
    end
```

When LPS is encoded:

```
    begin
        C = C + (A − Q);
        A = Q;
        Renormalize A and C;
    end
```

The probability estimation in QM-coder is accomplished by using a predetermined table of Q values. The value of Q of the LPS is updated each time a renormalization occurs during the encoding. The table consists of a preset ordered list of the Q values. For every renormalization, Q is updated by the next lower or next higher Q value in the table, depending on whether the renormalization takes place because of encoding of an LPS or MPS during the encoding process. An important issue in QM-coder is called the problem of *interval inversion*. This problem happens when the size of the subinterval assigned to MPS becomes smaller than the size of the LPS subinterval because of the result of approximation of A and C. This problem occurs when LPS actually occurs more frequently than the MPS due to some peculiar characteristics of the input bits. As a result, it is possible that the value of Q becomes of the order of 0.5 and as a result the size of the subinterval assigned to MPS can be as small as 0.25. In this situation, the problem is solved by reversing the assignment of the two subintervals whenever the LPS subinterval is greater than the MPS subinterval. This is known as the *conditional exchange*. The term *conditional* is used due to the fact that the subinterval reassignment takes place only when the LPS probability occupies more than half of the total interval A. Thus the condition for interval inversion is $Q > A - Q$. Since $Q \leq 0.5$, we get $0.5 \geq Q > A - Q$. As a result, both the subintervals Q and $A - Q$ are less than 0.75 and this necessitates renormalization of A and C. That's why the conditional exchange is performed only after the encoder detects that renormalization is needed.

We have demonstrated the situation of a *conditional exchange* as an example in Figure 15.5. We assume that the current value of A is 1.35, which is less than 1.5 as shown in Figure 15.5(a). Assuming that $Q = 0.5$, the subintervals $[0.0, 0.85)$ and $[0.85, 1.35)$ are assigned to MPS and LPS respectively as shown in Figure 15.5(a). Now if the value of A is changed to $A = A - Q = 0.85$ in the next coding cycle, the symbol orders are inverted and accordingly subintervals assigned to LPS and MPS are exchanged as shown in Figure 15.5(b). The subintervals assigned to LPS and MPS after conditional exchange are $[0.0, 0.35)$ and $[0.35, 0.85)$ respectively, as shown in Figure 15.5(b).

Incorporating the process of handling the *conditional exchange*, the encoding algorithm thus becomes as follows.

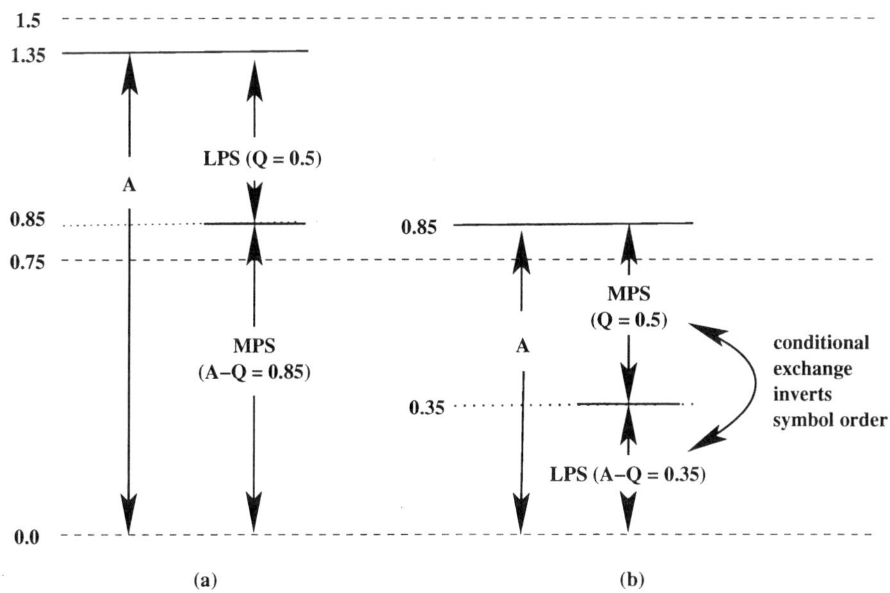

Fig. 15.5 Subinterval assignment in QM-coder, (a) without conditional exchange and (b) with conditional exchange.

QM-Coder: The Encoding Algorithm

When MPS is encoded:

begin
  C is unchanged;
  $A = A - Q$;
  if $(C < 0.75)$ **then**
    if $(A < Q)$ **then**
      $C = C + A$;
      $A = Q$;
    endif;
    Renormalize A and C;
  endif;
end

When LPS is encoded:

begin
  $A = A - Q$;
  if $(A \geq Q)$ **then**
    $C = C + A$;

```
            A = Q;
        endif;
        Renormalize A and C;
    end
```

15.6.3.1 The QM-Decoder The decoder decodes an MPS or LPS by determining which subinterval the value of the code stream belongs to. The QM-decoder is just the reverse of the encoder. For simplicity we ignore the *conditional exchange* situation here. The matching decoding algorithm is as follows.

QM-Decoder: The Decoding Algorithm

```
begin
    if (C ≥ Q) then
        (MPS is decoded)
        C = C − Q;
        A = A − Q;
    else
        (LPS is decoded)
        A = Q;
    endif;
    if (A < 0.75) then
        Renormalize A and C;
    endif;
end.
```

Another variation of Q-coder, called the MQ-coder, has been used for adaptive binary arithmetic coding in JPEG2000 encoding. The details of this algorithm will be discussed in Chapter 18.

15.7 SUMMARY

In this chapter, we have introduced readers to the fundamentals of data and image compression. We have discussed some fundamentals including information theory such as discrete memoryless model, entropy, noiseless source coding theorem, unique decipherability, etc., in order to aid readers in understanding the principles behind data compression. In this chapter, we have also presented some of the key source coding algorithms widely used in data and image compression. First we have described the run-length coding scheme with an example. We have described the popular Huffman coding scheme that is used in various image and data compression techniques. Arithmetic coding is an alternative approach for an efficient entropy encoding and it achieves compression efficiency very close to the entropy limit. We discussed the basic principles of arithmetic coding with an example. Binary arithmetic coding is a key algorithm for bilevel image compression. We discussed the QM-coder algorithm for implementation of an adaptive binary arithmetic coding, which has been adopted in the JBIG standard for bilevel image compression and also in a mode of JPEG standard. A variation of QM-coder called the MQ-coder is the basis of the entropy encoding of the new JPEG2000 standard for still picture compression.

REFERENCES

1. C. E. Shannon and W. Weaver, *The Mathematical Theory of Communication*, University of Illinois Press, Urbana, IL, 1949.

2. C. E. Shannon, "Certain Results in Coding Theory for Noisy Channels," *Information Control*, 1(1), September 1957, 6–25.

3. C. E. Shannon, "Coding Theorems for a Discrete Source with a Fidelity Criterion," *IRE National Convention Record*, Part 4, 1959, 142–163.

4. B. McMillan, "The Basic Theorems of Information Theory," *Ann. Math. Statist.*, Vol. 24, 1953, 196–219.

5. D. S. Hirschberg and D. A. Lelewer, "Efficient Decoding of Prefix Codes," *Comm. of the ACM*, 33(4), April 1990, 449–459.

6. R. M. Fano, *Transmission of Information*, MIT Press, Cambridge, MA, 1949.

7. D. A. Huffman, "A Method for the Construction of Minimum Redundancy codes," *Proc. IRE*, Vol. 40, 1952, 1098–1101.

8. P. Elias, "Universal Codeword Sets and Representations of the Integers," *IEEE Trans. on Info. Theory*, 21(2), March 1975, 194–203.

9. I. H. Witten, R. M. Neal, and J. G. Cleary, "Arithmetic Coding for Data Compression," *Communications of the ACM*, 30(6), June 1987.

10. T. Acharya and P. Tsai, *JPEG2000 Standard Image Compression: Concepts, Algorithms and VLSI Architectures*, Wiley, Hoboken, NJ, 2004.

11. R. Hunter and A. H. Robinson, "International Digital Facsimile Standard," *Proceedings of IEEE*, 68(7), 1980, 854–867.

12. I. H. Witten, R. M. Neal, and J. G. Cleary, "Arithmetic Coding for Data Compression," *Communications of the ACM*, 30(6), June 1987.

13. W. B. Pennebaker, J. L. Mitchell, G. G. Langdon, Jr., and R. B. Arps, "An Overview of the Basic Principles of the Q-Coder Adaptive Binary

Arithmetic Coder," *IBM Journal of Research and Development*, Vol. 32, November 1988, 717–726.

14. W. B. Pennebaker and J. L. Mitchell, *JPEG Still Image Data Compression Standard.* Chapman & Hall, New York, 1993.

16

JPEG: Still Image Compression Standard

16.1 INTRODUCTION

JPEG is the first international image compression standard for continuous-tone still images—both grayscale and color images [1, 2]. JPEG is the acronym for *Joint Photographic Experts Group.* This image compression standard is a result of collaborative efforts by the International Telecommunication Union (ITU), International Organization for Standardization (ISO), and International Electrotechnical Commission (IEC). The JPEG standard is officially referred to as *ISO/IEC IS* (International Standard) *10918-1*: *Digital Compression and Coding of Continuous-tone Still Images*, and also *ITU-T Recommendation T.81.* The goal of this standard is to support a variety of applications for compression of continuous-tone still images of most image sizes in any color space in order to achieve compression performance at or near the state-of-the-art with user-adjustable compression ratios and with very good to excellent reconstructed quality. Another goal of this standard is that it would have manageable computational complexity for widespread practical implementation. JPEG defines four modes of operations:

1. *Sequential lossless mode*: Compress the image in a single scan and the decoded image is an exact replica of the original image.

2. *Sequential DCT-based mode*: Compress the image in a single scan using DCT-based lossy compression technique. As a result, the decoded image is not an exact replica, but an approximation of the original image.

3. *Progressive DCT-based mode*: Compress the image in multiple scans and also decompress the image in multiple scans with each successive scan producing a better-quality image.

4. *Hierarchical mode*: Compress the image at multiple resolutions for display on different devices.

The three DCT-based modes (2, 3, and 4) in JPEG provide lossy compression because precision limitation to digitally compute DCT (and its inverse) and the quantization process introduce distortion in the reconstructed image. For sequential lossless mode of compression, predictive coding is used instead of the DCT-based transformation, and also there is no quantization involved in this mode. The hierarchical mode uses extensions of either the DCT-based coding or predictive coding techniques. The simplest form of the sequential DCT-based JPEG algorithm is called the *baseline JPEG* algorithm, which is based on Huffman coding for entropy encoding. The other form of sequential DCT-based JPEG algorithm is based on arithmetic coding for entropy encoding. The baseline JPEG algorithm is widely used in practice. We shall describe the JPEG lossless algorithm and the baseline JPEG algorithm in greater detail in this chapter.

People often mention *motion JPEG* for compression of moving pictures. This is not really a standard. Although it is not specifically defined as part of the standard, JPEG can be used to compress image sequences in video on the basis that video clips can be considered as a sequence of still image frames and each image frame can be compressed independently using the JPEG algorithm. This process of image sequence compression is popularly known as motion JPEG in the industry.

JPEG standard does not specify any inherent file format. It defines only the syntax of the compressed bitstream. This caused creation of a number of file formats to store the JPEG compressed images such as JFIF (JPEG File Interchange Format), JPEG extension to TIFF 6.0, FlashPix, etc. But none of them is considered to be an official international standard defined under the auspices of an international standards committee.

16.2 THE JPEG LOSSLESS CODING ALGORITHM

The lossless JPEG compression is based on the principles of predictive coding. Since the adjacent pixels in a typical image are highly correlated, it is possible to extract a great deal of information about a pixel from its neighboring pixel values. Predictive coding is a simple method for spatial redundancy reduction. In this method, a pixel value is predicted by a set of previously encoded adjacent pixels using a suitable prediction model. For an ideal prediction model, the predicted value of the pixel can be equal to the actual value. But that is not the case in reality. Using an effective prediction model, we can predict the pixel value, which is very close to its actual value and hence error of prediction can be very small.

A practical approach to the prediction model is to take a linear combination of the previously encoded immediate neighboring adjacent pixels. The reason for taking the previously encoded pixel values is that the same values will be available to the decoder when it decodes the pixels in the same order they were encoded by the encoder. The difference between the actual pixel value and the predicted value is called the *differential* or the *prediction error* value. The *prediction error* is then entropy encoded using a variable-length encoding technique to generate compressed image. This method is popularly known as *Differential Pulse Code Modulation* (DPCM).

In the lossless JPEG algorithm, the value of a pixel in any pixel location in the image is first predicted by using one or more of the previously encoded adjacent pixels A, B, and C as shown in Figure 16.1(a) to predict pixel X. It then encodes the difference between the pixel and its predicted value, usually called the *prediction error* or *prediction residual*, by either Huffman coding or arithmetic coding.

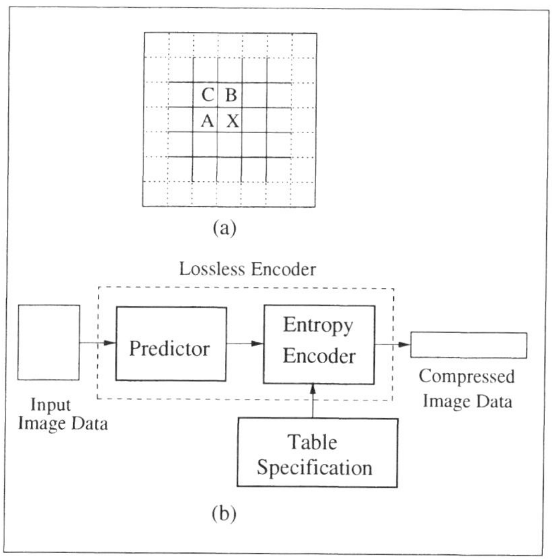

Fig. 16.1 (a) Three-pixel prediction neighborhood, (b) encoder diagram in lossless mode.

There are eight possible options for prediction as shown in Table 16.1. The *No prediction* option 0 in Table 16.1 is available only for differential coding in the JPEG hierarchical mode. We briefly discuss the essence of the hierarchical mode of coding later in this chapter. Options 1 to 3 are one-dimensional predictors and options 4 to 7 are two-dimensional predictors. Depending on the nature of the image, one predictor may yield better compression results compared to any other predictor. However, experimental results for various kinds of images show that on the average their performances are relatively close to each other [3]. The chosen option for prediction is indicated in the header of the compressed file so that both the encoder and decoder use the same function for prediction.

Table 16.1 Prediction functions in lossless JPEG.

Option	Prediction Function	Type of Prediction
0	No prediction	Differential Coding
1	$X_p = A$	1D Horizontal Prediction
2	$X_p = B$	1D Vertical Prediction
3	$X_p = C$	1D Diagonal Prediction
4	$X_p = A + B - C$	2D Prediction
5	$X_p = A + \frac{1}{2}(B - C)$	2-D Prediction
6	$X_p = B + \frac{1}{2}(A - C)$	2-D Prediction
7	$X_p = \frac{1}{2}(A + B)$	2-D Prediction

In lossless mode, the standard allows precision (P) of the input source image to be 2 bits to 16 bits wide. Since there is no previously encoded pixel known to the encoder when it encodes the very first pixels in the very first row of the image, it is handled differently. For a given input precision P and a point transform parameter P_t, the predicted value for the first pixel in the first line is 2^{P-P_t-1}. By default, we can assume $P_t = 0$. For details of the point transform parameter, the reader is advised to consult the JPEG standard [1].

For all other pixels (except the first one) in the first line, we use option 1 for prediction function. Except for the first line, option 2 is used to predict the very first pixel in all other lines. For all other pixels, we select one of the eight options for the prediction function from Table 16.1. Once a predictor is selected, it is used for all other pixels in the block.

In the lossless JPEG standard, the prediction error values are computed modulo 2^{16} in order to take consideration of the full precision allowed in this mode. These error values are not directly encoded using Huffman codes. They are first represented as a pair of symbols ($CATEGORY$, $MAGNITUDE$). The first symbol $CATEGORY$ represents the category of the error value. The second symbol $MAGNITUDE$ represents the variable-length integer (VLI) for the prediction error value. The category represents the number of bits to encode the $MAGNITUDE$ in terms of VLI. All the possible prediction error values modulo 2^{16} and their corresponding categories are shown in Table 16.2. Only the $CATEGORY$ in the symbol pair for each prediction error value is Huffman coded. The codeword for the symbol pair ($CATEGORY$, $MAGNITUDE$) is formed in two steps. First it assigns the Huffman code of the $CATEGORY$. This Huffman code is then appended with additional

Table 16.2 Categories of prediction error values.

Category	Prediction Error Value
0	0
1	-1, +1
2	-3, -2, +2, +3
3	-7, ..., -4, +4, ..., +7
4	-15, ..., -8, +8, ..., +15
5	-31, ..., -16, +16, ..., +31
6	-63, ..., -32, +32, ..., +63
7	-127, ..., -64, +64, ..., +127
8	-255, ..., -128, +128, ..., +255
9	-511, ..., -256, +256, ..., +511
10	-1023, ..., -512, +512, ..., +1023
11	-2047, ..., -1024, +1024, ..., +2047
12	-4095, ..., -2048, +2048, ..., +4095
13	-8191, ..., -4096, +4096, ..., +8191
14	-16383, ..., -8192, +8192, ..., +16383
15	-32767, ..., -16384, +16384, ..., +32767
16	+32768

$CATEGORY$ number of bits to represent the $MAGNITUDE$ in VLI. If the prediction error value is positive, the $MAGNITUDE$ is directly binary represented by a VLI using $CATEGORY$ number of bits and hence it starts with bit 1. If the error value is negative, the VLI is one's complement of its absolute value and hence it starts with bit 0. For example, the prediction error value 25 is represented by the pair (5, 25) because the number 25 belongs to category 5 in Table 16.2 and hence 25 is represented by a 5-bit VLI. If the Huffman code for category 5 is 011, then the binary codeword for the error value 25 will be 01111001. The first three bits correspond to the Huffman code 011 for category 5 and the next 5 bits, 11001, is the VLI for 25. Similarly, the prediction error value -25 will be represented as 01100110 where the last 5 bits, 00110, is the 1's complement of 11001 to represent -25, and since -25 belongs to the same category 5, the first three bits of the codeword corresponds to the Huffman code of category 5. Use of the category table greatly simplifies the Huffman coder. Without this categorization, we would need to use a Huffman table with 2^{16} entries for all the 2^{16} possible symbols of prediction error values, which definitely complicates the implementation of the Huffman coder both in software and hardware, if it is not rendered impossible for all practical purposes.

Detailed information for implementation of the JPEG lossless coding for both the Huffman coding mode and the arithmetic coding mode can be found in Annex H of the JPEG standard [1].

16.3 BASELINE JPEG COMPRESSION

The baseline JPEG compression algorithm is widely used among the four modes in JPEG family. This is defined for compression of continuous-tone images with 1 to 4 components. Number of components for grayscale images is 1, whereas a color image can have up to four color components. The baseline JPEG allows only 8-bit samples within each component of the source image. An example of a four-component color image is a CMYK (Cyan, Magenta, Yellow, and Black) image, which is used in many applications such as printing, scanning, etc. A color image for display has three color components, RGB (Red, Green, and Blue), though. In a typical color image, the spatial intercomponent correlation between the red, green, and blue color components is significant. In order to achieve good compression performance, correlation between the color components is first reduced by converting the RGB image into a decorrelated color space. In baseline JPEG, a three-color RGB image is first transformed into a luminance–chrominance (L–C) color space such as YC_bC_r, YUV, CIELAB, etc. The advantage of converting the image into luminance–chrominance color space is that the luminance and chrominance components are very much decorrelated between each other. Moreover, the chrominance channels contain much redundant information and can easily be subsampled without sacrificing any visual quality for the reconstructed image.

16.3.1 Color Space Conversion

In this book, we consider color space conversion from RGB to YC_bC_r and vice versa only. There are several ways to convert from RGB to YC_bC_r color space. In this book, we adopt the CCIR (International Radio Consultative Committee) Recommendation 601-1. This is the typical method for color conversion used in baseline JPEG compression. According to the CCIR 601-1 Recommendation, the transformation from RGB to YC_bC_r is done based on the following mathematical expression:

$$\begin{pmatrix} Y \\ Cb \\ Cr \end{pmatrix} = \begin{pmatrix} 0.299000 & 0.587000 & 0.114000 \\ -0.168736 & -0.331264 & 0.500002 \\ 0.500000 & -0.418688 & -0.081312 \end{pmatrix} \begin{pmatrix} R \\ G \\ B \end{pmatrix}.$$

Color space conversion from RGB to YC_bC_r using the above transformation may result in negative numbers for C_b and C_r, while Y is always positive. In order to represent C_b and C_r in unsigned 8-bit integers, they are level shifted by adding 128 to each sample followed by rounding and saturating the value in the range [0, 255]. Hence the above transformation can be expressed as

$$\begin{pmatrix} Y \\ Cb \\ Cr \end{pmatrix} = \begin{pmatrix} 0.29900 & 0.58700 & 0.11400 \\ -0.16874 & -0.33126 & 0.50000 \\ 0.50000 & -0.41869 & -0.08131 \end{pmatrix} \begin{pmatrix} R \\ G \\ B \end{pmatrix} + \begin{pmatrix} 0 \\ 128 \\ 128 \end{pmatrix}$$

in order to produce 8-bit unsigned integers for each of the components in the YC_bC_r domain. Accordingly, the inverse transformation from YC_bC_r to RGB is done as

$$\begin{pmatrix} R \\ G \\ B \end{pmatrix} = \begin{pmatrix} 1.0 & 0.0 & 1.40210 \\ 1.0 & -0.34414 & -0.71414 \\ 1.0 & 1.77180 & 0.0 \end{pmatrix} \begin{pmatrix} Y \\ Cb \\ Cr \end{pmatrix} - \begin{pmatrix} 0 \\ 128 \\ 128 \end{pmatrix}.$$

After the color space conversion, most of the spatial information of the image is contained in the luminance component (Y). The chrominance components (C_b and C_r) contain mostly redundant color information and we lose little information by subsampling these components both horizontally and/or vertically. We can subsample the chrominance components by simply throwing away every other sample in each row and/or in each column if desired. If we subsample the redundant chrominance components both horizontally and vertically, the amount of data required to represent the color image is reduced to half because the size of each chrominance component (C_b and C_r) is one-fourth of the original size. This color format is called the 4:2:0 color subsampling format. Baseline JPEG also supports 4:2:2 and 4:4:4 color formats. Each chrominance component, in 4:2:2 color format, has the same vertical resolution as the luminance component, but the horizontal resolution is halved by dropping alternate samples in each row. In 4:4:4 format, both the chrominance components C_b and C_r have identical vertical and horizontal resolution as the luminance component. Hence no subsampling is done in 4:4:4 format. The subsampling operation to generate in 4:2:0 or 4:2:2 color format is the first lossy step. For a grayscale image there is only one component and obviously no color transformation is required.

16.3.2 Source Image Data Arrangement

In the previous section, we have seen that the dimension of each of the color components Y, C_b, and C_r could be different depending on the color subsampling format. Each color component is divided into 8×8 nonoverlapping blocks, and we can form what is called a *minimum coded unit* (MCU) in JPEG by selecting one or more data blocks from each of the color components. The standard defines the arrangement of the data blocks in interleaved or noninterleaved scanning order of the color components. In noninterleaved scan, the data blocks in each color component are stored and processed separately in raster scan order, left-to-right and top-to-bottom. In interleaved order, data blocks from all the color components appear in each MCU. Definition of the MCUs for 4:4:4, 4:2:2, and 4:2:0 formats of YC_bC_r images in interleaved scan is shown in Figure 16.2.

Each dot in Figure 16.2 represents a 8×8 data block. In 4:4:4 format interleaved scan, each MCU consists of a data block from each of the Y, C_b, and C_r components as shown in Figure 16.2(a). The order of processing these blocks is in the scan order from left to right and top to bottom. For example,

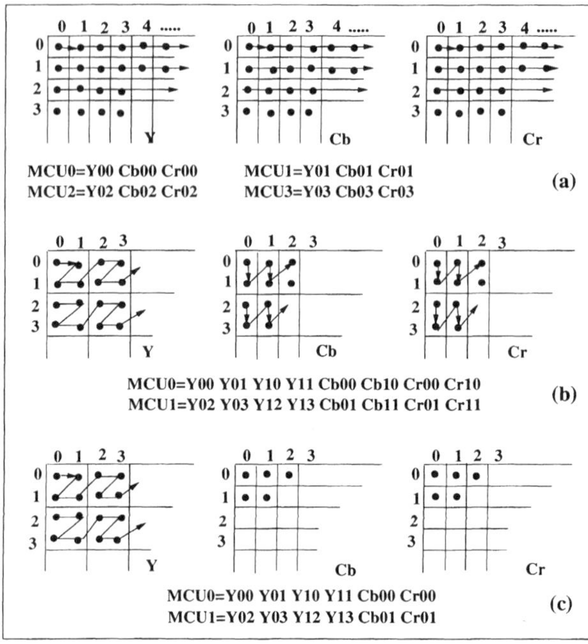

Fig. 16.2 (a) YC_bC_r 4:4:4, (b) YC_bC_r 4:2:2, (c) YC_bC_r 4:2:0.

the first MCU consists of the first data blocks Y00 from the Y component followed by the first data blocks Cb00 from the C_b component followed by Cr00 from the C_r component as shown in Figure 16.2(a). The next MCU consists of Y01, Cb01, and Cr01 respectively. After all the MCUs consisting of the 8 × 8 data blocks from the first row, as shown in Figure 16.2(a), are encoded the second row of 8 × 8 blocks are scanned in a similar fashion. This procedure is continued until the last 8 × 8 block in the raster scan is encoded. In 4:2:2 format, each MCU consists of a 2 × 2 unit of four data blocks from the Y component, a 2 × 1 unit of two data blocks from each of the C_b and C_r components, and the corresponding order of processing is shown in Figure 16.2(b). In 4:2:0 format, each MCU consists of 2 × 2 units of four data blocks from the Y component, one from each of the C_b and C_r components, and the corresponding order of processing is shown in Figure 16.2(c).

16.3.3 The Baseline Compression Algorithm

The baseline JPEG algorithm follows the principles of block-based transform coding. Block diagram of the baseline JPEG algorithm for a grayscale image with a single component is shown in Figure 16.3. For a color image, the same algorithm is applied in each 8 × 8 data block based on the source image data arrangement as described in the previous section.

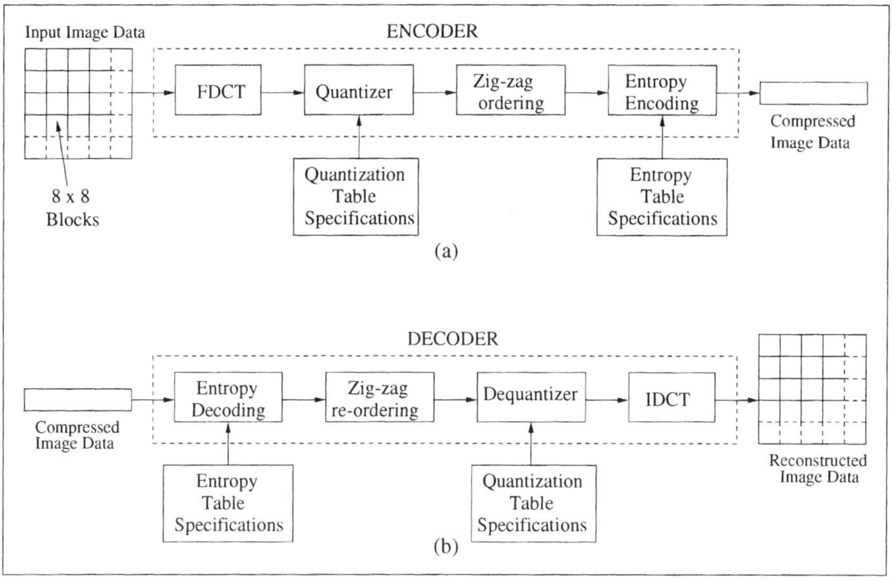

Fig. 16.3 Baseline JPEG: (a) compression, (b) decompression.

The image component is first divided into nonoverlapping 8 × 8 blocks in the raster scan order left-to-right and top-to-bottom as shown in Figure 16.3(a). Each block is then encoded separately by the ENCODER shown in the broken box in Figure 16.3(a). The first step is to level shift each pixel in the block to convert into a signed integer by subtracting 128 from each pixel. Each level-shifted pixel in the 8 × 8 block is then transformed into frequency domain via forward discrete cosine transform (FDCT). The FDCT of an 8 × 8 block of pixels $f(x, y)$ for $(x, y = 0, 1, \cdots, 7)$ is defined by:

$$F(u,v) = \frac{1}{4}C(u)C(v)\sum_{x=0}^{7}\sum_{y=0}^{7} f(x,y)\cos\left[\frac{\pi(2x+1)u}{16}\right]\cos\left[\frac{\pi(2y+1)u}{16}\right]$$

for $u = 0, 1, \ldots, 7$ and $v = 0, 1, \ldots, 7$, where

$$C(k) = \begin{cases} \frac{1}{\sqrt{2}} & \text{for } k = 0 \\ 1 & \text{otherwise.} \end{cases}$$

We discuss discrete cosine transform in great detail in Chapter 4.

16.3.4 Coding the DCT Coefficients

The transformed 8 × 8 block now consists of 64 DCT coefficients. The first coefficient $F(0, 0)$ is the DC component of the block and the other 63 coefficients are AC components $AC_{u,v} = F(u, v)$ of the block as shown in Figure 16.4.

The DC component $F(0,0)$ is essentially the sum of the 64 pixels in the input 8×8 pixel block multiplied by the scaling factor $\frac{1}{4}C(u)C(v) = \frac{1}{8}$ as shown in the expression for $F(u,v)$.

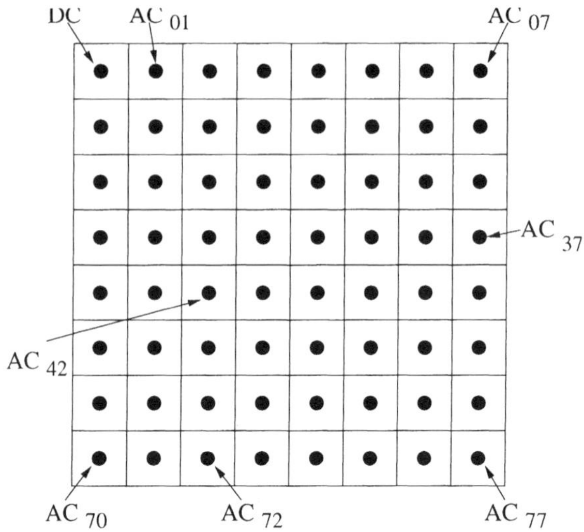

Fig. 16.4 DC and AC components of the transformed block.

The next step in the compression process is to quantize the transformed coefficients. This step is primarily responsible for losing the information and hence introduces distortion in the reconstructed image. That's why baseline JPEG is a lossy compression. Each of the 64 DCT coefficients are uniformly quantized. The 64 quantization step-size parameters for uniform quantization of the 64 DCT coefficients form an 8×8 *quantization matrix*. Each element in the quantization matrix is an integer between 1 and 255. Each DCT coefficient $F(u,v)$ is divided by the corresponding quantizer step-size parameter $Q(u,v)$ in the quantization matrix and rounded to the nearest integer as

$$F_q(u,v) = Round\left(\frac{F(u,v)}{Q(u,v)}\right).$$

The standard does not define any fixed quantization matrix. It is the prerogative of the user to select a quantization matrix. There are two quantization matrices provided in Annex K of the JPEG standard for reference, but not as a requirement. These two quantization matrices are shown in Tables 16.3 and 16.4 respectively.

Table 16.3 is the *luminance quantization matrix* for quantizing the transformed coefficients of the luminance component of an image. Table 16.4 is the *chrominance quantization matrix* for quantizing the transformed coefficients of the chrominance components of the image. These two quantization tables have been designed based on the psychovisual experiments by Lohscheller [4]

Table 16.3 Luminance quantization matrix

16	11	10	16	24	40	51	61
12	12	14	19	26	58	60	55
14	13	16	24	40	57	69	56
14	17	22	29	51	87	80	62
18	22	37	56	68	109	103	77
24	35	55	64	81	104	113	92
49	64	78	87	103	121	120	101
72	92	95	98	112	100	103	99

Table 16.4 Chrominance quantization matrix

17	18	24	47	99	99	99	99
18	21	26	66	99	99	99	99
24	26	56	99	99	99	99	99
47	66	99	99	99	99	99	99
99	99	99	99	99	99	99	99
99	99	99	99	99	99	99	99
99	99	99	99	99	99	99	99
99	99	99	99	99	99	99	99

to determine the visibility thresholds for 2D basis functions. These tables may not be suitable for all kinds of images, but they provide reasonably good results for most of the natural images with 8-bit precision for luminance and chrominance samples. If the elements in these tables are divided by 2, we get perceptually lossless compression—the reconstructed image is indistinguishable from the original one by human eyes. If the quantization tables are designed based on the perceptual masking properties of human eyes, many of the small DCT coefficients and mainly high-frequency samples are zeroed out to aid significant compression. This is done by using larger quantization step-size parameters for higher-frequency AC components as shown in Tables 16.3 and 16.4. Quality of the reconstructed image and the achieved compression can be controlled by a user by selecting a quality factor Q_JPEG to tune the elements in the quantization tables as proposed by the *Independent JPEG Group* (IJG) and implemented in their software [5]. The value of Q_JPEG may vary from 1 to 100. The quantization matrices in Tables 16.3 and 16.4 have been set for $Q\_JPEG = 50$. For other Q_JPEG values, each element in both the tables is simply scaled by the factor alpha (α) as defined in [5], where

$$\alpha = \begin{cases} \frac{50}{Q\_JPEG} & \text{if } 1 \le Q\_JPEG \le 50 \\ 2 - \frac{Q\_JPEG}{50} & \text{if } 50 \le Q\_JPEG \le 100, \end{cases}$$

subject to the condition that the minimum value of the scaled quantization matrix elements $\alpha Q(u,v)$ is 1. For the best reconstructed quality, Q_JPEG is set to 100.

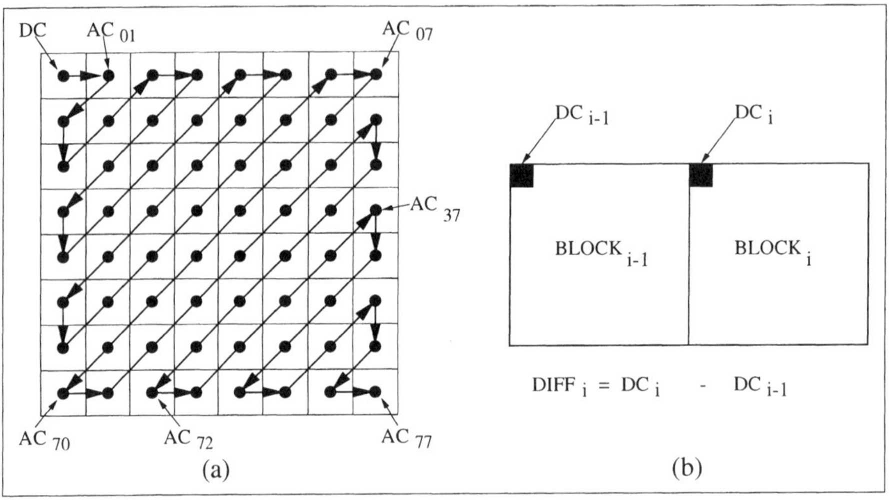

Fig. 16.5 (a) Zig-zag ordering of AC coefficients; (b) differential coding of DC.

After quantization of the DCT coefficients, the quantized DC coefficient is encoded by differential encoding. The DC coefficient DC_i of the current block is subtracted by the DC coefficient DC_{i-1} of the previous block and the difference $DIFF = DC_i - DC_{i-1}$ is encoded as shown in Figure 16.5(b). This is done to exploit the spatial correlation between the DC values of the adjacent blocks. Encoding of the AC coefficients is not straightforward. Instead of encoding each AC coefficient in the block, only the significant (nonzero) coefficients are encoded by an efficient manner such that the runs of zeros preceding a nonzero value are embedded into the encoding. Usually there are few significant low-frequency AC coefficients in the whole 8×8 block and most of the higher-frequency coefficients are quantized to 0's. In order to exploit this property, the AC coefficients are ordered in a particular irregular order sequence as shown in Figure 16.5(a). This irregular ordering of the AC coefficients is called *zig-zag ordering*. This is done to keep the low-frequency coefficients together and form long runs of 0's corresponding to the higher-frequency quantized coefficients. This zig-zag sequence is then broken into runs of zeros ending in a nonzero value. Before we explain the entropy encoding procedure, let us show the results of level shifting, DCT, quantization, and zig-zag ordering with an example 8 × 8 block extracted from a natural image.

Example: A Sample 8 × 8 Data Block

110	110	118	118	121	126	131	131
108	111	125	122	120	125	134	135
106	119	129	127	125	127	138	144
110	126	130	133	133	131	141	148
115	116	119	120	122	125	137	139
115	106	99	110	107	116	130	127
110	91	82	101	99	104	120	118
103	76	70	95	92	91	107	106

The 8 × 8 Data Block After Level Shifting

−18	−18	−10	−10	−7	−2	3	3
−20	−17	−3	−6	−8	−3	6	7
−22	−9	1	−1	−3	−1	10	16
−18	−2	2	5	5	3	13	20
−13	−12	−9	−8	−6	−3	9	11
−13	−22	−29	−18	−21	−12	2	−1
−18	−37	−46	−27	29	−24	−8	−10
−25	−52	−58	−33	−36	−37	−21	−22

DCT Coefficients of the above 8 × 8 Block

−89.00	−63.47	18.21	−6.85	7.50	13.45	−7.00	0.13
74.14	−2.90	−19.93	−21.04	−17.88	−10.81	8.29	5.26
−63.65	3.10	5.08	14.82	10.12	9.33	1.31	−0.62
3.73	2.85	6.67	8.99	−3.38	1.54	1.04	−0.62
2.50	0.57	−4.46	0.52	3.00	−2.89	−0.32	1.33
7.52	−1.80	−0.63	−0.10	0.41	−3.21	−2.74	−2.07
−3.40	0.43	0.81	0.28	−0.40	−0.19	−0.58	−1.09
−2.26	−0.88	1.73	0.23	−0.21	−0.12	1.23	1.61

Results of DCT Coefficients Quantized by Luminance Quantization Matrix

−6	−6	2	0	0	0	0	0
6	0	−1	−1	−1	0	0	0
−5	0	0	1	0	0	0	0
0	0	0	0	0	0	0	0
0	0	0	0	0	0	0	0
0	0	0	0	0	0	0	0
0	0	0	0	0	0	0	0
0	0	0	0	0	0	0	0

After the DC coefficient is differentially encoded, the AC coefficients are ordered in the zig-zag sequence and the sequence is subsequently broken into a number of runs of zeros ending in a nonzero coefficient. The entropy encoding procedure for differentially encoded DC coefficient is identical to the entropy encoding of the prediction error values that we explained for lossless JPEG. For 8-bit images in baseline JPEG, the DCT coefficients fall in the range [−1023, +1023]. Since the DC coefficient is differentially encoded, the differential values of DC fall in the range [−2047, +2047]. Assuming that the DC coefficient of the previous block is −4 as an example, the differential DC value of the present block is −2. From Table 16.2, we find that this belongs to category 2 and hence −2 is described as (2, 01). If the Huffman code of category 2 is 011, then −2 is coded as 01101, where the last two bits 01 represent the variable-length integer (VLI) code of −2. There are two Huffman tables (Tables K.3 and K.4) for encoding the DC coefficients in Annex K of the baseline JPEG standard for reference. But the user can choose any table and add them as part of the header of the compressed file [1]. Table K.3 is supplied for coding the luminance DC differences as a reference. Table K.4 is supplied for chrominance DC differences.

After zig-zag ordering of the AC coefficients in the example, the resulting sequence is −6 6 −5 0 2 0 −1 0 0 0 0 0 −1 0 0 −1 1 0 0 0 0 0 0 0 0 0 0 0 0 0 0 0 0 0 0 0 0 0 0 0 0 0 0 0 0 0 0 0 0 0 0 0 0 0 0 0 0 0 0 0 0 0 0. This sequence of AC coefficients can be mapped into an *intermediate sequence* of a combination of two symbols, $symbol_1$ and $symbol_2$: $symbol_1$ is represented by a pair $(RUNLENGTH, CATEGORY)$, where $RUNLENGTH$ is the number of consecutive zeros preceding the nonzero AC coefficient being encoded and $CATEGORY$ is the number of bits to represent the VLI code of this nonzero AC coefficient; $symbol_2$ is a single piece of information designated $(AMPLITUDE)$ that is encoded by the VLI code of the nonzero AC coefficient. Accordingly, the zig-zag sequence in the example can be compactly represented as:

(0, 3)(−6), (0, 3)(6), (0, 3)(−5), (1, 2)(2), (1, 1)(−1), (5, 1)(−1), (2, 1)(−1), (0, 1)(1), (0,0).

The first significant (nonzero) AC coefficient in the zig-zag sequence is −6. It is represented as (0, 3)(−6) because it precedes with no run of zeros (i.e., $RUNLENGTH = 0$) and the $AMPLITUDE = -6$ belongs to $CATEGORY = 3$. Similarly, the following two nonzero coefficients 6 and −5 are represented as (0, 3)(6) and (0, 3)(−5) respectively. The next significant coefficient 2 is represented by (1, 2)(2) because it precedes a 0 coefficient (i.e., $RUNLENGTH = 1$) and $AMPLITUDE = 2$ belongs to $CATEGORY = 2$. Similarly, the next significant symbol is represented as (1, 1)(−1). The following significant coefficient −1 is represented as (5, 1)(−1) because it precedes five 0's (i.e., $RUNLENGTH = 5$) and $AMPLITUDE = -1$ belongs to $CATEGORY = 1$. Following the same procedure, the next two nonzero coefficients −1 and 1 are represented by (2, 1)(−1) and (0, 1)(1) respectively. There are no other nonzero coefficients in the remainder of the zig-zag sequence. It is represented by a special symbol (0,0) to indicate that the remaining elements in the zig-zag block are all zeros. Each ($RUNLENGTH$, $CATEGORY$) pair is encoded using a Huffman code and the corresponding $AMPLITUDE$ is encoded by the VLI code.

There are two special symbols in encoding the zig-zag sequence of AC coefficients—(0,0) and (15, 0). The first special symbol is (0,0), and it is referred to as EOB (end-of-block), to indicate that the remaining elements in the zig-zag block are zeros. The other special symbol is (15, 0) and it is also referred to as ZRL (zero-run-length) to indicate a run of 16 zeros. Maximum length of a run of zeros allowed in baseline JPEG is 16. If there are more than 16 zeros, then the run is broken into the number of runs of zeros of length 16. For example, consider 57 zeros before a nonzero coefficient, say −29. This will be represented by (15, 0) (15, 0) (15, 0), (9, 5)(−29). The first three (15, 0) pairs represent 48 zeros and (9, 5)(−29) represents 9 zeros followed by the coefficient −29 which belongs to category 5.

The baseline JPEG allows a maximum of four Huffman tables–two for encoding AC coefficients and two for encoding DC coefficients. In luminance–chrominance image data, usually two Huffman tables (one for AC and one for DC) are used for encoding luminance data and similarly two for encoding chrominance data. The Huffman tables used during the compression process are stored as header information in the compressed image file in order to uniquely decode the coefficients during the decompression process. There are two Huffman tables (Table K.5 and K.6) for encoding the AC coefficients and two others (Table K.3 and K.4) for encoding the DC coefficients in Annex K of the baseline JPEG standard for reference. The users can choose any table of their choice and store it as part of the header of the compressed file [1]. Tables K.3 and K.5 are recommended for luminance DC differences

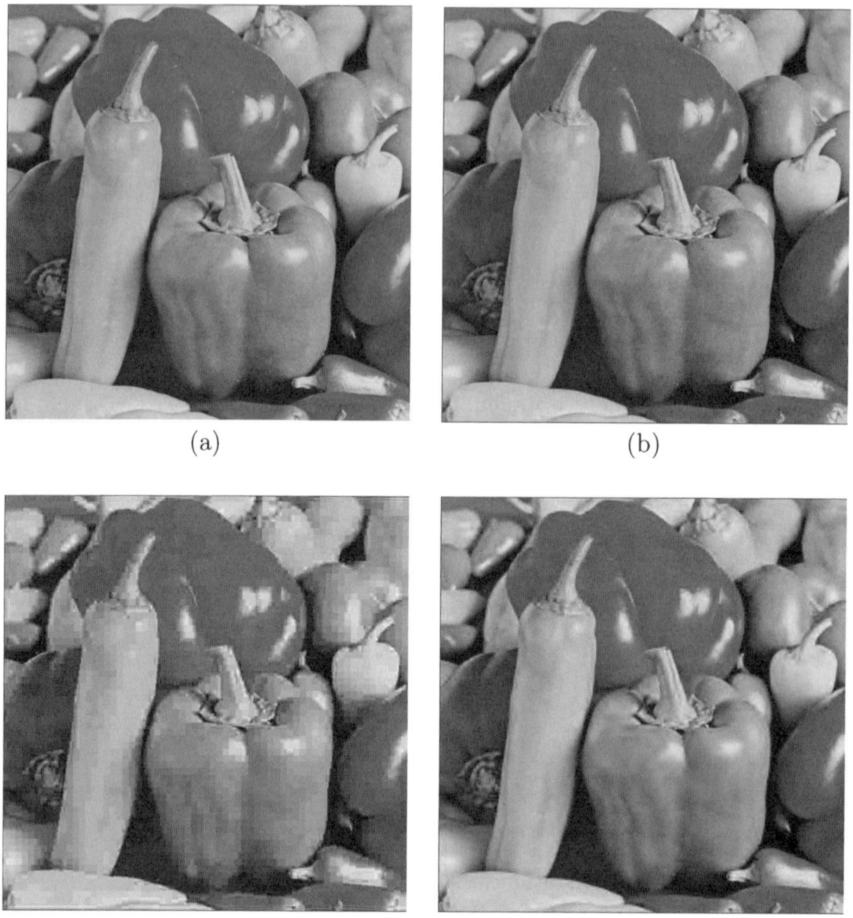

(a) (b) (c) (d)

Fig. 16.6 (a) Original Peppers image, (b) compressed with baseline JPEG using quality factor 75 (1.57 bit/pixel), (c) compressed with baseline JPEG using quality factor 10 (0.24 bit/pixel), and (d) compressed with the new JPEG2000 standard using the same bit rate (0.24 bit/pixel).

and AC coefficients. Tables K.4 and K.6 are recommended for corresponding chrominance channels.

Let us now allocate the variable-length codes in the last example. The codewords for (0, 0), (0, 1), (0, 3), (1, 1), (1, 2), (2, 1), and (5, 1) from Table K.5 are 1010, 00, 100, 1100, 11011, 11100, and 1111010 respectively. VLI codes for the nonzero AC coefficients 1, -1, 2, -5, 6 and -6 are 1, 0, 10, 010, 110, and 001 respectively. Codeword for the differential DC value is 01101. The compressed bitstream for the 8×8 block is shown below, and it requires only 52 bits as opposed to 512 bits required by the original 8×8 block of

8-bit pixels,

01101 100001 100110 100010 1101110 11000 11110100 111000 001 1010

where the first five bits, 01101, represent the DC coefficient and the other 47 bits represent the AC coefficients. Hence, we achieved approximately 10:1 compression using baseline JPEG to compress the block as shown above.

Decompression is the inverse process to decode the compressed bitstream in order to properly reconstruct the image. The inverse functions in the decompression process are obvious and the corresponding block diagram of the baseline decompression algorithm is shown in Figure 16.3(b).

We show a picture of the famous "Peppers" image in Figure 16.6(a). The color version of Figure 16.6 is provided in the color figures page. The image is compressed using the baseline JPEG algorithm with quality factor $Q\_JPEG = 75$ and the reconstructed image is perceptually almost identical to the original image. This is shown in Figure 16.6(b). When we compress the same image with quality factor $Q\_JPEG = 10$, we can see prominent artifacts in the image as shown in Figure 16.6(c). The nature of artifacts that is caused by lossy JPEG compression/decompression is called *blocking* artifacts. This happens because of the discontinuities created at the 8×8 block boundaries, since the blocks are compressed and decompressed independently. The new JPEG2000 standard solves this problem by using *discrete wavelet transform* (DWT) over the whole image [2]. Figure 16.6(d) shows the result of the JPEG2000 standard compressing the image with the same bit-rate (0.24 bits per pixel). We discuss details of this new standard in Chapters 17 and 18.

16.4 SUMMARY

In this chapter, we have described the JPEG standard for still image compression. JPEG is essentially the first international standard for gray level and color image compression. JPEG has four different modes of algorithms. We have discussed both the lossy and lossless compression algorithms. We described the prediction-based lossless JPEG algorithm and the DCT based lossy compression as well. Baseline JPEG is a lossy compression algorithm and the most widely used algorithm among all different modes in the JPEG standard for still image compression. We have discussed the principles and algorithms for the baseline JPEG standard in great detail in this chapter. We have presented some results of the baseline JPEG algorithm and compared its performance with the discrete wavelet transformed (DWT) based new JPEG2000 standard for image compression. We describe the new JPEG2000 standard in great details in Chapters 17 and 18.

REFERENCES

1. W. B. Pennebaker and J. L. Mitchell, *JPEG Still Image Data Compression Standard.* Chapman & Hall, New York, 1993.

2. T. Acharya and P. Tsai, *JPEG2000 Standard for Image Compression: Concepts, Algorithms and VLSI Architectures*, Wiley, Hoboken, NJ, 2004.

3. T. Acharya and A. Mukherjee, "High-Speed Parallel VLSI Architectures for Image Decorrelation," *International Journal of Pattern Recognition and Artificial Intelligence*, 9(2), 1995, 343–365.

4. H. Lohscheller, "A Subjectively Adapted Image Communication System," *IEEE Transactions on Communications*, COM-32, Vol. 12, 1984, 1316–1322.

5. The independent JPEG Group, "The Sixth Public Release of the Independent JPEG Group's Free JPEG Software," *C source code of JPEG Encoder research 6b*, March 1998 (ftp://ftp.uu.net/graphics/jpeg/).

17

JPEG2000 Standard For Image Compression

17.1 INTRODUCTION

JPEG2000 is the new international standard for image compression [1] developed jointly by the *International Organization for Standardization* (ISO), and the *International Electrotechnical Commission* (IEC) and also recommended by *International Telecommunications Union* (ITU).

Although JPEG (actually baseline JPEG) has been very successful in the marketplace for more than a decade, it lacks many features desired by interactive multimedia applications, its usage in current communications (wired or wireless) environments, and Internet applications platforms. A fundamental shift in the image compression approach came after the *discrete wavelet transform* (DWT) became popular [2]–[6]. Exploiting the interesting features in DWT, many scalable image compression algorithms were proposed in the literature [7]–[13].

The JPEG2000 standard is effective in wide application areas such as Internet, digital photography, digital library, image archival, compound documents, image databases, color reprography (photocopying, printing, scanning, facsimile), graphics, medical imaging, multispectral imaging such as remotely sensed imagery, satellite imagery, mobile multimedia communication, 3G cellular telephony, client-server networking, e-commerce, etc.

17.2 WHY JPEG2000?

The underlying philosophy behind development of the JPEG2000 standard was to compress an image once and decode the compressed bitstream in many ways to meet different applications requirements. Some of the salient features offered by the JPEG2000 standard that are effective in vast areas of applications are as follows:

- *Superior low bit-rate performance*: It offers superior performance in terms of visual quality and PSNR (peak signal-to-noise ratio) at very low bit-rates (below 0.25 bit/pixel) compared to the baseline JPEG. For equivalent visual quality JPEG2000 achieves more compression compared to JPEG. This has been demonstrated in Fig.16.6 in Chapter 16, and its color version is provided in the color figures page.

- *Continuous tone and bi-level image compression*: The JPEG2000 standard is capable of compressing and decompressing both the continuous-tone (gray scale and color) and bi-level images. The JBIG2 standard was defined to compress the bi-level images and it uses the same MQ-coder that is used to entropy encode the wavelet coefficients of the grayscale or color image components.

- *Large dynamic range of the pixels*: The JPEG2000 standard-compliant systems can compress and decompress images with various dynamic ranges for each color component. Although the desired dynamic range for each component in the requirement document is 1 to 16 bits, the system is allowed to have a maximum of 38 bits precision based on the bitstream syntax.

- *Large images and large numbers of image components*: The JPEG2000 standard allows the maximum size of an image to be $(2^{32}-1) \times (2^{32}-1)$ and the maximum number of components in an image to be 2^{14}.

- *Lossless and lossy compression*: The single unified compression architecture can provide both the lossless and the lossy mode of image compression. As a result, the same technology is applicable in varying applications areas ranging from medical imagery requiring lossless compression to digital transmission of images through communication networks.

- *Fixed size can be preassigned*: The JPEG2000 standard allows users to select a desired size of the compressed file. This is possible because of the bit-plane coding of the architecture and controlling the bit-rate through the rate control. The compression can continue bit-plane by bit-plane in all the code-blocks until the desired compressed size is achieved and the compression process can terminate.

- *Progressive transmission by pixel accuracy and resolution*: Using the JPEG2000 standard, it is possible to organize the code-stream in a progressive manner in terms of *pixel accuracy* (i.e., visual quality or SNR) of images that allows reconstruction of images with increasing pixel accuracy as more and more compressed bits are received and decoded. This is possible by progressively decoding most significant bit-plane to less significant bit-planes until all the bit-planes are reconstructed. The code-stream can also be organized as progressive in resolution such that the higher-resolution images are generated as more compressed data are received and decoded.

- *Region of interest (ROI) coding*: The user may desire certain parts of an image that are of greater importance to be encoded with higher fidelity compared to the rest of the image. During decompression the quality of the image also can be adjusted depending on the degree of interest in each region of interest. For example, a medical practitioner may find a certain region (or number of regions) in the radiograph to be more informative than the other parts.

- *Random access and compressed domain processing*: By randomly extracting the code-blocks from the compressed bitstream, it is possible to manipulate certain areas (or regions of interest) of the image. Some of the examples of compressed-domain processing could be cropping, flipping, rotation, translation, scaling, feature extraction, etc.

- *Robustness to bit-errors (error resiliency)*: Robustness to bit-errors is highly desirable for transmission of images over noisy communications channels. The JPEG2000 standard facilitates this by coding small size independent code-blocks and including resynchronization markers in the syntax of the compressed bitstream. There are also provisions to detect and correct errors within each code-block.

- *Sequential buildup capability*: The JPEG2000–compliant system can be designed to encode an image from top to bottom in a single sequential pass without the need to buffer an entire image, and hence is suitable for low-memory on-chip VLSI implementation. The line-based implementation of DWT and tiling of the images facilitates this feature.

- *Metadata*: The extended file syntax format allows inclusion of metadata information to describe the data (image) into the compressed bitstream. For example, the JPX file format, defined in JPEG2000 Part 2: Extensions, allows any legal ICC (International Color Consortium) profile to be embedded in the file.

In Figure 17.1, we have demonstrated the results of some of the capabilities of the JPEG2000 technology. The input image shown in Figure 17.1(a) is actually a color image, although we have shown the gray scale version of the

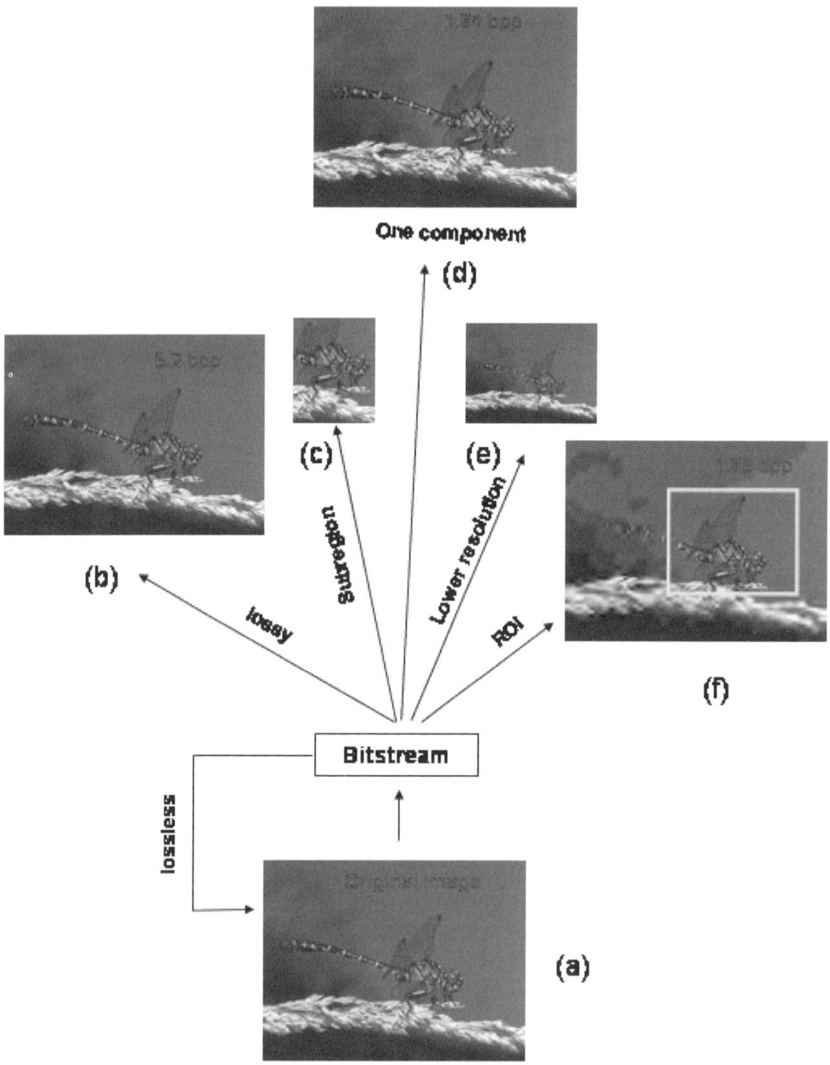

Fig. 17.1 Example of capabilities of JPEG2000 technology.

image here. We applied three levels of DWT to decompose the input image and generated the compressed bitstream with ROI encoding. The bitstream was generated by compressing the image losslessly. From the same bitstream, we decoded the image progressively until we reconstructed the original image as shown by the lossless arrow. While decoding in progressive manner, the reconstructed image is visually lossless at 5.2 bits per pixel or above as shown in Figure 17.1(b). In Figure 17.1(c), we show the random-access capability. We have accessed the compressed bits only for the code-blocks forming the subregion (or cropped version of the image) and decoded the result as shown in Figure 17.1(c). When we decode only one component (in this example we decoded the G component), we get a grayscale image as shown in Figure 17.1(d). After decoding the bitstream progressively at two levels of resolution, we generate a 2:1 downscaled (horizontally and vertically) version of the image as shown in Figure 17.1(e). After decoding to 1.89 bits per pixel, we losslessly reconstructed the ROI portion of the image, but introducing artifacts in the rest of the image as shown in Figure 17.1(f). As a result, we can conclude that an image can be compressed once and the compressed bitstream can be decoded in many different ways to suit the desired requirement. For further details the readers can refer to [1].

17.3 PARTS OF THE JPEG2000 STANDARD

As of writing this book, the standard has 11 parts (because Part 7 has been abandoned) with each part adding new features to the core standard in Part 1. The 11 parts and their features are as follows:

- Part 1—Core Coding System [1, 14] specifies the basic feature set and code-stream syntax for JPEG2000.
- Part 2—Extensions [15] to Part 1. This part adds a lot more features to the core coding system.
- Part 3—Motion JPEG2000 [16] specifies a file format (MJ2) that contains an image sequence encoded with the JPEG2000 core coding algorithm for motion video.
- Part 4—Conformance Testing [17] is now published as an International Standard (ISO/IEC 15444-4:2002). It specifies compliance-testing procedures for encoding/decoding using Part 1 of JPEG2000.
- Part 5—Reference Software [18]. In this part, two software source packages (using Java and C programming languages) are provided for the purpose of testing and validation for JPEG2000 systems implemented by the developers.
- Part 6—Compound Image File Format [19] specifies another file format (JPM) for the purpose of storing compound images. The ITU-T

T.44|ISO 16485 [20] multilayer *mixed raster content* (MRC) model is used to represent a compound image in Part 6 of JPEG2000.

- *Part 7*—This part has been abandoned.
- Part 8—Secure JPEG2000 (JPSEC). This part deals with security aspects for JPEG2000 applications such as encryption, watermarking, etc.
- Part 9—Interactivity Tools, APIs, and Protocols (JPIP). This part defines an interactive network protocol, and it specifies tools for efficient exchange of JPEG2000 images and related metadata.
- Part 10—3D and Floating Point Data (JP3D). This part is developed with the concern of three-dimensional data such as 3D medical image reconstruction, as an example.
- Part 11—Wireless (JPWL). This part is developed for wireless multimedia applications. The main concerns for JPWL are error protection, detection, and correction for JPEG2000 in an error-prone wireless environment.
- Part 12—ISO Base *media file format* has a common text with ISO/IEC 14496-12 for MPEG-4.

Parts 8 to 11 (and possible additional parts) are still under development as of writing this book.

17.4 OVERVIEW OF THE JPEG2000 PART 1 ENCODING SYSTEM

Like JPEG, the JPEG2000 standard is also written from the decoder point of view. This means that the decoder is specified quite precisely from marker segments to bitstream syntax in the JPEG2000 standard document. The detail of the specification of the decoder is sufficient to dictate the functionalities of the encoder. However, it is very difficult for a beginner to understand the standard document. Once the encoder system is well understood, it becomes easier to comprehend the decoder system described in the standard document. The whole compression system is simply divided into three phases. We call them (1) image preprocessing, (2) compression, and (3) compressed bitstream formation.

17.5 IMAGE PREPROCESSING

The image preprocessing phase consists of three optional major functions: first *tiling*, then *DC level shifting*, followed by the *multicomponent transformation*.

17.5.1 Tiling

The first preprocessing operation is *tiling*. In this step, the input source image is (optionally) partitioned into a number of rectangular non-overlapping blocks if the image is very large. Each of these blocks is called a *tile*. All the tiles have exactly the same dimension except the tiles at the image boundary if the dimension of the image is not an integer multiple of the dimension of the tiles. The tile sizes can be arbitrary up to the size of the original image. For an image with multiple components, each tile also consists of these components. For a gray scale image, the tile has a single component. Since the tiles are compressed independently, visible artifacts may be created at the tile boundaries when it is heavily quantized for very-low-bit-rate compression as typical in any block transform coding. Smaller tiles create more boundary artifacts and also degrade the compression efficiency compared to the larger tiles. Obviously, no tiling offers the best visual quality. On the other hand, if the tile size is too large, it requires larger memory buffers for implementation either by software or hardware.

17.5.2 DC Level Shifting

Originally, the pixels in the image are stored in unsigned integers. For mathematical computation, it is essential to convert the samples into two's complement representation before any transformation or mathematical computation starts in the image. The purpose of DC level shifting (optional) is to ensure that the input image samples have a dynamic range that is approximately centered around the zero. The DC level shifting is performed on image samples that are represented by unsigned integers only. All samples $I_i(x, y)$ in the i th component of the image (or tile) are level shifted by subtracting the same quantity $2^{S_{siz}^i - 1}$ to produce the DC level shifted sample $I_i'(x, y)$ as follows,

$$I_i'(x, y) \leftarrow I_i(x, y) - 2^{S_{siz}^i - 1}$$

where S_{siz}^i is the precision of image samples signaled in the SIZ (image and tile size) marker segment in compressed bitstream.

For images whose samples are represented by signed integers, such as CT (computed tomography) images, the dynamic range is already centered about zero, and no DC level shifting is required.

17.5.3 Multicomponent Transformations

The multicomponent transform is effective in reducing the correlations (if any) among the multiple components in a multicomponent image. This results in reduction in redundancy and increase in compression performance. Actually, the standard does not consider the components as color planes and in that sense the standard itself is colorblind. However, it defines an optional multicomponent transformation in the first three components only. These first three components can be interpreted as three color planes (R, G, B) for ease of understanding. That's why they are often called multicomponent color transformation as well. However, they do not necessarily represent Red-Green-Blue data of a color image. In general, each component can have different bit-depth (precision of each pixel in a component) and different dimension. However, the condition of application of the multicomponent transform is that the first three components should have identical bit-depth and identical dimension as well.

The JPEG2000 Part 1 standard supports two different transformations: (1) reversible color transform (RCT), and (2) irreversible color transform (ICT). The RCT can be applied for both lossless and lossy compression of images. However, ICT is applied only in lossy compression.

17.5.3.1 Reversible Color Transformation For lossless compression of an image, only the reversible color transform (RCT) is allowed because the pixels can be exactly reconstructed by the inverse RCT. Although it has been defined for lossless image compression, the standard allows it for lossy compression as well. In case of lossy compression, the errors are introduced by the transformation and/or quantization steps only, not by the RCT. The forward RCT and inverse RCT are given by:

Forward RCT:

$$\left.\begin{aligned} Y_r &= \lfloor \tfrac{R+2G+B}{4} \rfloor \\ U_r &= B - G \\ V_r &= R - G \end{aligned}\right\} \tag{17.1}$$

Inverse RCT:

$$\left.\begin{aligned} G &= Y_r - \lfloor \tfrac{U_r+V_r}{4} \rfloor \\ R &= V_r + G \\ B &= U_r + G \end{aligned}\right\} \tag{17.2}$$

17.5.3.2 Irreversible Color Transformation The irreversible color transformation (ICT) is the same as the luminance–chrominance color transformation used in baseline JPEG, presented in Chapter 16. ICT is applied for lossy compression only. Y is the luminance component of the image representing intensity of the pixels (light) and Cb and Cr are the two chrominance components representing the color information in each pixel. In baseline JPEG, the chrominance components can be subsampled to reduce the amount of data to start with. However, in the JPEG2000 standard, this subsampling is not allowed.

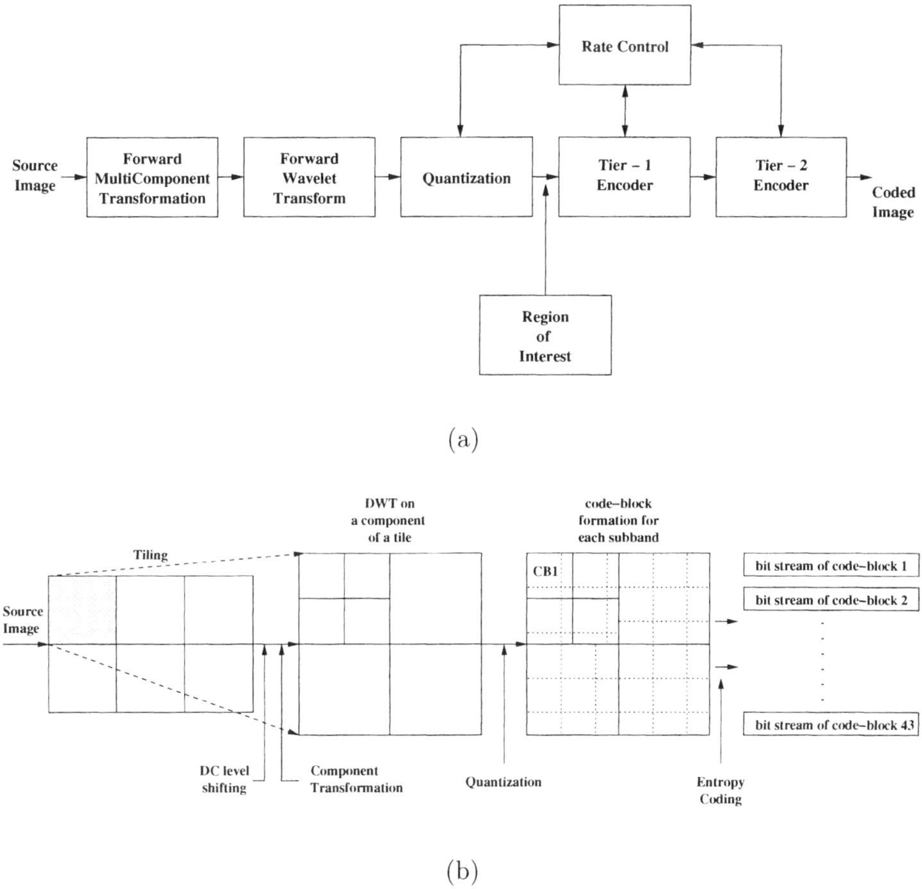

Fig. 17.2 (a) Block diagram of the JPEG2000 encoder algorithm; (b) dataflow.

17.6 COMPRESSION

After the optional preprocessing phase, as described in the previous section, the compression phase actually generates the compressed code. The computational block diagram of the functionalities of the compression system is shown in Figure 17.2(a). The data flow of the compression system is shown in Figure 17.2(b). As shown in Figure 17.2(b), each preprocessed component is independently compressed and transmitted as shown in Figure 17.2(a).

The compression phase is mainly divided into three sequential steps: (1) *discrete wavelet transform* (DWT), (2) *quantization*, and (3) *entropy encoding*. After preprocessing, each component is independently analyzed by a suitable discrete wavelet transform (DWT). The DWT essentially decomposes each component into a number of subbands in different resolution levels. Each subband is then independently quantized by a quantization parameter, in case of lossy compression. The quantized subbands are then divided into a number of smaller *code-blocks* of equal size, except for the code-blocks at the boundary of each subband. Typical size of the code-blocks is usually 32 × 32 or 64 × 64 for better memory handling. Each code-block is then entropy encoded independently to produce compressed bitstreams as shown in the dataflow diagram in Figure 17.2(b).

17.6.1 Discrete Wavelet Transformation

We discussed the DWT in greater detain in Chapter 5. The maximum number of levels of decomposition allowed in Part 1 is 32. In Part 1 of the JPEG2000 standard, only power of 2 dyadic decomposition in multiple levels of resolution is allowed.

17.6.1.1 Discrete Wavelet Transformation for Lossy Compression For lossy compression, the default wavelet filter used in the JPEG2000 standard is the Daubechies (9, 7) *biorthogonal spline* filter. By (9, 7) we indicate that the analysis filter is formed by a 9-tap low-pass FIR filter and a 7-tap high-pass FIR filter. Both filters are symmetric. The analysis filter coefficients (for forward transformation) are as follows:

- 9-tap low-pass filter: $[h_{-4}, h_{-3}, h_{-2}, h_{-1}, h_0, h_1, h_2, h_3, h_4]$

$$
\begin{aligned}
h_4 = h_{-4} &= +0.026748757410810 \\
h_3 = h_{-3} &= -0.016864118442875 \\
h_2 = h_{-2} &= -0.078223266528988 \\
h_1 = h_{-1} &= +0.266864118442872 \\
h_0 &= +0.602949018236358
\end{aligned}
$$

- 7-tap high-pass filter: $[g_{-3}, g_{-2}, g_{-1}, g_0, g_1, g_2, g_3]$

$$
\begin{aligned}
g_3 = g_{-3} &= +0.0912717631142495 \\
g_2 = g_{-2} &= -0.057543526228500 \\
g_1 = g_{-1} &= -0.591271763114247 \\
g_0 &= +1.115087052456994
\end{aligned}
$$

For the synthesis filter pair used for inverse transformation, the low-pass FIR filter has seven filter coefficients and the high-pass FIR filter has nine coefficients. The corresponding synthesis filter coefficients are as follows:

- 7-tap low-pass filter: $[h'_{-3}, h'_{-2}, h'_{-1}, h'_0, h'_1, h'_2, h'_3]$

$$
h'_3 = h'_{-3} = -0.0912717631142495
$$

$$
\begin{aligned}
h'_2 = h'_{-2} &= -0.057543526228500 \\
h'_1 = h'_{-1} &= +0.591271763114247 \\
h'_0 &= +1.115087052456994
\end{aligned}
$$

- 9-tap high-pass filter: $[g'_{-4}, g'_{-3}, g'_{-2}, g'_{-1}, g'_0, g'_1, g'_2, g'_3, g'_4]$

$$
\begin{aligned}
g'_4 = g'_{-4} &= +0.026748757410810 \\
g'_3 = g'_{-3} &= +0.016864118442875 \\
g'_2 = g'_{-2} &= -0.078223266528988 \\
g'_1 = g'_{-1} &= -0.266864118442872 \\
g'_0 &= +0.602949018236358
\end{aligned}
$$

17.6.1.2 Reversible Wavelet Transform for Lossless Compression For lossless compression, the default wavelet filter used in the JPEG2000 standard is the Le Gall (5, 3) spline filter [21]. Although this is the default filter for lossless transformation, it can be applied in lossy compression as well. However, experimentally it has been observed that the (9, 7) filter produces better visual quality and compression efficiency in lossy mode than the (5, 3) filter. The analysis filter coefficients for the (5, 3) filter are as follows:

- 5-tap low-pass filter: $[h_{-2}, h_{-1}, h_0, h_1, h_2]$

$$
\begin{aligned}
h_2 = h_{-2} &= -1/8 \\
h_1 = h_{-1} &= 1/4 \\
h_0 &= 3/4
\end{aligned}
$$

- 3-tap high-pass filter: $[g_{-1}, g_0, g_1]$

$$
\begin{aligned}
g_1 = g_{-1} &= -1/2 \\
g_0 &= 1
\end{aligned}
$$

The corresponding synthesis filter coefficients are as follows:

- 3-tap low-pass filter: $[h'_{-1}, h'_0, h'_1]$

$$
\begin{aligned}
h'_1 = h'_{-1} &= 1/2 \\
h'_0 &= 1
\end{aligned}
$$

- 5-tap high-pass filter: $[g'_{-2}, g'_{-1}, g'_0, g'_1, g'_2]$

$$
\begin{aligned}
g'_2 = g'_{-2} &= -1/8 \\
g'_1 = g'_{-1} &= -1/4 \\
g'_0 &= 3/4
\end{aligned}
$$

17.6.1.3 Boundary Handling Like a convolution, filtering is applied to the input samples by multiplying the filter coefficients with the input samples and accumulating the results. Since these filters are not causal, they cause discontinuities at the tile boundaries and create visible artifacts at the image boundaries as well. This introduces the dilemma of what to do at the boundaries. In order to reduce discontinuities in tile boundaries or reduce artifacts at image boundaries, the input samples should be first extended *periodically* at both sides of the input boundaries before applying the one-dimensional filtering both during row-wise and column-wise computation. By symmetrical/mirror extension of the data around the boundaries, one is able to deal with the noncausal nature of the filters and avoid edge effects. The number of additional samples needed to extend the boundaries of the input data is dependent on filter length. The general idea of period extension of the finite-length signal boundaries is explained by the following two examples.

Example 1: Consider the finite-length input signal $ABCDEFGH$. For an FIR filter of odd length, the signal can be extended periodically as

$$\cdots FG\underline{HGFEDCB}\overline{ABCDEFGH}\underline{GFEDCBA}BC\cdots$$

The two underlined sequences demonstrate the symmetry of extension with respect to the first sample (A) and the last sample (H) of the input signal as axis, and hence the boundary samples (A and H) are not included in the extension. The overlined sequence is the original input signal. This is called "whole-sample" symmetric (WSS) extension. The (9, 7) and (5, 3) filter kernels in Part 1 of the standard are odd-length filters and the boundary handling is done using the whole-sample symmetric extension.

Example 2: For an FIR filter of even length, the signal can be extended periodically as

$$\cdots FGH\underline{HGFEDCBA}\overline{ABCDEFGH}\underline{HGFEDCBA}ABC\cdots$$

The two underlined sequences demonstrate the *mirror* symmetry of the input signal at both of the boundaries and the overlined sequence is the original input signal. This is called "half-sample" symmetric (HSS) extension, in which the boundary samples (A and H) are also included in the extension because of the mirror symmetry. The even-length filters are allowed in the Part 2 extension of the standard and the boundary handling is accomplished by the half-sample symmetric extension.

17.6.2 Quantization

After the DWT, all the subbands are quantized in lossy compression mode in order to reduce the precision of the subbands to aid in achieving compression. Quantization of DWT subbands is one of the main sources of information loss in the encoder. Coarser quantization results in more compression and hence in reducing the reconstruction fidelity of the image because of greater loss of information. Quantization is not performed in case of lossless encoding. In Part 1 of the standard, the quantization is performed by uniform scalar quantization with dead-zone about the origin. In dead-zone scalar quantizer with step-size $\triangle_b$, the width of the dead-zone (i.e., the central quantization bin around the origin) is $2\triangle_b$ as shown in Figure 17.3. The standard supports separate quantization step-sizes for each subband. The quantization step size ($\triangle_b$) for a subband (b) is calculated based on the dynamic range of the subband values. The formula of uniform scalar quantization with a dead-zone is

$$q_b(i,j) = sign(y_b(i,j)) \left\lfloor \frac{|y_b(i,j)|}{\triangle_b} \right\rfloor, \tag{17.3}$$

where $y_b(i,j)$ is a DWT coefficient in subband b and $\triangle_b$ is the quantization step size for the subband b. All the resulting qunantized DWT coefficients $q_b(i,j)$ are signed integers.

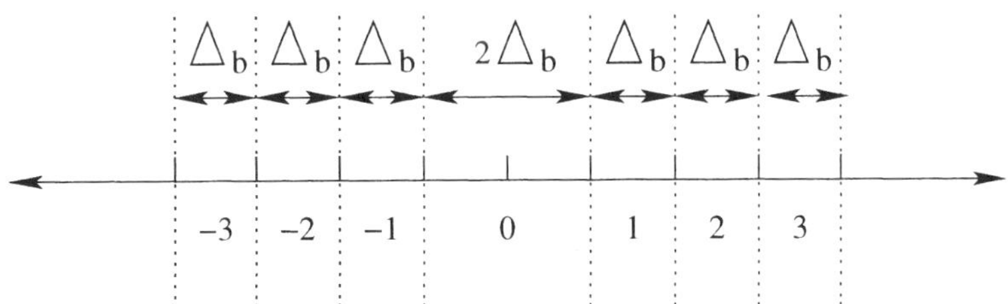

Fig. 17.3 Dead-zone quantization about the origin.

All the computations up to the quantization step are carried out in two's complement form. After the quantization, the quantized DWT coefficients are converted into sign-magnitude represented prior to entropy coding because of the inherent characteristics of the entropy encoding process, which will be described in greater detail in Chapter 18.

17.6.3 Region of Interest Coding

The *region of interest* (ROI) coding is a unique feature of the JPEG2000 standard. It allows different regions of an image to be coded with different fidelity criteria. These regions can have arbitrary shapes and be disjoint to each other. In Figure 17.4, we show an example of ROI coding. We compressed the ROI portion of the Zebra image losslessly and introduced losses in the non-ROI (background) part of the image. The reconstructed image after decompression is shown in Figure 17.4(a). We indicate the ROI by a circle around the head of the Zebra in Figure 17.4(a). In Figure 17.4(b), we pictorially show the difference between the original image and the reconstructed

(a) (b)

Fig. 17.4 (a) Reconstructed image with circular shape ROI; (b) difference between original image and reconstructed image.

image after ROI coding and decoding. The values of difference of the original and the reconstructed pixels in the ROI region (i.e., inside the circle) are all zeros (black) and they are nonzero (white) in the non-ROI parts of the image. This shows the capability of the JPEG2000 standard in how we can compress different regions of an image with different degrees of fidelity.

The ROI method defined in the JPEG2000 Part 1 standard is called the MAXSHIFT method [22]. The MAXSHIFT method is an extension of the scaling-based ROI coding method [23]. During ROI coding, a binary mask is generated in the wavelet domain for distinction of the ROI from the background as shown in Figure 17.5(a). In the scaling-based ROI coding, the bits associated with the wavelet coefficients corresponding to an ROI (as indicated by the ROI mask) are scaled (shifted) to higher bit-planes than the bits associated with the non-ROI portion of the image. This is shown by a block diagram in Figure 17.5(b). During the encoding process, the most significant ROI bit-planes are encoded and transmitted progressively before encoding the bit-planes associated with the non-ROI background region. As a result, during the decoding process, the most significant bit-planes of ROI can be decoded before the background region progressively in order to produce high fidelity in the ROI portions of the image compared to its background. In this method, the encoding can stop at any point and still the ROI portion of the reconstructed image will have higher quality than the non-ROI portion. In scaling-based ROI, the scaling parameter and the shape information needs to be transmitted along with the compressed bitstream. This is used in the Part 2 extension of the standard.

In JPEG2000 Part 1, the MAXSHIFT technique is applied instead of the more general scaling-based technique. The MAXSHIFT allows arbitrary-shaped regions to be encoded without requiring to transmit the shape information along with the compressed bitstream. As a result, there is no need for shape coding or decoding in the MAXSHIFT technique. The basic principle of

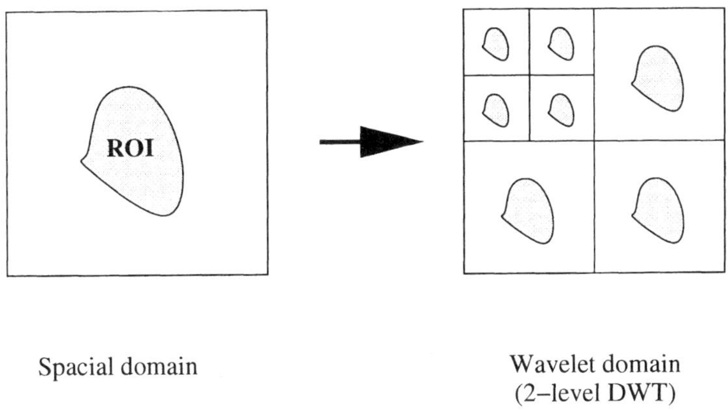

(a)

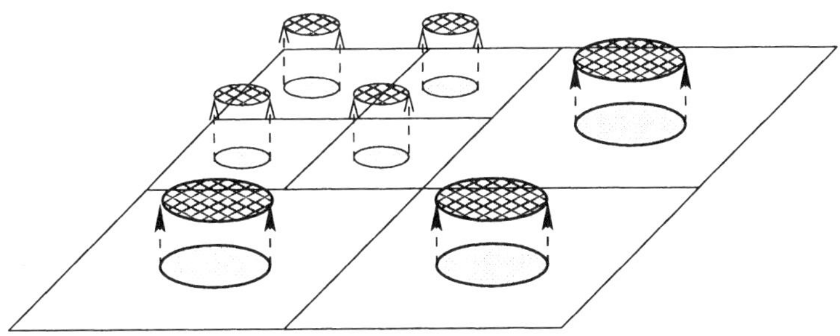

(b)

Fig. 17.5 (a) ROI mask; (b) scaling of ROI coefficients.

the MAXSHIFT method is to find the minimum value (V_{min}) in the ROI and the maximum value in the background (both in wavelet transformed domain) and then scale (shift) the wavelet coefficients in ROI in such a manner that the smallest coefficient in the ROI is always greater than the largest coefficient in the background. Then the bit-planes are encoded in the order of the most significant bit (MSB) plane first to the least significant bit (LSB) plane last. Figure 17.6 shows an example where the LSB plane of ROI is shifted above the MSB plane of the background region. During the decompression process, the wavelet coefficients that are larger than V_{min} are identified as the ROI coefficients without requiring any shape information or the binary mask that was used during the encoding process. The ROI coefficients are now shifted down relative to V_{min} in order to represent it with original bits of precision.

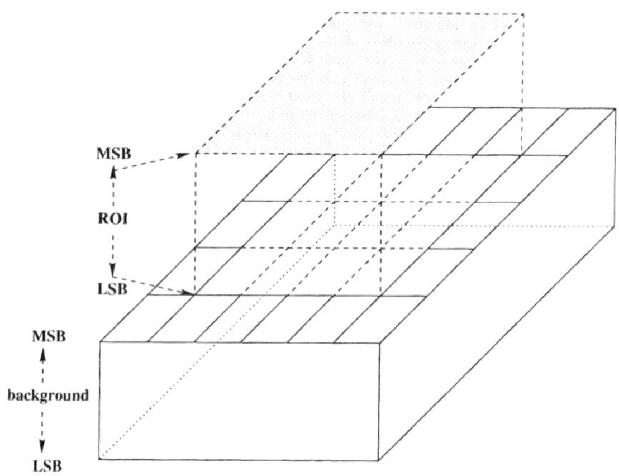

Fig. 17.6 MAXSHIFT.

In JPEG2000, due to the sign-magnitude representation of the quantized wavelet coefficients required in the bit-plane coding, there is an implementation precision for number of bit-planes. Scaling the ROI coefficients up may cause an overflow problem when it goes beyond this implementation precision. Therefore, instead of shifting ROI up to higher bit-planes, the coefficients of background are downscaled by a specified value s, which is stored in the RGN marker segment in the bitstream header. The decoder can deduce the shape information based on this shift value s and magnitude of the coefficients. By choosing an appropriate value of s, we can decide how many bit-planes to truncate in the background in order to achieve overall bit-rate without sacrificing the visual quality of the ROI.

17.6.4 Rate Control

Although the key encoding modules of JPEG2000 such as wavelet transformation, quantization, and entropy coding (bit-plane coding and binary arithmetic coding) are clearly specified, some implementation issues are left up to the prerogative of the individual developers. Rate control is one such open issue in JPEG2000 standard. Rate control is a process by which the bit-rates (sometimes called coding rates) are allocated in each code-block in each subband in order to achieve the overall target encoding bit-rate for the whole image while minimizing the distortion (errors) introduced in the reconstructed image due to quantization and truncation of codes to achieve the desired code rate [24]. It can also be treated in another way. Given the allowed distortion in the MSE (mean square energy) sense, the rate control can dictate the optimum encoding rate while achieving the maximum given MSE.

The JPEG2000 encoder generates a number of independent bitstreams by encoding the code-blocks. Accordingly a rate-distortion optimization algorithm generates the truncation points for these bitstreams in an optimal way in order to minimize the distortion according to a target bit rate. After the image is completely compressed, the rate-distortion optimization algorithm is applied once at the end using all the rate and rate-distortion slope information of each coding unit. This is the so-called post-compression rate-distortion (PCRD) algorithm [24, 25].

The bit-rate control is purely an encoder issue, and remains an open issue for the JPEG2000 standard. It is up to the prerogative of the developers how they want to accomplish the rate-distortion optimization in a computationally efficient way without incurring too much computation and/or hardware cost.

17.6.5 Entropy Encoding

Physically the data are compressed by the entropy encoding of the quantized wavelet coefficients in each code-block in each subband. We have devoted a complete chapter (Chapter 18) to discussion of entropy encoding. Here we just summarize the entropy-encoding scheme at the top level for the sake of completeness of this chapter. The entropy coding and generation of compressed bitstream in JPEG2000 is divided into two coding steps: *Tier-1* and *Tier-2* coding.

17.6.5.1 Tier-1 Coding In Tier-1 coding, the code-blocks are encoded independently. If the precision of the elements in the code-block is p, then the code-block is decomposed into p bit-planes and they are encoded from the most significant bit-plane to the least significant bit-plane sequentially. Each bit-plane is first encoded by a *fractional bit-plane coding* (BPC) mechanism to generate intermediate data in the form of a *context* and a binary *decision* value for each bit position. In JPEG2000 the *embedded block coding with optimized truncation* (EBCOT) algorithm [25] has been adopted for the BPC. EBCOT encodes each bit-plane in three coding passes, with a part of a bit-plane being coded in each coding pass without any overlapping with the other two coding passes. That is the reason why the BPC is also called *fractional* bit-plane coding. The three coding passes in the order in which they are performed on each bit-plane are *significant propagation pass*, *magnitude refinement pass*, and *cleanup pass*. The algorithm is very complex and we have devoted a complete chapter on this with a number of examples to aid the reader in better understanding of the algorithm. The details of these coding passes and the EBCOT algorithm are dealt with in Chapter 18.

The binary decision values generated by the EBCOT are encoded using a variation of *binary arithmetic coding* (BAC) to generate compressed codes for each code-block. The variation of the binary arithmetic coder is a *context adaptive BAC* called the MQ-coder, which is the same coder used in the JBIG2 standard to compress bi-level images [26]. The *context* information generated by EBCOT is used to select the estimated probability value from a lookup table and this probability value is used by the MQ-coder to adjust the intervals and generate the compressed codes. JPEG2000 standard uses a predefined lookup table with 47 entries for only 19 possible different contexts for each bit type depending on the coding passes. This facilitates rapid probability adaptation in the MQ-coder and produces compact bitstreams.

The working principles and detail flowchart (algorithm) for implementation of the MQ-coder are presented in Chapter 18.

17.7 TIER-2 CODING AND BITSTREAM FORMATION

After the compressed bits for each code-block are generated by Tier-1 coding, the Tier-2 coding engine efficiently represents the layer and block summary information for each code-block. A *layer* consists of consecutive bit-plane coding passes from each code-block in a tile, including all the subbands of all the components in the tile. The block summary information consists of length of compressed code words of the code-block, the most significant magnitude bit-plane at which any sample in the code-block is nonzero, as well as the truncation point between the bitstream layers, among others. The decoder receives this information in an encoded manner in the form of two tag trees. This encoding helps to represent this information in a very compact form without incurring too much overhead in the final compressed file. The encoding process is popularly known as *Tag Tree coding*.

17.8 SUMMARY

In this chapter, we have presented an overview of the JPEG2000 standard for image compression. We have discussed different salient features of the new JPEG2000 standard and how they influence vast applications areas. We have introduced different parts of the standard. The core coding system in JPEG2000 has been defined in Part 1 of the standard. We have dealt in great length with the underlying principles and algorithms for the Part 1 core coding system for JPEG2000 standard. The whole compression system can be divided into three phases—image preprocessing, compression, and compressed bitstream formation. In this chapter, we first discussed the concepts behind the preprocessing functionalities, including tiling of the input image, DC level shifting, and multicomponent transformation, before the actual compression takes place. We discussed the implementation issues of the discrete wavelet transform supported by the JPEG2000 Part 1 standard including the symmetric extension at the boundary of the signals both for lossless and lossy compression. Theoretical foundation of the discrete wavelet transform and its implementation issues were elaborated on earlier in Chapter 5. In lossy compression mode, a dead-zone scalar quantization technique is applied on the wavelet coefficients. The concept of region of interest coding allows one to encode different regions of the input image with different fidelity. We saw that the region of interesting coding can be achieved in terms of simply scaling the wavelet coefficients, and demonstrated some examples. The entropy coding and the generation of compressed bitstream in JPEG2000 are divided into two coding steps: *Tier-1* and *Tier-2* coding. We introduced the Tier-1 coding step in entropy encoding based on a fractional bit-plane coding scheme (EBCOT) and an adaptive binary arithmetic coding (MQ-coder) to generate the compressed codes. The details of the algorithms for both the Tier-1 coding and the Tier-2 coding are presented in greater detail in Chapter 18.

REFERENCES

1. T. Acharya and P. Tsai, *JPEG2000 Standard Image Compression: Concepts, Algorithms and VLSI Architectures*, Wiley, Hoboken, NJ, 2004.

2. S. Mallat, "A Theory for Multiresolution Signal Decomposition: The Wavelet Representation," *IEEE Trans. on Pattern Analysis and Machine Intelligence*, 11(7), July 1989, 674–693.

3. I. Daubechies, "The Wavelet Transform, Time-Frequency Localization and Signal Analysis," *IEEE Trans. on Inform. Theory*, 36(5), September 1990, 961–1005.

4. Y. Meyers, *Wavelet: Algorithms and Applications*. SIAM, Philadelphia, 1993 (Translated by Robert D. Ryan).

5. W. Sweldens, "The Lifting Scheme: A Custom-Design Construction of Biorthogonal Wavelets," *Applied and Computational Harmonic Analysis*, 3(15), 1996, 186–200.

6. I. Daubechies and W. Sweldens, "Factoring Wavelet Transforms into Lifting Schemes," *Journal of Fourier Analysis and Applications*, 4, 1998, 247–269.

7. R. A. DeVore, B. Jawerth, and B. J. Lucier, "Image Compression Through Wavelet Transform Coding," *IEEE Trans. on Information Theory*, 38(2), March 1992, 719–746.

8. M. Antonini, M. Barlaud, P. Mathieu, and I. Daubechies, "Image Coding Using Wavelet Transform," *IEEE Trans. on Image Processing*, 1(2), April 1992, 205–220.

9. A. S. Lewis and G. Knowles, "Image Compression Using the 2-D Wavelet Transform," *IEEE. Trans. on Image Processing*, 1(2), April 1992, 244–250.

10. J. M. Shapiro, "Embedded Image Coding Using Zerotrees of Wavelet Coefficients," *IEEE Trans. on Signal Processing*, 41(12), December 1993, 3445–3462.

11. A. Said and W. A. Peralman, "A New Fast and Efficient Image Codec Based on Set Partitioning in Hierarchical Trees," *IEEE Trans. on Circuits and Systems for Video Technology*, 6(3), June 1996, 243–250.

12. A. Said and W. A. Peralman, "An Image Multiresolution Representation for Lossless and Lossy Compression," *IEEE Trans. on Image Processing*, 5, September 1996, 1303–1310.

13. P. Chen, T. Acharya, and H. Jafarkhani, "A Pipelined VLSI Architecture for Adaptive Image Compression," *International Journal of Robotics and Automation*, 14(3), 1999, 115–123.

14. ISO/IEC 15444-1, "Information Technology—JPEG2000 Image Coding System—Part 1: Core Coding System," 2000.

15. ISO/IEC 15444-2, Final Committee Draft, "Information Technology—JPEG2000 Image Coding System—Part 2: Extensions," 2000.

16. ISO/IEC 15444-3, "Information Technology—JPEG2000 Image Coding System—Part 3: Motion JPEG2000," 2002.

17. ISO/IEC 15444-4, "Information Technology—JPEG2000 Image Coding System—Part 4: Conformance Testing," 2002.

18. ISO/IEC 15444-5, "Information Technology—JPEG2000 Image Coding System—Part 5: Reference Software," 2003.

19. ISO/IEC 15444-6, Final Committee Draft, "Information Technology—JPEG2000 Image Coding System—Part 6: Compound Image File Format," 2001.

20. ITU-T T.44|ISO/IEC 16485, "Information Technology—Mixed Raster Content (MRC)," 2000.

21. D. L. Gall and A. Tabatabai, "Subband Coding of Digital Images Using Symmetric Short Kernel Filters and Arithmetic Coding Techniques," *Proceedings of IEEE International Conference Acoustics, Speech, and Signal Processing*, Vol. 2, pp. 761–764, New York, April 1988.

22. D. Nister and C. Christopoulos, "Lossless Region of Interest with Embedded Wavelet Image Coding," *Signal Processing*, Vol. 78, No. 1, pp. 1–17, 1999.

23. E. Atsumi and N. Farvardin, " Loss/Lossless Region-of-Interest Image Coding Based on Set Partitioning in Hierarchical Tree," *IEEE International Conference Image Processing*, 87–91, Chicago, October 1998.

24. T. Kim, H. Kim, P. Tsai, and T. Acharya, "Memory Efficient Progressing Rate-Distortion Algorithm for JPEG2000," *IEEE Transactions on Circuits and Systems for Video Technology*, 15(1), January 2005, 181-187.

25. D. S. Taubman, "High Performance Scalable Image Compression with EBCOT," *IEEE Transaction Image Processing*, 9(7), 1158–1170, July 2000.

26. ISO/IEC 14492-1, "Lossy/Lossless Coding of Bi-level Images," 2000.

18

Coding Algorithms in JPEG2000 Standard

18.1 INTRODUCTION

As shown in Section 17.6, Figure 17.2(a), after the DWT and quantization, the encoding phase in JPEG2000 is divided into two steps—*Tier-1* coding and *Tier-2* coding. In Tier-1 coding, each code-block is entropy encoded independently. In Tier-2 coding, the information of the compressed codewords generated in the Tier-1 coding step are encoded using a Tag Tree coding mechanism, which will be discussed in great detail in this chapter.

Entropy coding in JPEG2000 [1, 2] is combination of fractional bit-plane coding (BPC) [3] and binary arithmetic coding (BAC) [4], which is known as Tier-1 coding in JPEG2000. In this chapter, we will explain this new paradigm of fractional bit-plane coding technique. The implementation of the MQ-coder for binary arithmetic coding follows the same principle of QM-coder described in Chapter 15.

18.2 PARTITIONING DATA FOR CODING

During entropy encoding, each wavelet subband is further divided into a number of code-blocks. At this stage all the elements in all the subbands are represented in sign and magnitude representation of integers instead of two's complement. Dimension of the code-blocks is always a power of 2 with the minimum height and width being 4 and maximum height and width being 1024. Further restriction in dimension of a code-block is that if height of a code-block is 2^x and width of the code-block 2^y, then $x+y$ is limited to be less than or equal to 12. Typical choice of code-block size is 64×64 or 32×32. It has been found experimentally that the compression performance degrades when the code-block size is chosen below 16×16. It should be noted that the profile-0 of JPEG2000 Part 1 amendments further restricts the code-block size to be either 32×32 or 64×64.

During the coding phase, each code-block is decomposed into a number of bit-planes. If the precision of the subband is P bits, then each code-block in the subband is decomposed into P number of bit-planes. *Bit-plane coding* (BPC) is applied on each bit-plane of the code-blocks to generate intermediate data in the form of a *context* and a *binary decision value.* The intermediate data is input to the *binary arithmetic coding* (BAC) step to generate the final compressed bitstream.

18.3 TIER-1 CODING IN JPEG2000

In JPEG2000, the *Embedded Block Coding with Optimized Truncation (EBCOT)* algorithm by David S. Taubman [1, 3] has been adapted to implement the BPC. This algorithm has been built to exploit the symmetries and redundancies within and across the bit-planes so as to minimize the statistics to be maintained and minimize the coded bitstream that BAC would generate. EBCOT encodes each bit-plane in three passes, with a part of the bit-plane being coded in each of these passes without any overlapping with the other two passes. That is the reason why this bit-plane coding is also called **fractional** bit-plane coding. The three passes in the order they are performed on each bit-plane are:

1. **Significance Propagation Pass (SPP)**: Bit positions that have a magnitude of 1 for the first time (i.e., the most significant bit of the corresponding sample coefficients) are coded in this pass.

2. **Magnitude Refinement Pass (MRP)**: Bit positions that have not been coded in SPP and that have had magnitude of 1 in previous bit-planes (i.e., the current bit is not the most significant bit of the corresponding sample coefficient) are coded in this pass.

3. **Cleanup pass (CUP)**: Bit positions that have not been coded in either of the two earlier passes are coded in this pass. This pass also incorporates a form of run-length coding to help in coding a string of zeros.

18.3.1 Fractional Bit-Plane Coding

In order to make it easy for readers to understand this complex algorithm, we first provide the definition of terms used to describe the algorithm, followed by the explanation of four basic *coding operations* and three *coding passes*. Then we provide a simple example to illuminate the detailed process of the BPC coder.

18.3.1.1 Definition of Terms

- **Code-block (y):** A code-block is a two-dimensional array that consists of integers (wavelet coefficients with or without quantization). Each code-block has width and height that specify its size. Each integer of the code-block can be either positive, zero, or negative. Each of the elements of a code-block are associated with σ, σ', and η to indicate their coded states (see σ, σ', and η for detailed descriptions).

- **Sign array (χ):** χ is a two-dimensional array representing the signs of the elements of a code-block. It has the exact same dimensions as the code-block. Each element $\chi[m,n]$ represents the sign information of the corresponding element $y[m,n]$ in the code-block as follows.

$$\chi[m,n] = \begin{cases} 1 & \text{if } y[m,n] < 0 \\ 0 & \text{otherwise} \end{cases}$$

  When referenced to $\chi[m,n]$ during encoding or decoding, m or n may be out of range of a code-block, such as $m = -1$. This will happen when we are working on the boundary of the code-block. In those cases, $\chi[m,n]$ is always set to equal zero.

- **Magnitude array (v):** v is a two-dimensional array of unsigned integers. Dimension of this array is exactly the same as the dimension of the code-block. Each element of v represents the absolute value of the integer at the corresponding location in the code-block, that is, $v[m,n] = |y[m,n]|$, where $y[m,n]$ is the integer element at spatial location (m,n) in the code-block. The notation $v^P[m,n]$ is used to denote the Pth bit of $v[m,n]$.

- **Bit-plane:** The magnitude array v can be conceptually viewed as a three-dimensional array, with the bit sequence of the integers in the third dimension. Each particular order of bits of every element of v constitutes a single bit-plane. We can also view v as a one-dimensional array consisting of a number of bit-planes.

  Example: Consider (2, 0) denotes a 2×1 array v with only two elements. The two bit-planes for this array will be (1, 0) and (0, 0).

  We say one bit-plane is more significant than the other if its bits are more significant than those of the other. Not all bit-planes of v are going to be coded because a bit sequence may have as many leading zeros as possible for a particular nonnegative value. We code only the most significant bit-plane containing at least one 1 and all the other subsequent less significant bit-planes, regardless of whether they have 1 or not. In other words, a bit-plane that entirely consists of 0 is ignored unless there is a higher significant bit-plane that contains at least one 1. We call those uncoded bit-planes "leading-zero bit-planes." A nonnegative integer P is frequently used to refer to a bit-plane to be coded. We use $v^P[m, n]$ to represent the bit at spatial location (m, n) of the bit-plane P of the code-block.

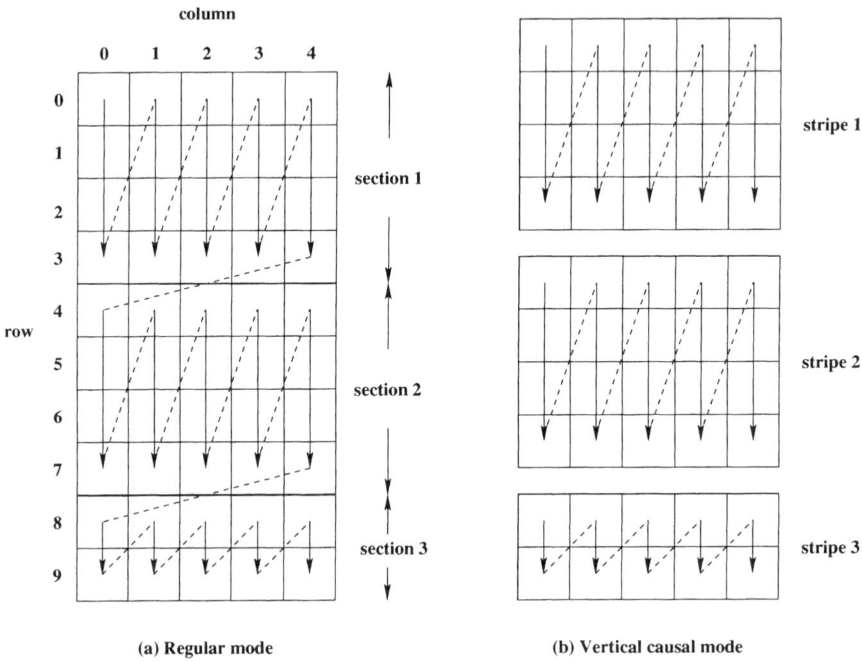

Fig. 18.1 Scan pattern with a 5 × 10 code-block: (a) regular mode; (b) vertical causal mode.

- **Scan pattern:** Scan pattern defines the order of encoding or decoding the bit-planes of a code-block. The scan pattern of a code-block can be conceptually divided into sections (or stripes), each with four consecutive rows starting from the first row of a code-block. If the total number of rows of a code-block is not a multiple of 4, all the sections will have four consecutive rows except the very last section. The scan starts from the first section and down to the last section until all elements of a code-block are encoded or decoded. Each section is scanned from the first row of the first column. The next location to be scanned will be the next row on the same column. After a column in the section is completely coded, start scanning at the first row of the next column in the same section. Continue the coding process until all columns in a section are coded. This same process is then applied to the next adjacent section until all of them are coded. An example of scan pattern for a 5 $\times$ 10 code-block is shown in Figure 18.1. Figure 18.1(a) shows the regular mode of the scan pattern. The JPEG2000 Part 1 also specified another scanning mode, **vertical causal** mode. In vertical causal mode, every section (sometimes referred to as stripe), that is, 4 rows by N columns, will be considered as a standalone module. In other words, under the vertical causal mode, all the information of neighbors from the same code-block but different sections will not be used in the current section. Figure 18.1(b) shows the vertical causal mode for the same example shown in Figure 18.1(a).

- **State variables σ, σ' and η:** Three two-dimensional "binary" arrays, σ, σ', and η, are created to indicate the coding states of each element in the code-block during the coding process. These arrays have the exact the same dimension as the code-block. Initially, each of the elements of these arrays are set to 0 (i.e., $\sigma[m,n] = 0$, $\sigma'[m,n] = 0$, and $\eta[m,n] = 0$, for all m and n). Once the coding process starts, the values of the two variables $\sigma[m,n]$ and $\sigma'[m,n]$ may change to 1 depending on certain conditions, but are never changed back to 0 until the entire code-block is encoded. On the other hand, the values of $\eta[m,n]$ are reset to 0 right after completion of coding of each bit-plane. Interpretations of these three variables are as follows:

  1. When $\sigma[m,n] = 1$, it indicates that the first nonzero bit of $v[m,n]$ at row m and column n has been coded; otherwise it is equal to 0. When referenced to $\sigma[m,n]$ during encoding or decoding, m or n may be out of range or have an invalid value, such as $m = -1$. In this case, $\sigma[m,n]$ is equal to 0.
  2. When $\sigma'[m,n] = 1$, it indicates that a *magnitude refinement coding* operation (defined in the next section) has been applied to $v[m,n]$; otherwise, it is equal to zero.
  3. When $\eta[m,n] = 1$, it indicates that *zero coding* operation (defined in the next section) has been applied to $v^P[m,n]$ in the *significant propagation pass*; otherwise, it is equal to 0.

- **Preferred neighborhood:** An element $y[m,n]$ in the code-block is said to be in a preferred neighborhood if at least one of its eight adjacent neighbors has σ value equal to 1.

- **Zero coding tables:** There are three zero coding tables for the purpose of *zero coding* operations as shown in Tables 18.1–18.3. The *context* information generated in zero coding operation is based on the values of the *significance* states (σ) of the eight neighbors of the element being encoded. In Figure 18.2, we show the eight neighbors of an element (say **X**) used in these tables to form the "context." For example, **X** is an element in the LL subband, and the two horizonal neighbors have significance state value of 1 (i.e., $\sum H = 2$). The context value 8 will be used as shown in Table 18.1.

$\mathbf{D}_0$	$\mathbf{V}_0$	$\mathbf{D}_1$
$\mathbf{H}_0$	$\mathbf{X}$	$\mathbf{H}_1$
$\mathbf{D}_3$	$\mathbf{V}_1$	$\mathbf{D}_2$

Fig. 18.2 Neighborhood for *zero coding* context generation.

Table 18.1 Zero Coding context table for code-blocks from LL and LH subbands

LL and LH Subbands			Context Label
$\sum H$	$\sum V$	$\sum D$	**CX**
2	x	x	8
1	≥ 1	x	7
1	0	≥ 1	6
1	0	0	5
0	2	x	4
0	1	x	3
0	0	≥ 2	2
0	0	1	1
0	0	0	0

Note: "x" in the table denotes "do not care."

Table 18.2 Zero coding context table for code-blocks from HL subbands

HL Subbands			Context Label
$\sum H$	$\sum V$	$\sum D$	**CX**
x	2	x	8
≥ 1	1	x	7
0	1	≥ 1	6
0	1	0	5
2	0	x	4
1	0	x	3
0	0	≥ 2	2
0	0	1	1
0	0	0	0

18.3.1.2 Coding Operations There are four possible coding operations used in EBCOT to generate the values of *context* (CX) and *decision* (D) as in-

Table 18.3 Zero coding context table for code-blocks from HH subbands

HH Subbands		Context Label
$\sum(H+V)$	$\sum D$	**CX**
x	≥ 3	8
≥ 1	2	7
0	2	6
≥ 2	1	5
1	1	4
0	1	3
≥ 2	0	2
1	0	1
0	0	0

termediate data before the BAC. CX is a nonnegative integer while D is a binary value, 0 or 1. There are 19 different context values (0–18) used in these four coding operations. The index of the current bit-plane is assumed to be P. Exactly when or where these operations are applied is subject to current coding pass, the location of the current element, and the status of the state variables.

- **Zero coding (ZC):** For zero coding operation, the decision bit D is equal to $v^P[m,n]$ and CX is selected from one of the three "zero coding context tables" related to the relevant subband (LH, HL or HH) the code-block belongs to. There are nine entries in each context table. They are derived using the values of the significance states of the eight surrounding neighbors of the current coefficient bit, $v^P[m,n]$. As shown in the tables, each entry depends on how many and which neighbors of $v^P[m,n]$ are significant.

- **Sign coding (SC):** The D and CX for the sign coding are determined by a horizontal reference value H and a vertical reference value V. Suppose that the current scan location is (m,n); the values of H and V are obtained by the following equations.

  $$H = \min[1, \max(-1, \sigma[m,n-1] \times (1-2\chi[m,n-1]) + \sigma[m,n+1] \times (1-2\chi[m,n+1]))]$$

  $$V = \min[1, \max(-1, \sigma[m-1,n] \times (1-2\chi[m-1,n]) + \sigma[m+1,n] \times (1-2\chi[m+1,n]))]$$

  The reference values of H and V indicate three possible situations as follows:

  1. **0** indicates that both neighbors are insignificant, or both neighbors are significant but have opposite signs.
  2. **1** indicates that one or both neighbors are significant with positive sign.

3. **−1** indicates that one or both neighbors are significant with negative sign.

The neighbors mean the two adjacent horizontal locations of the current scan location for H and the two vertical locations of the current scan location for V. Significant at a location means the value of the state variable σ at that location is 1 while insignificant means the value of σ is 0.

As shown in Table 18.4, H and V are used together to determine the context (CX) and a binary value $\hat{\chi}$, which in terms is used to calculate the value of D as $D = \hat{\chi} \otimes \chi[m,n]$, where $\otimes$ represents an Exclusive-OR operation.

Table 18.4 Reference Table for Sign Coding

H	V	$\hat{\chi}$	**CX**
1	1	0	13
1	0	0	12
1	−1	0	11
0	1	0	10
0	0	0	9
0	−1	1	10
−1	1	1	11
−1	0	1	12
−1	−1	1	13

- **Magnitude refinement coding (MRC):** For magnitude refinement coding, D at position (m,n) in the Pth bit-plane is simply equal to the bit value $v^P[m,n]$. The value of CX is determined by $\sigma'[m,n]$ and the sum of its eight adjacent values of the state variable σ is as follows:
  - If $\sigma' = 1$ at the current position, which indicates that it is not the first magnitude refinement for this element, then CX $= 16$.
  - When $\sigma' = 0$ at the current position and the sum of the values of σ of its eight adjacent neighbors is also 0, then CX $= 14$.
  - When $\sigma' = 0$ at the current position and the sum of the values of σ of its eight adjacent neighbors is greater than 0, then CX $= 15$.

  In Table 18.5, we summarize the logic for generation of the context values (CX) as described above.

- **Run-length coding (RLC):** Unlike the other three coding operations, run-length coding may code from one to four consecutive bits in the current scan pattern stripe. Exactly how many bits are encoded depends

Table 18.5 Reference Table for Magnitude Refinement Coding

$\sigma'[m,n]$	$\sigma[m-1,n]$ + $\sigma[m+1,n]$ + $\sigma[m-1,n-1]$ + $\sigma[m-1,n+1]$ + $\sigma[m+1,n-1]$ + $\sigma[m+1,n+1]$	**CX**
1	x	16
0	≥ 1	15
0	0	14

on where the first 1 bit (if any) is located in the four consecutive bits. If all of them are 0's, then all four bits are coded. If one (or more) of these four bits is 1, then the first 1 in the scan pattern and any preceding 0's in between the current scan location are coded. For example, suppose that 0101 are four consecutive bits along the scan pattern of a bit-plane. If the current location is at the first 0 and we are going to apply run-length coding, then the first two bits, 01, are coded and the next location will be at the second 0.

A run-length coding operation may generate either one D or three D's, depending on whether the four consecutive bits are all 0's or not. The first D is equal to 0 if all four bits are equal to 0; otherwise it is equal to 1. For both cases, CX is equal to a unique run-length context value 17. In other words, a (CX, D) pair with values (17, 0) indicates four consecutive 0 bits, and a (CX, D) pair with values (17, 1) means there is at least one 1 bit in the current scan pattern stripe.

In the case that at least one of the four bits in the current scan pattern is 1, two more D's are used with a "UNIFORM" context value 18 to indicate the location of the first 1 bit in the 4-bit scan pattern. Since height of the scan pattern is four, a zero-based index with two bits is sufficient to indicate the location of the first 1 bit from the top. The first and the second D's with a UNIFORM context represent the most significant and the least significant bits of these two bits representing the distance.

Continuing with the example for coding 01, the values of the first and the second D's are 0 and 1, respectively. The corresponding (CX, D) pairs will be (18, 0) and (18, 1).

Table 18.6 shows the summary of all 19 different contexts used in the four different coding operations, and their corresponding initial index values for the probability estimation lookup table used in BAC (discussed in the next section).

18.3.1.3 Coding passes There are three coding passes—*significance propagation pass* (SPP), *magnitude refinement pass* (MRP), and *cleanup pass* (CUP).

Table 18.6 Nineteen different contexts and their initial index for BAC probability estimation lookup table

Operation	**Context** CX	**Initial index** I(CX)
Zero Coding	0	**4**
	1	0
	2	0
	3	0
	4	0
	5	0
	6	0
	7	0
	8	0
Sign Coding	9	0
	10	0
	11	0
	12	0
	13	0
Magnitude Refinement Coding	14	0
	15	0
	16	0
Run-Length Coding	17	**3**
UNIFORM	18	**46**

Three different coding passes are applied to each bit-plane of a code-block except the first bit-plane (the most significant bit-plane), which is applied only with the cleanup pass. Even if we do apply the SPP and MRP to the most significant bit-plane, due to the initial condition there will be no bits coded in those two passes. After each coding pass completes a run of scan pattern in the current bit-plane, the next coding pass restarts the scan pattern from the beginning. The first bit-plane is only encoded by the cleanup pass. The remaining bit-planes are coded in the order of significance propagation pass, magnitude refinement pass, and cleanup pass. They are described below.

- **Cleanup pass (CUP):** CUP applies to every bit-plane of a code-block after completion of MRP, except the first bit-plane, which does not need the MRP.

  In each position (m, n) follow in the scan pattern, CUP first checks where $\sigma[m, n]$ and $\eta[m, n]$ are both 0's. If any one of them is not 0, then proceed to the next bit position in the bit-plane. If they are both 0's, then check whether to apply run-length coding (RLC) or zero coding (ZC), but not both. RLC is applied when all of the following three conditions are true:

1. m is a multiple of four, including $m = 0$.
2. $\sigma = 0$ for the four consecutive locations on the same column, starting from current scan position.
3. $\sigma = 0$ for all the adjacent neighbors of the four consecutive bits in the column.

If any one of the above conditions is false, then zero coding (ZC) is applied to the current location. Depending on whether run-length coding or zero coding is applied in the current location, the number of bits coded may vary. The next bit to be coded is the bit after the last coded bit. Note that run-length coding should not be applied to the last section with fewer than four rows in a scan pattern because there would not be four consecutive bits available in the same column.

After completion of the run-length coding or zero coding, we need to check whether we need to apply *sign coding* (SC) before we move on to code the next bit. Suppose the last coded position is (i, j). If $v^P[i, j] = 1$, which indicates this bit is the most significant bit of the current sample, the cleanup pass applies sign coding and assigns σ value of the last coded location to be 1 (i.e., $\sigma[i, j] = 1$) right after run-length or zero coding is done. Otherwise, no sign coding is needed.

Continue coding the bits along the scan pattern until all of the bits of the bit-plane are coded. After completion of the cleanup pass for a bit-plane, reset $\eta[m, n] = 0$ for all m and n in the bit-plane before moving into the next bit-plane. Figure 18.3 shows the flowchart of the cleanup pass we just described.

- **Significance propagation pass (SPP):** This is the first pass applied to every bit-plane of a code-block, except the first bit-plane. Significance propagation pass first applies zero coding if the current scan position (m, n) is in a preferred neighborhood and $\sigma[m, n] = 0$. If zero coding cannot be applied, then proceed to the next bit position. If the zero coding is applied, $\eta[m, n]$ is set to 1. After zero coding is completed, we need to check whether sign coding is needed at the current bit position (m, n). If $v^P[m, n] = 1$, then sign coding is applied and we set $\sigma[m, n] = 1$.

  Continue coding the bits along the scan pattern until all of the bits of the bit-plane are coded. Figure 18.4 shows the flowchart of the significance propagation pass.

- **Magnitude refinement pass (MRP):** This is the second pass applied to every bit-plane of a code-block, except the first bit-plane, which does not need magnitude refinement pass.

  If the state variables $\sigma[m, n] = 1$ and $\eta[m, n] = 0$, then we apply magnitude refinement coding (MRC) to the current scan position (m, n) and

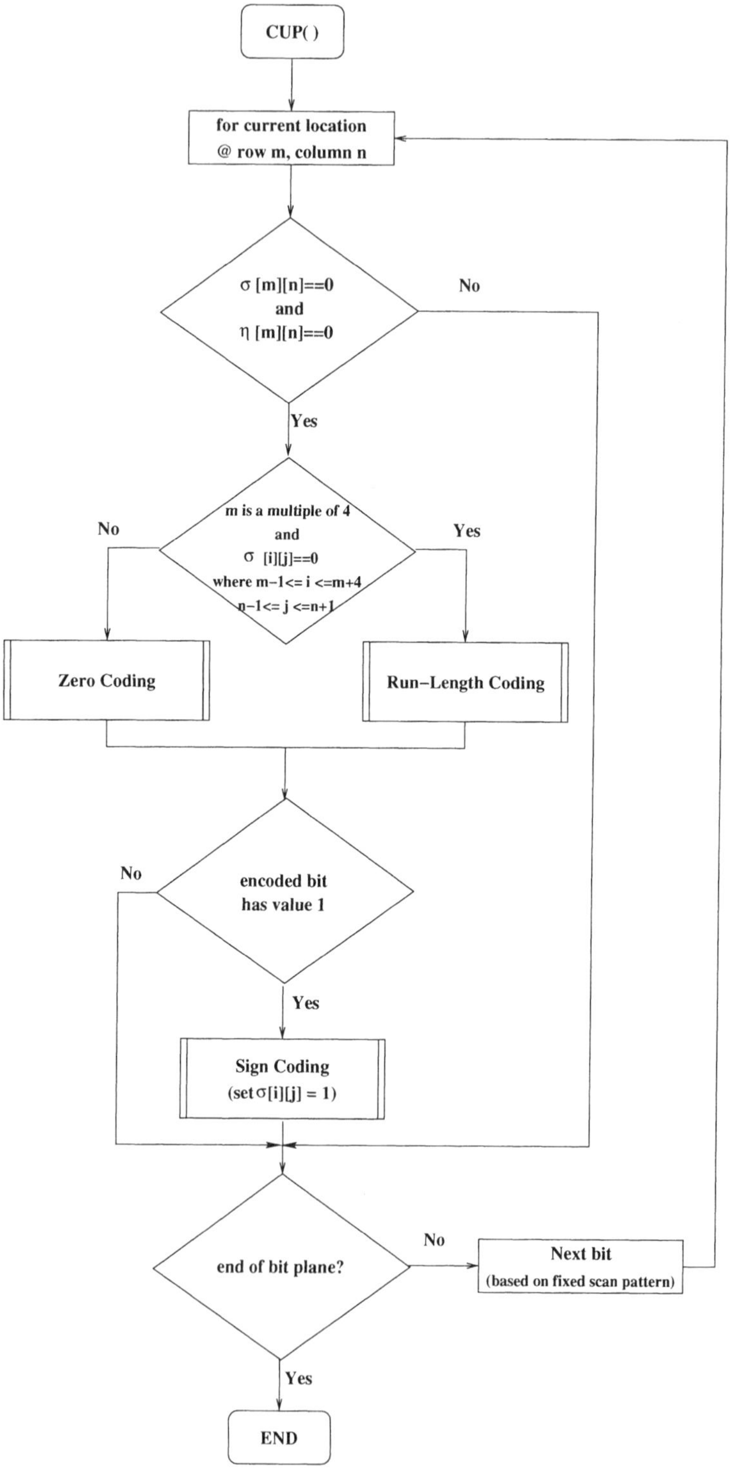

Fig. 18.3 Flowchart of cleanup pass.

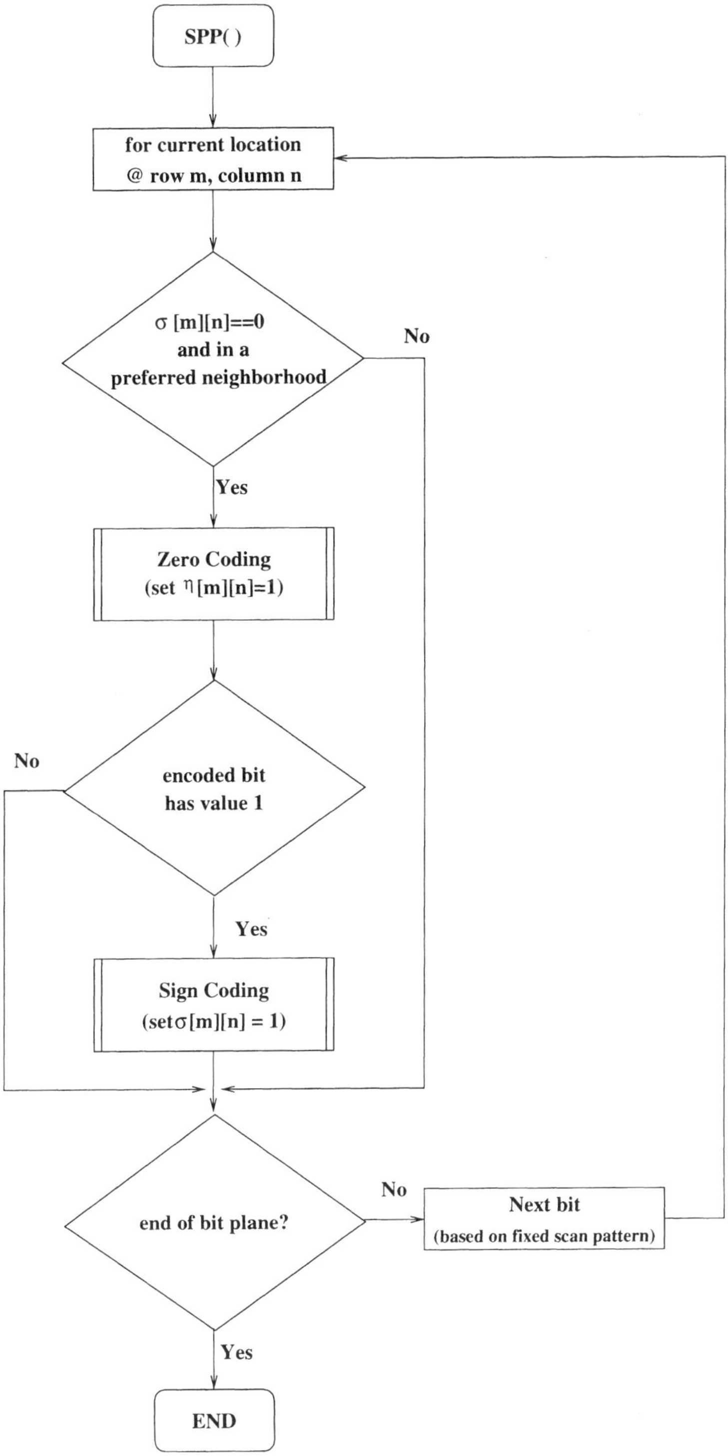

Fig. 18.4 Flowchart of significance propagation pass.

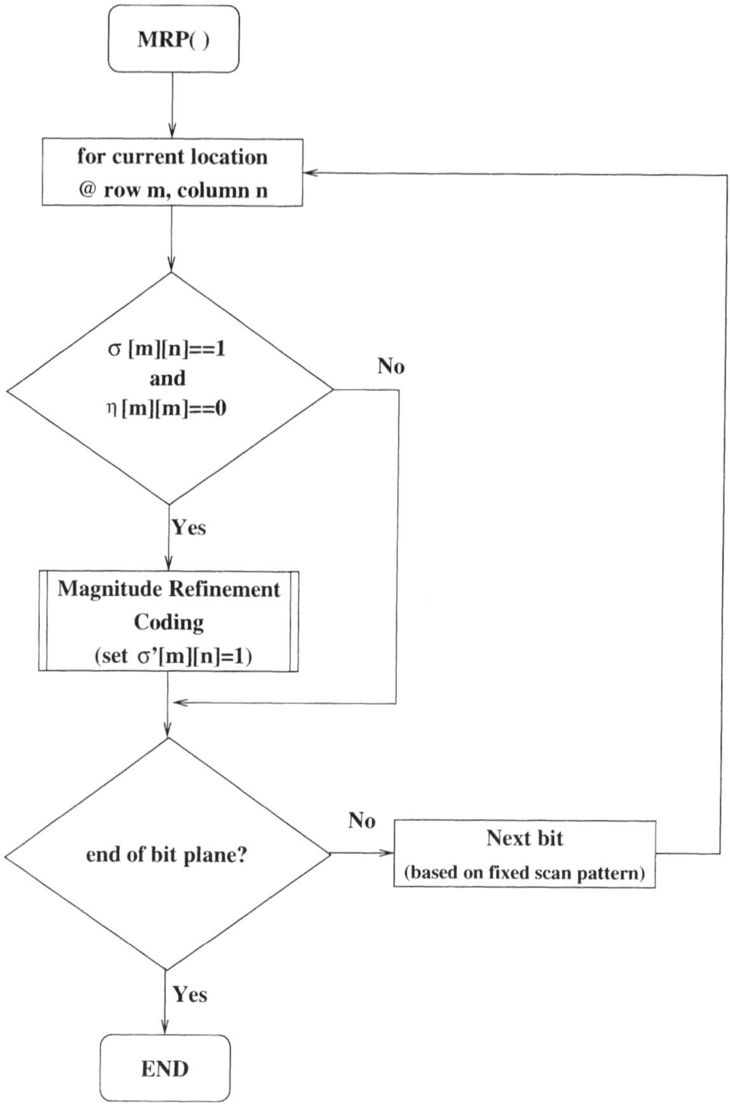

Fig. 18.5 Flowchart of magnitude refinement pass.

set $\sigma'[m, n] = 1$. Continue coding the bits along the scan pattern until all of the bits of the bit-plane are coded. Figure 18.5 shows the flowchart of the magnitude refinement pass.

- **Selective binary arithmetic coding—bypass mode:** Instead of applying the binary arithmetic coding (MQ-coder) on symbols (the contexts and decision bits) generated during all three coding passes, the bypass mode allows bypassing MQ-coder for the SPP and MRP after the four most significant bit-planes are coded. In other words, only those symbols generated in the CUP will be coded with the MQ-coder, and raw decision bits and sign bits will be coded during the SPP and MRP, if the bypass mode is selected.

18.3.1.4 JPEG2000 Bit-Plane Coding: Encoder and Decoder Algorithms As we discussed earlier, the quantized wavelet coefficients in each subband are converted into sign-magnitude represented before the entropy encoding starts. For each input code-block, we can first initialize the two-dimensional arrays υ and χ, where the value of $\upsilon[m, n]$ is the magnitude and $\chi[m, n]$ is the sign information of the element at position (m, n) in the code-block. The number of bit-planes in the code-block to be encoded (P) is determined by searching the largest value in array υ. Initially, all elements in two-dimensional arrays σ, σ', and η are set to 0's.

The first bit-plane to be coded is the most significant bit-plane. As mentioned at the definition of bit-plane, the leading-zero bit-planes consisting entirely of zeros are ignored. A more (higher) significant bit-plane is always coded before coding a less (lower) significant bit-plane. If $P = 0$, we don't need to do any coding and the output is empty. If $P \geq 1$, then we apply the cleanup pass only to the first bit-plane. For the remaining bit-planes, we first apply the significance propagation pass, then the magnitude refinement pass, and then the cleanup pass. Figure 18.6 shows the top-level flowchart of the fractional bit-plane coder.

The procedure of the decoder is essentially the same as for the encoder. For the sake of completeness, we also show the encoding and decoding procedures via the flowcharts as shown in Figures 18.7 and 18.8 respectively for $P \geq 1$.

18.3.2 Examples of BPC Encoder

In this section we present an example detailing all the operations step by step and the output generated by encoding a 4×4 code-block.

- Input of the encoder: An input 4×4 code-block is shown below:

3	0	0	5
-3	7	2	1
-4	-1	-2	3
0	6	0	2

- The magnitude array (υ) is shown below:

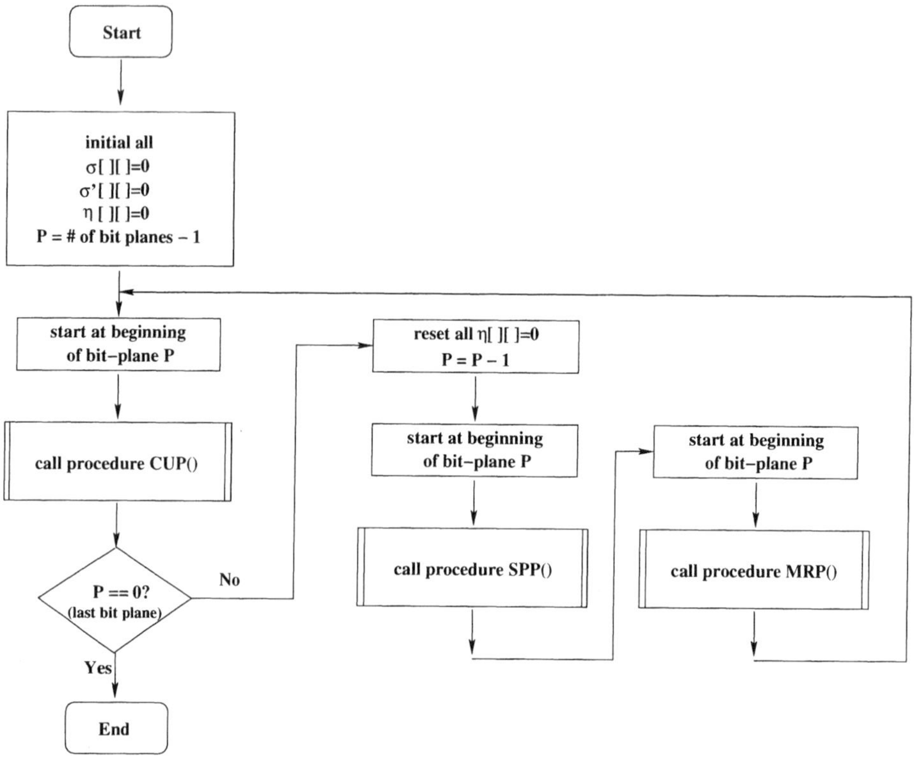

Fig. 18.6 Top-level flowchart of fractional bit-plane coder for $P > 0$.

3	0	0	5
3	7	2	1
4	1	2	3
0	6	0	2

- The sign array (χ) is shown below:

0	0	0	0
1	0	0	0
1	1	1	0
0	0	0	0

- The three bit-planes are shown below as bit-planes of the magnitude array υ^2, υ^1, and υ^0 respectively:

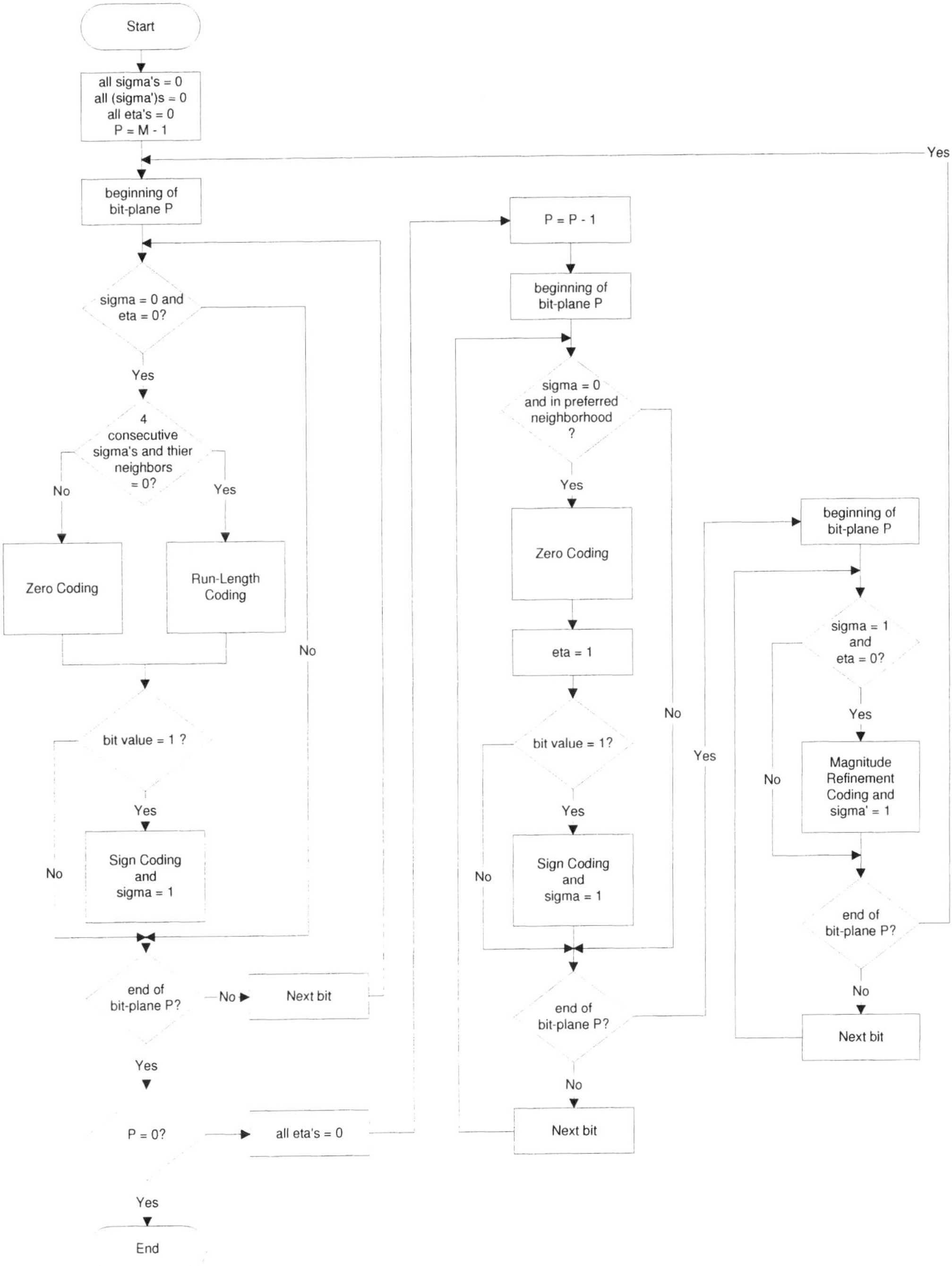

Fig. 18.7 Flowchart of fractional bit-plane encoder for $P > 0$.

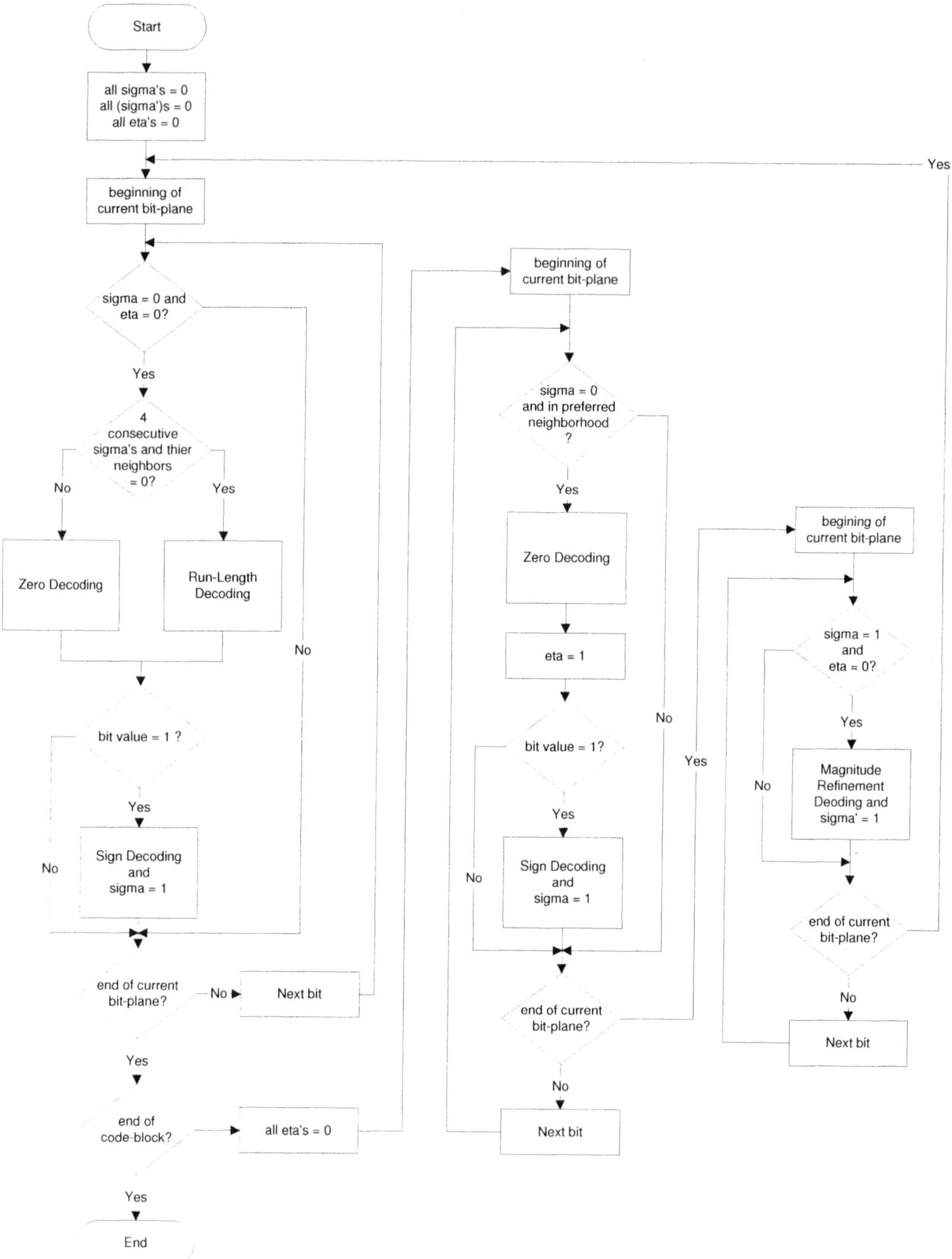

Fig. 18.8 Flowchart of fractional bit-plane decoder for $P > 0$.

v^2

0	0	0	1
0	1	0	0
1	0	0	0
0	1	0	0

v^1

1	0	0	0
1	1	1	0
0	0	1	1
0	1	0	1

v^0

1	0	0	1
1	1	0	1
0	1	0	1
0	0	0	0

- The coding sequence for the code-block and the output (CX, D) generated are shown here. We show the sequence of coding operations (ZC, RLC, SC, or MRC) at bit position (*row*, *column*) for the bit-plane P. The resulting CX and D are shown after each colon in each step. For example, the first operation in the sequence below indicates a run-length coding (RLC) applied in CUP at position (0, 0) in the first bit-plane. The output are CX $= 17$ and $D = 1$. The (CX, D) pair with (17, 1) indicates there is a 1 bit in the current four rows of scanned samples. Therefore, two pairs of (CX, D) are generated with values (18, 1) and (18, 0). The two decision bits indicate that the first 1 bit is at the zero-based location $(10)_2 = 2$ as shown in first column of v^2. A sign coding (SC) is followed after the RLC, and since all the state variables have initial value zeros, a context 9 and $\hat{\chi} = 0$ are selected from Table 18.4. The decision bit $D = \chi(0,0) \otimes \hat{\chi} = 1 \otimes 0 = 1$ is generated as shown in step 4 of CUP for bit-plane 2 below. After each pass, the temporal contents of the state variables σ, η, and σ' in the current bit-plane are also listed.

CUP for Bit-Plane 2

1. RLC for (0, 0): CX=17, D=1
2. RLC for (2, 0): 18, 1
3. RLC for (2, 0): 18, 0
4. SC for (2, 0): 9, 1
5. ZC for (3, 0): 3, 0
6. ZC for (0, 1): 0, 0
7. ZC for (1, 1): 1, 1
8. SC for (1, 1): 9, 0
9. ZC for (2, 1): 7, 0
10. ZC for (3, 1): 1, 1
11. SC for (3, 1): 9, 0
12. ZC for (0, 2): 1, 0
13. ZC for (1, 2): 5, 0
14. ZC for (2, 2): 2, 0

15. ZC for (3, 2): 5, 0
16. RLC for (0, 3): 17, 1
17. RLC for (0, 3): 18, 0
18. RLC for (0, 3): 18, 0
19. SC for (0, 3): 9, 0
20. ZC for (1, 3): 3, 0
21. ZC for (2, 3): 0, 0
22. ZC for (3, 3): 0, 0

υ^2

0	0	0	1
0	1	0	0
1	0	0	0
0	1	0	0

σ

0	0	0	1
0	1	0	0
1	0	0	0
0	1	0	0

η

0	0	0	0
0	0	0	0
0	0	0	0
0	0	0	0

σ'

0	0	0	0
0	0	0	0
0	0	0	0
0	0	0	0

SPP for Bit-Plane 1

23. ZC for (0, 0): CX=1, $D = 1$
24. SC for (0, 0): 9, 0
25. ZC for (1, 0): 7, 1
26. SC for (1, 0): 12, 1
27. ZC for (3, 0): 7, 0
28. ZC for (0, 1): 7, 0
29. ZC for (2, 1): 7, 0
30. ZC for (0, 2): 6, 0
31. ZC for (1, 2): 6, 1
32. SC for (1, 2): 12, 0
33. ZC for (2, 2): 3, 1
34. SC for (2, 2): 10, 1
35. ZC for (3, 2): 7, 0
36. ZC for (1, 3): 7, 0
37. ZC for (2, 3): 6, 1
38. SC for (2, 3): 12, 1
39. ZC for (3, 3): 3, 1
40. SC for (3, 3): 10, 0

υ^1				σ				η				σ'			
1	0	0	0	1	0	0	1	1	1	1	0	0	0	0	0
1	1	1	0	1	1	1	0	1	0	1	1	0	0	0	0
0	0	1	1	1	0	1	1	0	1	1	1	0	0	0	0
0	1	0	1	0	1	0	1	1	0	1	1	0	0	0	0

MRP for Bit-Plane 1

41. MRC for (2, 0): CX=15, D=0
42. MRC for (1, 1): 15, 1
43. MRC for (3, 1): 14, 1
44. MRC for (0, 3): 14, 0

υ^1				σ				η				σ'			
1	0	0	0	1	0	0	1	1	1	1	0	0	0	0	1
1	1	1	0	1	1	1	0	1	0	1	1	0	1	0	0
0	0	1	1	1	0	1	1	0	1	1	1	1	0	0	0
0	1	0	1	0	1	0	1	1	0	1	1	0	1	0	0

CUP for Bit-Plane 1

(Note: This pass does not generate any CX or D.)

υ^1				σ				η				σ'			
1	0	0	0	1	0	0	1	1	1	1	0	0	0	0	1
1	1	1	0	1	1	1	0	1	0	1	1	0	1	0	0
0	0	1	1	1	0	1	1	0	1	1	1	1	0	0	0
0	1	0	1	0	1	0	1	1	0	1	1	0	1	0	0

SPP for Bit-Plane 0

45. ZC for (3, 0): CX=7, D= 0
46. ZC for (0, 1): 7, 0
47. ZC for (2, 1): 8, 1
48. SC for (2, 1): 11, 0
49. ZC for (0, 2): 7, 0
50. ZC for (3, 2): 8, 0

51. ZC for (1, 3): 7, 1
52. SC for (1, 3): 13, 0

υ^0				σ				η				σ'			
1	0	0	1	1	0	0	1	0	1	1	0	0	0	0	1
1	1	0	1	1	1	1	1	0	0	0	1	0	1	0	0
0	1	0	1	1	1	1	1	0	1	0	0	1	0	0	0
0	0	0	0	0	1	0	1	1	0	1	0	0	1	0	0

MRP for Bit-Plane 0

53. MRC for (0, 0): CX=15, D= 1
54. MRC for (1, 0): 15, 1
55. MRC for (2, 0): 16, 0
56. MRC for (1, 1): 16, 1
57. MRC for (3, 1): 16, 0
58. MRC for (1, 2): 15, 0
59. MRC for (2, 2): 15, 0
60. MRC for (0, 3): 16, 1
61. MRC for (2, 3): 15, 1
62. MRC for (3, 3): 15, 0

υ^0				σ				η				σ'			
1	0	0	1	1	0	0	1	0	1	1	0	1	0	0	1
1	1	0	1	1	1	1	1	0	0	0	1	1	1	1	0
0	1	0	1	1	1	1	1	0	1	0	0	1	0	1	1
0	0	0	0	0	1	0	1	1	0	1	0	0	1	0	1

CUP for Bit-Plane 0

(Note: This pass does not generate any CX or D.)

υ^0				σ				η				σ'			
1	0	0	1	1	0	0	1	0	1	1	0	1	0	0	1
1	1	0	1	1	1	1	1	0	0	0	1	1	1	1	0
0	1	0	1	1	1	1	1	0	1	0	0	1	0	1	1
0	0	0	0	0	1	0	1	1	0	1	0	0	1	0	1

18.3.3 Binary Arithmetic Coding—MQ-Coder

As explained in the previous section, the fractional bit-plane coding (EBCOT) produces a sequence of symbols, pairs of context and decision (CX, D), in each coding pass. The context-based adaptive binary arithmetic MQ-coder that is used in JBIG2 [5] is adapted in JPEG2000 standard to encode these symbols. The probability values (Qe) and probability estimation/mapping process are provided by the standard as a lookup table with four fields (or four functions), which is defined in Table 18.7. We discussed the principles of arithmetic coding, binary arithmetic coding (BAC), and the implementation procedure of an adaptive version of BAC (the QM-coder, used in JPEG) in Chapter 15. In JPEG2000, the binary arithmetic coder is called the MQ-coder, which is a variation of the QM-coder. Here we present the implementation procedures of the MQ-coder based on the informative materials provided by the standard. Besides the probability table, the Qe-table (Table 18.7), we need two more lookup tables, I(CX) and MPS(CX). This is because there could be 19 different contexts generated by the bit-plane coder, and we need to keep track of the state and the index of the Qe-table for each context. The I(CX) is used to keep track of the index of the Qe-table and the initial values are provided by the standard (as shown in Table 18.6). The MPS(CX) specifies the sense (0 or 1) of the more probable symbol of context CX, and all MPS(CX) are initialized with value zero. Table 18.7 can be viewed as four lookup tables, Qe(I(CX)), NMPS(I(CX)), NLPS(I(CX)), and SWITCH(I(CX)) respectively. The I(CX) is the current index for the context CX. The Qe(I(CX)) provides the probability value, NMPS(I(CX))/NLPS(I(CX)) indicates the next index for a MPS/LPS renormalization, and SWITCH(I(CX)) is a flag used to indicate whether a change of the sense of MPS(CX) is needed. The same tables and initial values will be used in both encoder and decoder. We will see more details on how these variables are used in the following implementation subsections.

18.4 TIER-2 CODING IN JPEG2000

In the JPEG2000 standard [1, 2], the Tier-2 coding engine is responsible for efficiently representing layer and block summary information for each code-block, including:

- The bitstream layers to which the code-block contributes the compressed codewords; this is also known as the "inclusion information."
- The length of these codewords.
- The most significant magnitude bit-plane at which any sample in the code-block is nonzero, also known as the zero bit-planes information.

Table 18.7 BAC Qe-value and probability estimation lookup table

Index	Qe	NMPS	NLPS	SWITCH
0	0x5601	1	1	1
1	0x3401	2	6	0
2	0x1801	3	9	0
3	0x0AC1	4	12	0
4	0x0521	5	29	0
5	0x0221	38	33	0
6	0x5601	7	6	1
7	0x5401	8	14	0
8	0x4801	9	14	0
9	0x3801	10	14	0
10	0x3001	11	17	0
11	0x2401	12	18	0
12	0x1C01	13	20	0
13	0x1601	29	21	0
14	0x5601	15	14	1
15	0x5401	16	14	0
16	0x5101	17	15	0
17	0x4801	18	16	0
18	0x3801	19	17	0
19	0x3401	20	18	0
20	0x3001	21	19	0
21	0x2801	22	19	0
22	0x2401	23	20	0
23	0x2201	24	21	0
24	0x1C01	25	22	0
25	0x1801	26	23	0
26	0x1601	27	24	0
27	0x1401	28	25	0
28	0x1201	29	26	0
29	0x1101	30	27	0
30	0x0AC1	31	28	0
31	0x09C1	32	29	0
32	0x08A1	33	30	0
33	0x0521	34	31	0
34	0x0441	35	32	0
35	0x02A1	36	33	0
36	0x0221	37	34	0
37	0x0141	38	35	0
38	0x0111	39	36	0
39	0x0085	40	37	0
40	0x0049	41	38	0
41	0x0025	42	39	0
42	0x0015	43	40	0
43	0x0009	44	41	0
44	0x0005	45	42	0
45	0x0001	45	43	0
46	0x5601	46	46	0

- The truncation points between the bitstream layers, that is, the number of coding passes information.

This information is known to the encoder. The decoder receives this information in an encoded format, which combines two Tag Trees (one for the inclusion information and the other for the zero bit-planes information) in the encoding procedure.

Tag Tree is a particular type of quad-tree data structure, which provides the framework for efficiently representing information in the Tier-2 coding engine of JPEG2000. Size of the header included in the compressed file in the JPEG2000 standard is much larger than in the JPEG standard and contains lots of important information. The Tag Tree coding mechanism helps in representing the layer and block summary information for each code-block to be included in the header of the compressed file. In this section, we first discuss the basic Tag Tree compression technique. Then we discuss the bitstream formation methodology and how the Tag Trees are integrated in Tier-2 coding in detail.

18.4.1 Bitstream Formation

Before we discuss how Tag Tree encoding can be used to encode the packet header information, we need to explain some terminology about bitstream formation and the progression order defined in JPEG2000.

18.4.1.1 Definition of Terms

Packet: Compressed data representing a component, specific tile, layer, resolution level, and precinct.

Layer: The coded data (bitstream) of each code-block is distributed across one or more layers in the code-stream. Each layer consists of some number of consecutive bit-plane coding passes from each code-block in the tile, including all subbands of all components for that tile.

Resolution: Partition of DWT subbands in one tile.

There are (N_L +1) resolutions for N_L levels DWT decomposition.

r = 0 : LL(N_L) subband only

r = 1 : HL(N_L), LH(N_L), HH(N_L)

⋮

r : HL(N_L- r+1), LH(N_L- r+1), HH(N_L- r+1)

⋮

r = N_L : HL1, LH1, HH1

Precinct: Partition in each resolution (formed in DWT domain). Power of 2 in size (line up with code-block size boundary). Don't cause block (tile) artifacts. Figure 18.9 shows an example for a precinct from a two-level DWT with three resolutions.

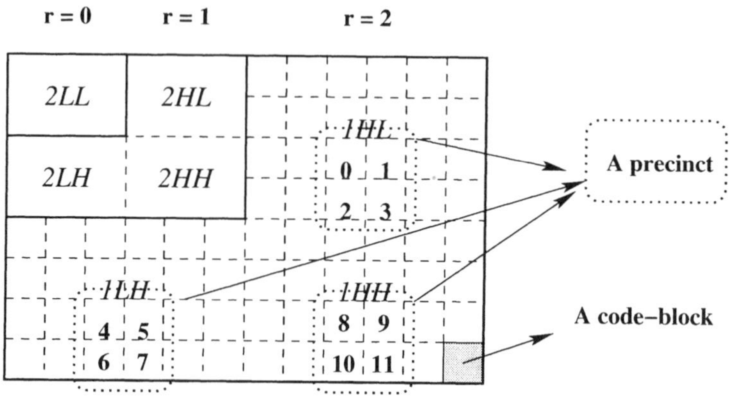

Fig. 18.9 A precinct from a two-level DWT with three resolutions.

Component: A color image may have several components from a specified color space. A component is a two-dimensional array of samples from an image.

18.4.1.2 Progression Order The standard allows five different progression orders, which are specified in the coding style default (COD) markers unless otherwise overridden by the progression order default (POD) marker. Five different possible progression orders defined by the standard are listed below.

1. Layer–resolution–component–position progressive

2. Resolution–layer–component–position progressive

3. Resolution–position–component–layer progressive

4. Position–component–resolution–layer progressive

5. Component–position–resolution–layer progressive

The standard has the layer–resolution–component–position progressive order as the default order.

Figure 18.10(a) shows an example of a single-component image applied with two-level DWT decomposition and partitioned into 16 code-blocks. For the sake of discussion and simplicity, as shown in Figure 18.10(b), we assume all code-blocks (CB1–CB16) have four bit-planes (BP4 denotes the MSB plane); the letters *S*, *M*, and *C* stand for the bitstream generated during significance propagation pass, magnitude refinement pass, and cleanup pass,

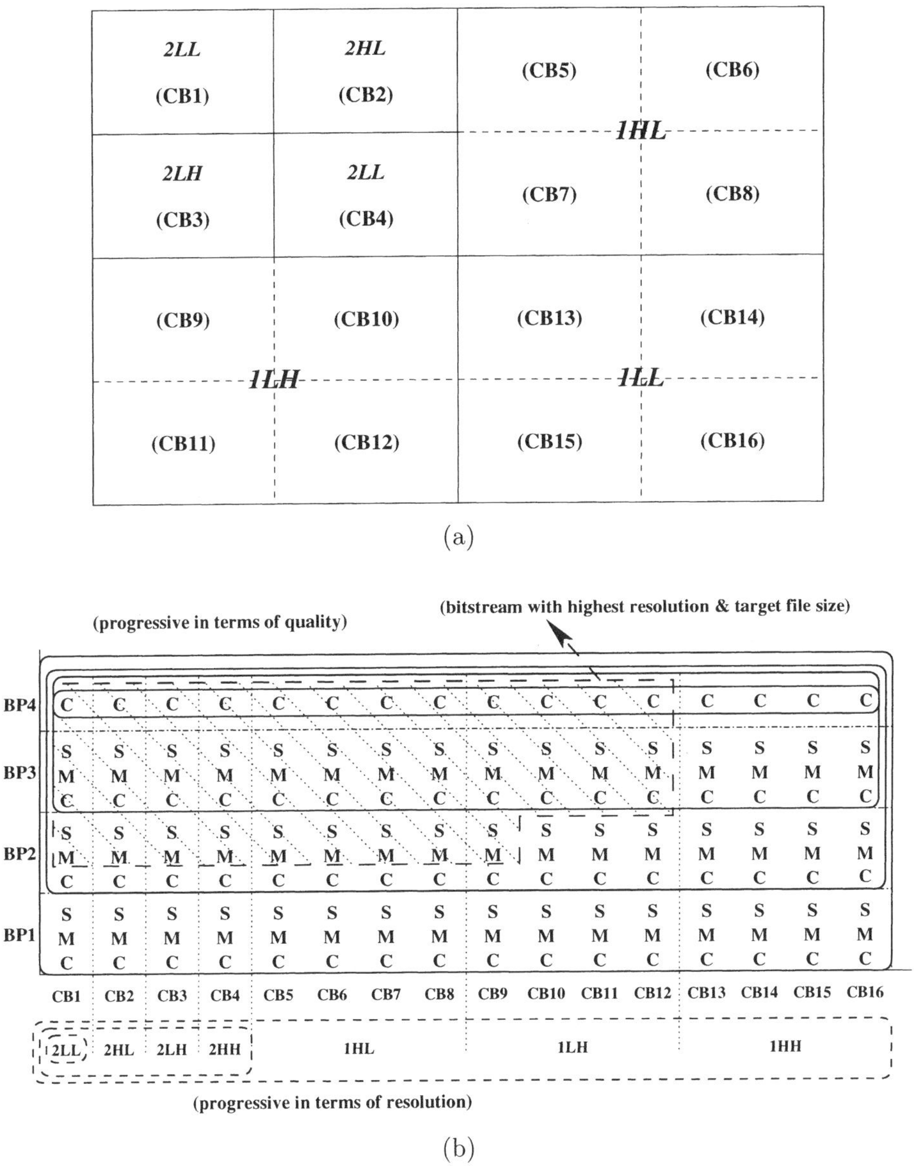

Fig. 18.10 (a) A two-level DWT with 16 code-blocks, (b) bitstream formations. (Note: The letters S, M, and C in (b) stand for the bitstream generated during significance propagation pass, magnitude refinement pass, and cleanup pass, respectively.)

respectively. Conceptually, the bottom-nested dash-line boxes stand for progressive in terms of resolution where the bitstreams generated from the lowest-resolution code-block (CB1) were sent out first, followed by the bitstreams generated from resolution 1 (CB2, CB3, and CB4), and then the bitstreams generated from resolution 2 (CB5–CB16). The progressive in terms of quality can be done by first sending out all the bitstreams generated from all the code-blocks for the MSB plane (BP4), followed by the bitstreams generated from all the code-blocks for the next bit-plane (BP2), and so on (as shown in Figure 18.10(b) with nested solid-line boxes). On the other hand, in order to meet a certain bandwidth or target file size with the highest resolution, we can even send out the bitstreams that are corresponding to the shading area as shown in Figure 18.10(b).

18.4.2 Packet Header Information Coding

All the compressed bitstreams from a specific tile, layer, resolution, component, and precinct are stored in a contiguous segment called a *packet*. The packet header information appears in the bitstream immediately preceding the packet data, unless one of the PPM (main packed packet header marker) or PPT (tile-part packed packet header) marker segments has been used. Both the packet header information and packet data are constructed based on the order of contribution from the LL, HL, LH, and HH subbands. The packet header contains the following information:

- **Zero-length packet**, which is encoded using one bit; the value 0 denotes a zero-length packet, and the value 1 indicates a nonzero-length packet.
- **Inclusion information**, which is encoded with a separate Tag Tree. The value in this Tag Tree is the number of the layer (a zero-based index) in which the code-block is first included.
- Number of (leading) **zero bit-planes**, which is used to identify the actual number of bit-planes for coefficients from the code-block. A second Tag Tree is used to encode this information.
- Number of **coding passes** for each code-block in this packet. This number is encoded using the codewords shown in Table 18.8.
- **Length** (in bytes) of the bitstream from a given code-block, which is encoded either in a single-codeword segment or a multiple-codeword segment. This is the case when a termination occurs between coding passes that are included in the packet.

The codewords (as shown in Table 18.8) for the number of coding passes for each code-block are variable-length codes. The number 37 through 164 has a 9-bit 1111 11111 as prefix followed by 7 bits as the offset from 37. The number 6 through 36 has a 4-bit 1111 as prefix followed by 5 bits as the offset from 6.

Table 18.8 Codewords for the number of coding passes

# of Coding Passes	Codeword
1	0
2	10
3	1100
4	1101
5	1110
6–36	**1111** *0000 0*–**1111** *1111 0*
37–164	**1111 11111** *0000 000*–**1111 11111** *1111 111*

The single-codeword segment is used to encode the number of bytes contributed to a packet by a code-block. The number of bits needed to represent the length of bytes can be derived using

$$bits = LBlocks + \lfloor log_2(number\ of\ coding\ passes) \rfloor,$$

where $LBlocks$ is a code-block-wise state variable with initial value *three*. The value of $LBlocks$ can be increased if there are $\boldsymbol{k}$ 1's (i.e., '$\overbrace{11111....1}^{k}$') followed by a bit 0 (this is also known as the code-block length indicator); the value of $LBlocks$ is increased by k. If $k = 0$, then the bit 0 is used as a delimiter, which means no increase for the value of $LBlocks$. There is no restriction on number of bits used to represent the code length in the packet.

18.5 SUMMARY

In this chapter, we have dealt with details of coding algorithms and bitstream formation in JPEG2000 standard. We have discussed the fractional bit-plane coding (EBCOT) and the MQ-coder for implementation of the adaptive binary arithmetic coding scheme proposed in JPEG2000 standard. We described the terminology and their underlying concepts behind these algorithms. The EBCOT has three coding passes and four coding operations. We discussed in great detail the concepts behind these coding operations and how they are used in each coding pass in order to encode the bit-planes of the DWT coefficients. We took an example and showed how to generate the context and decision information by the EBCOT algorithm. This context and decision pair is then used by the MQ-coder to generate compressed codewords. We described the flowcharts for all the modules in both the EBCOT and the MQ-coder algorithms and their implementation issues. We have discussed the Tier-2 encoding scheme using the Tag Tree coding mechanism. We have discussed the basic definitions for the Tag Tree with an example and showed how the two-dimensional integer arrays can be mapped into a Tag Tree. This Tag Tree encoding is the basis for encoding the code-block and layer information in the compressed file header. We compressed a small component of size 80 × 60, generated the code-block information, and displayed the results as an example. We also showed the different progression orders achievable by the JPEG2000 standard in order to influence different areas of applications. In summary, we have detailed entropy encoding (Tier-1 and Tier-2 encoding) in this chapter. For further detail of the JPEG2000 standard and implementation issues both in software and hardware, the readers are suggested to refer the book by Acharya et al. [1].

REFERENCES

1. T. Acharya and P. Tsai, *JPEG2000 Standard Image Compression: Concepts, Algorithms and VLSI Architectures*, Wiley, Hoboken, NJ, 2004.

2. ISO/IEC 15444-1, "Information Technology—JPEG2000 Image Coding System, Part 1: Core Coding System," 2000.

3. D. S. Taubman, "High-Performance Scalable Image Compression with EBCOT," *IEEE Transactions on Image Processing*, 9(7), July 2000, 1158–1170.

4. W. B. Pennebaker, J. L. Mitchell, G. G. Langdon, Jr., and R. B. Arps, "An Overview of the Basic Principles of the Q-Coder Adaptive Binary Arithmetic Coder," *IBM Journal of Research and Development*, 32(6), November 1988, 717–726.

5. ISO/IEC 14492-1, "Lossy/Lossless Coding of Bi-level Images," 2000.

Index

About the Authors

Dr. Tinku Acharya is the Chief Technology Officer of Avisere Inc., Tucson, Arizona, and Managing Director of Avisere Technology Pvt. Ltd., Kolkata, India. He is also an Adjunct Professor in the Department of Electrical Engineering, Arizona State University, Tempe, Arizona. He received his Bachelor's (Honors) in Physics, and Bachelor's and Master's in Computer Science and Engineering from the University of Calcutta, India in 1984, 1987, and 1989 respectively. He received his Ph.D. in Computer Science from the University of Central Florida, Orlando, in 1994.

Before joining Avisere, Dr. Acharya served in Intel Corporation from June 1996 to June 2002, where he led several R&D teams in numerous projects toward development of algorithms and architectures in image and video processing, multimedia computing, PC-based digital camera, high-performance reprographics architecture for color photocopiers, biometrics, multimedia architecture for 3G cellular mobile telephony, analysis of next-generation microprocessor architecture, etc. Before joining Intel Corporation, he was a consulting engineer at AT&T Bell Laboratories (1995–1996) in New Jersey, a research faculty member at the Institute of Systems Research, University of Maryland at College Park (1994–1995), and held visiting faculty positions at Indian Institute of Technology (IIT), Kharagpur (on several occasions during 1998–2003). He also served as Systems Analyst in National Informatics Center, Planning Commission, Government of India (1988–1990). He has held many other positions in industry and research laboratories. He collaborated in research and development with Palo Alto Research Center (PARC) in Xerox Corporation, Eastman Kodak Corporation, and many other institutions and research laboratories worldwide.

Dr. Acharya is an inventor of 83 U.S. patents and 15 European patents in the areas of electronic imaging, data compression, multimedia computing, biometrics, and their VLSI architectures and algorithms, and more than 50 patents are currently pending in the U.S. Patent Office. He has contributed to over 70 refereed technical papers published in international journals and conferences. He is author of 3 books: (1) *JPEG2000 Standard for Image Compression: Concepts, Algorithms, and VLSI Architectures* (John Wiley & Sons, Hoboken, NJ, 2004), (2) *Information Technology: Principles and Applications* (Prentice-Hall India, New Delhi, 2004), and (3) *Data Mining: Multimedia, Soft Computing and Bioinformatics* (John Wiley & Sons, Hoboken, NJ, 2003). His pioneering works won him international acclamation. He has been awarded the *Most Prolific Inventor* in Intel Corporation Worldwide in 1999 and *Most Prolific Inventor* in Intel Corporation Arizona site for five consecutive years (1997–2001).

Dr. Acharya is a Life Fellow of the Institution of Electronics and Telecommunication Engineers (FIETE), and Senior Member of IEEE. He served on the U.S. National Body of JPEG2000 standards committee (1998–2002). His current research interests are in image processing, computer vision for enterprise applications, biometrics, multimedia computing, multimedia data mining, and VLSI architectures and algorithms.

Dr. Ajoy K. Ray is currently a Senior Corporate Research Scientist of Avisere Inc., Tucson, Arizona. He is also a Professor in the Department of Electronics and Electrical Communication Engineering, and Chairman of Nehru Museum of Science and Technology at Indian Institute of Technology (IIT), Kharagpur, India. He received his Bachelor's in Electronics and Telecommunication engineering from BE College, University of Calcutta, Master's and Ph.D. in Automation and Computer Vision from Indian Institute of Technology, Kharagpur in 1975, 1977, and 1984 respectively.

During 1989 to 1990, Dr. Ray served as a visiting Research Scientist at the University of Southampton, UK. He has completed 15 sponsored research projects from various national and international agencies in the areas of image processing and machine intelligence. He published more than 80 refereed technical papers in international journals and conferences. He is author of 3 books (1) *Advanced Microprocessors: Principles and Applications* (TATA-McGraw Hill, India, 2000), (2) *Intel Microprocessors* (McGraw Hill International, Singapore, 2001), and (3) *Information Technology: Principles and Applications* (Prentice Hall of India, New Delhi, 2004) as well as many other booklets.

Dr. Ray is a Fellow of the Institution of Electronics and Telecommunication Engineers (FIETE), Institution of Engineers (India) and member of many professional bodies. His current research interest is in diverse areas of image processing, machine intelligence, soft computing, computer vision, etc.